ADVANCES IN CHEMICAL ENGINEERING

Volume 9

CONTRIBUTORS TO THIS VOLUME

Renato G. Bautista
Norman Epstein
Kishan B. Mathur
R. E. Peck
W. C. Reynolds
D. T. Wasan

ADVANCES IN
CHEMICAL ENGINEERING

Edited by

THOMAS B. DREW

Department of Chemical Engineering
Massachusetts Institute of Technology
Cambridge, Massachusetts

GILES R. COKELET

Department of Chemical Engineering
Montana State University
Bozeman, Montana

JOHN W. HOOPES, Jr.

Imperial Chemical Industries America, Inc.
Wilmington, Delaware

THEODORE VERMEULEN

Department of Chemical Engineering
University of California
Berkeley, California

Volume 9

Academic Press • New York • London 1974

A Subsidiary of Harcourt Brace Jovanovich, Publishers

TP
145
.D7
1974
v.9

COPYRIGHT © 1974, BY ACADEMIC PRESS, INC.
ALL RIGHTS RESERVED.
NO PART OF THIS PUBLICATION MAY BE REPRODUCED OR
TRANSMITTED IN ANY FORM OR BY ANY MEANS, ELECTRONIC
OR MECHANICAL, INCLUDING PHOTOCOPY, RECORDING, OR ANY
INFORMATION STORAGE AND RETRIEVAL SYSTEM, WITHOUT
PERMISSION IN WRITING FROM THE PUBLISHER.

ACADEMIC PRESS, INC.
111 Fifth Avenue, New York, New York 10003

United Kingdom Edition published by
ACADEMIC PRESS, INC. (LONDON) LTD.
24/28 Oval Road, London NW1

LIBRARY OF CONGRESS CATALOG CARD NUMBER: 56-6600

PRINTED IN THE UNITED STATES OF AMERICA

CONTENTS

LIST OF CONTRIBUTORS vii
PREFACE ix
CONTENTS OF PREVIOUS VOLUMES xi

Hydrometallurgy

RENATO G. BAUTISTA

I. Introduction 1
II. Raw Material Preparation 3
III. Leaching 7
IV. Separation and Concentration Processes 51
V. Metal Reduction from Aqueous Solutions 72
VI. Hydrometallurgical Operations 83
VII. Summary 98
References 99

Dynamics of Spouted Beds

KISHAN B. MATHUR AND NORMAN EPSTEIN

I. The Phenomenon of Spouting 111
II. Location of Spouting in the Gas–Solids Contacting Spectrum . . . 115
III. The Onset of Spouting 117
IV. Flow Patterns 140
V. Bed Structure 163
VI. Spouting Stability 173
Nomenclature 187
References 188

Recent Advances in the Computation of Turbulent Flows

W. C. REYNOLDS

I. Background and Overview 193
II. Mean-Velocity Field Closure 200
III. Mean Turbulent Energy Closure 216
IV. Mean Reynolds-Stress Closure 231
V. Opportunities and Outlook 236
Nomenclature 244
References 245

Drying of Solid Particles and Sheets

R. E. Peck and D. T. Wasan

I.	Introduction	247
II.	Estimation of Heat- and Mass-Transfer Coefficients	248
III.	Moisture Movement through Porous Solids	252
IV.	Drying of Porous Solids—Batch Operations	258
V.	Drying Porous Solids—Continuous Operations	279
VI.	Summary	288
	Nomenclature	289
	References	290

Author Index 295
Subject Index 306

CONTRIBUTORS TO VOLUME 9

RENATO G. BAUTISTA, *Department of Chemical Engineering and Ames Laboratory, USAEC, Iowa State University, Ames, Iowa*

NORMAN EPSTEIN, *Department of Chemical Engineering, University of British Columbia, Vancouver, Canada*

KISHAN B. MATHUR, *Department of Chemical Engineering, University of British Columbia, Vancouver, Canada*

R. E. PECK, *Department of Chemical Engineering, Illinois Institute of Technology, Chicago, Illinois*

W. C. REYNOLDS, *Department of Mechanical Engineering, Stanford University, Stanford, California*

D. T. WASAN, *Department of Chemical Engineering, Illinois Institute of Technology, Chicago, Illinois*

PREFACE

The aim of this Volume 9, like that of each of its predecessors in the series, is perhaps best expressed by quoting the first two paragraphs of our instructions to our authors:

> Ideally, a chapter in *Advances in Chemical Engineering* is a short monograph in which the author summarizes the current state of knowledge of his topic for the benefit of professional colleagues in engineering who by reason of their normal duties have not been able to make a study of the subject in depth. They want an authoritative account not cloaked in unintelligible specialized terminology. They will read it, not only for general information, but also because as sophisticated engineers they know that major progress in science and engineering is made by those who see connections between matters others have imagined unrelated: they may spot in the author's specialty a method or an idea with an analog useful in theirs. Many readers will not have ready access to large university libraries and many with such access are by hypothesis too inexpert to assess the validity of journal articles on the author's topic—they expect such assessment in the chapter.
>
> Typically, one expects a chapter to be a critical review and an evaluation of the results and opinions which various workers have presented in journal articles or books. The author is expected to point out discrepancies in previous work and, if he cannot resolve them, to suggest the nature of further studies needed for that purpose. Except where it may be necessary to introduce them to justify his evaluations and conclusions, an article in *Advances in Chemical Engineering* is not ordinarily the appropriate place of first publication for new experimental or theoretical results of the author. In exceptional cases, especially when the space required for intelligible presentation would exceed that normally available in a journal article, the Editors will consider chapters that are essentially reports of previously unpublished work by the author.

The Editors hope that the four chapters herein published satisfy the criteria set forth above.

Thomas B. Drew
Giles R. Cokelet
John W. Hoopes, Jr.
Theodore Vermeulen

CONTENTS OF PREVIOUS VOLUMES

Volume 1

Boiling of Liquids
  J. W. Westwater
Non-Newtonian Technology: Fluid Mechanics, Mixing, and Heat Transfer
  A. B. Metzner
Theory of Diffusion
  R. Byron Bird
Turbulence in Thermal and Material Transport
  J. B. Opfell and B. H. Sage
Mechanically Aided Liquid Extraction
  Robert E. Treybal
The Automatic Computer in the Control and Planning of Manufacturing Operations
  Robert W. Schrage
Ionizing Radiation Applied to Chemical Processes and to Food and Drug Processing
  Ernest J. Henley and Nathaniel F. Barr
AUTHOR INDEX—SUBJECT INDEX

Volume 2

Boiling of Liquids
  J. W. Westwater
Automatic Process Control
  Ernest F. Johnson
Treatment and Disposal of Wastes in Nuclear Chemical Technology
  Bernard Manowitz
High Vacuum Technology
  George A. Sofer and Harold C. Weingartner
Separation by Adsorption Methods
  Theodore Vermeulen
Mixing of Solids
  Sherman S. Weidenbaum
AUTHOR INDEX—SUBJECT INDEX

Volume 3

Crystallization from Solution
  C. S. Grove, Jr., Robert V. Jelinek, and Herbert M. Schoen
High Temperature Technology
  F. Alan Ferguson and Russell C. Phillips
Mixing and Agitation
  Daniel Hyman
Design of Packed Catalytic Reactors
  John Beek
Optimization Methods
  Douglass J. Wilde
AUTHOR INDEX—SUBJECT INDEX

Volume 4

Mass-Transfer and Interfacial Phenomena
  J. T. Davies
Drop Phenomena Affecting Liquid Extraction
  R. C. Kintner
Patterns of Flow in Chemical Process Vessels
  Octave Levenspiel and Kenneth B. Bischoff
Properties of Cocurrent Gas-Liquid Flow
  Donald S. Scott
A General Program for Computing Multistage Vapor-Liquid Processes
  D. N. Hanson and G. F. Somerville
AUTHOR INDEX—SUBJECT INDEX

Volume 5

Flame Processes—Theoretical and Experimental
  J. F. Wehner
Bifunctional Catalysts
  J. H. Sinfelt
Heat Conduction or Diffusion with Change of Phase
  S. G. Bankoff
The Flow of Liquids in Thin Films
  George D. Fulford
Segregation in Liquid-Liquid Dispersions and Its Effect on Chemical Reactions
  K. Rietema
AUTHOR INDEX—SUBJECT INDEX

Volume 6

Diffusion-Controlled Bubble Growth
  S. G. Bankoff
Evaporative Convection
  John C. Berg, Andreas Acrivos, and Michel Boudart
Dynamics of Microbial Cell Populations
  H. M. Tsuchiya, A. G. Fredrickson, and R. Aris
Direct Contact Heat Transfer between Immiscible Liquids
  Samuel Sideman
Hydrodynamic Resistance of Particles at Small Reynolds Numbers
  Howard Brenner
AUTHOR INDEX—SUBJECT INDEX

Volume 7

Ignition and Combustion of Solid Rocket Propellants
  Robert S. Brown, Ralph Anderson, and Larry J. Shannon
Gas-Liquid-Particle Operations in Chemical Reaction Engineering
  Knud Østergaard
Thermodynamics of Fluid-Phase Equilibria at High Pressures
  J. M. Prausnitz
The Burn-Out Phenomenon in Forced-Convection Boiling
  Robert V. Macbeth
Gas-Liquid Dispersions
  William Resnick and Benjamin Gal-Or
AUTHOR INDEX—SUBJECT INDEX

Volume 8

Electrostatic Phenomena with Particulates
  C. E. Lapple
Mathematical Modeling of Chemical Reactions
  J. R. Kittrell
Decomposition Procedures for the Solving of Large Scale Systems
  W. P. Ledet and D. M. Himmelblau
The Formation of Bubbles and Drops
  R. Kumar and N. R. Kuloor
AUTHOR INDEX—SUBJECT INDEX

HYDROMETALLURGY

Renato G. Bautista

Department of Chemical Engineering and Ames Laboratory, USAEC
Iowa State University
Ames, Iowa

I. Introduction 1
II. Raw Material Preparation 3
III. Leaching 7
    A. Atmospheric Pressure Leaching 11
    B. Elevated Pressure Leaching 34
    C. Leaching at Reduced Pressure 50
IV. Separation and Concentration Processes 51
    A. Resin Ion Exchange 52
    B. Solvent Extraction 61
V. Metal Reduction from Aqueous Solutions 72
    A. Displacement Reactions 74
    B. Electrolysis 78
    C. Chemical Reduction 81
VI. Hydrometallurgical Operations 83
    A. Copper 84
    B. Molybdenum 88
    C. Nickel 90
    D. Copper–Zinc 96
VII. Summary 98
    References 99

I. Introduction

The chemical processing of ores to recover the metal values in pure form is commonly referred to as extractive metallurgy; as presently practiced it is subdivided into hydrometallurgy and pyrometallurgy. Hydrometallurgy involves reacting the ore at low or moderate temperatures with a liquid solvent that will selectively dissolve the valuable metal or metals, separating the dissolved metals from each other by chemical means, concentrating and purifying the desired metal values, and finally preparing the pure metal, usually in powder form, suitably for processing into finished products. The analogous processing stages of pyrometallurgy are carried out at high temperatures, usually close to or at liquid-metal temperature. Roasting, smelting, and volatilization are examples of the unit processes of pyrometallurgy. The majority of production techniques for large-tonnage metals at present are primarily pyrometallurgical, although some hydrometallurgical steps may be involved.

Hydrometallurgical unit processes go back to the 1700s when the first large-scale leaching and precipitation of copper was carried out at Rio Tinto in Spain (L1, T1). The method essentially consists of leaching weathered piles of massive pyrites containing copper sulfides with water. The copper in solution is recovered by precipitation with iron. This method is still in use all over the world, with changes in terms of machinery and operating modes dictated by location, weather, and availability of iron or iron-bearing materials. An indication of the importance of this particular process is shown by the fact that an estimated 220,000 tons of copper were produced by this process in the United States in 1969 (R14).

The current greatly increased interest in hydrometallurgy is due to the fact that: (1) it is suitable for treatment of low-grade ores and complex ores; (2) it requires low capital investment; (3) it has low operating costs because of by-product recovery and recycling of some chemical reagents; (4) almost complete recovery of all the metal values from the ore is possible; (5) hydrometallurgical processes are amenable to automatic controls resulting in lower labor cost; and (6) the water and air pollution problems are typically less troublesome than those normally associated with the corresponding pyrometallurgical operations. The last thirty years have seen the development of processes for the production of uranium, rare earths, zirconium, hafnium, and beryllium needed for the nuclear energy program, and the development by the Sheritt Gordon Mines Limited of Canada of their hydrometallurical "smelter" and refinery for nickel (300,000,000 lb/year) and cobalt (1,000,000 lb/year), making Sheritt Gordon the largest producer of cobalt in the North American continent (B25, M4, M32). The innovation in the processing of nickel and cobalt ores was soon taken advantage of by several other companies whose ore deposits are not amenable to the standard pyrometallurgical process (W13): in 1958, the Freeport Nickel Company at Moa Bay, Cuba, (B25, C4, M34, W15) started the production of nickel and cobalt from lateritic ores; a 3,000,000-lb/year cobalt plant started operation in 1967 at Outokumpu Oy, Finland using pyrite as the ore (M33); a 50-million lb/year nickel and cobalt plant using the basic Sheritt Gordon process is presently under construction in the southern Philippines and will be on stream by 1975 (M35).

It is apparent that as the requirements for water and air pollution standards are made stricter, the traditional methods of producing the common metals from their ores will come under closer scrutiny. There is now active search for alternative processing methods for most of the smelting operations; the most likely candidate is the hydrometallurgical technique so successfully used in the base metals, uranium, cobalt, and nickel.

It is not the intention of the author in writing this chapter to make a complete listing of all the available literature in hydrometallurgy, but rather to make a definitive and exhaustive examination of the important advances made and of the more promising developments. Also important to this discussion are why and how hydrometallurgy can best be applied and the promise of its fulfillment in industrial operations in the near future. There are excellent reviews available in the specialized areas of leaching (F14, H7, M5, R11, S10), ion exchange (C16, E12, H25, H26), solvent extraction (B29, M14, R9, Z1), reduction of metals from aqueous solutions (E10, M25, S10), and in industrial applications (B38, C20, G12). There have been a number of symposiums particularly oriented toward hydrometallurgy (U1, W2) and extractive metallurgy (A1, C5, P13). Books on the hydrometallurgy (B25, H26, Q1, V1) and chemistry (B36) of specific metals have been on the market for many years.

The laboratories most active in this field include the many U.S. Bureau of Mines Research Centers that are located throughout the country, the U.S. Atomic Energy Commissions national laboratories, the Canadian Department of Energy, Mines and Resources, the atomic energy laboratories, and other agencies of various countries, the research centers and laboratories of metal companies throughout the world, and the several universities and research institutes engaged in research in this field.

II. Raw Material Preparation

The process of putting the desired metal values into solution can be done by *in situ* operation when the conditions are geologically and metallurgically favorable or after mining of the ore. The former has great appeal because it involves no movement of large-tonnage ore and gangue materials. A number of the nonmetallic mining operations are now using this technique, presently referred to as solution mining or chemical mining. It is sometimes necessary to fracture the massive ore deposit in place by use of conventional explosives. The use of a nuclear device in an underground explosion to extract copper from a low-grade deposit that is uneconomical to mine has been proposed (M31).

The conventional method of ore concentration after mining involves: (a) crushing and grinding to size, which is determined by the degree of liberation of the minerals from the bulk of the ore; and (b) beneficiation of the ore, either by physical means or by flotation technique, whenever applicable to produce a concentrate acceptable to the smelter. In a typical copper operation, the mill feed averages 0.70% copper and the flotation concentrate is anywhere from 25 to 30% copper.

A concentrate is perferable to as-mined ore as a raw material for leaching since the majority of unwanted materials have already been removed. The saving in terms of chemical reagents alone is tremendous, not to mention the decrease in volume of the material to be handled. In the recovery of copper values from mine waste dumps, the raw material is the as-mined below-grade ore which has been stockpiled in a specific area where topography is ideal for a dump leaching operation. Although this is very low grade in terms of copper content, the cost of mining is not figured in the operation since this tonnage needs to be moved to get to the ore body regardless of whether or not further metal recovery is planned. Where the ore is not amenable to concentration by physical or flotation methods, the as-mined ore is the feed to the leaching stage.

Some of the important but expensive rare metals are usually extracted as by-products of other metal separation processes. Selenium and tellurium are recoverable from copper refinery slime by pressure leaching (M40), scandium from uranium plant iron sludge (R15), uranium from gold cyanidation residues (G3), silver from aqueous chlorination process for the treatment of slimes, and gravity concentrates from gold ores (V2). A host of other processes are in use.

Because all ores contain more than one metallic element of value, it is quite possible that in the very near future, hydrometallurgical processes will be developed to extract a whole line of products which at present are being discarded with the waste material. The dwindling supply of naturally occurring ore deposits, the increased demand due to an expanding technological age, and the pressure for maintaining a clean environment will help accelerate the development of such ideal processes.

The recovery of the metal values from sources other than freshly mined ores is gaining a lot of interest. Old mine workings are further exploited for their metal values by flooding of the underground workings with leach solutions and recovering the metal by conventional separation processes. Copper and uranium have been recovered in this way. The mine waste dumps and the old tailings ponds are leached for more of their copper and gold content.

The waste streams in standard metallurgical processing are good potential sources of important elements. The U.S. Bureau of Mines has developed processes for the recovery of elemental sulfur from stack gases discharged by base metal smelters (G4) and for the recovery and production of alumina from waste solutions of mining operations (G6). A potential of 1,750,000 tons of sulfur per year and an estimated 2000 tons of alumina per day are recoverable just from 14 copper mines included in the study.

With some ore deposits, it has been found advantageous to go one or more steps beyond beneficiation before the leaching stage. This usually involves heating the ore in an oxidizing or reducing atmosphere to effect some favorable chemical change in order that the solubility of the desired metal values in a given dissolution media will be enhanced. At times, this also has the effect of increasing the selectivity of the leaching process.

The oxidizing atmosphere during the pretreatment of the ore helps to break up chemically the naturally occurring stable bonds in the solid, resulting in better dissolution rates. In a process developed for the recovery of scandium from uranium plant iron sludges (R15), the calcination of the sludge at 250°C was found to be very effective in the removal of organic materials and appreciably decreased the consumption of acid.

In a reducing atmosphere, some of the metallic compounds are reduced to metals and others to the lower oxides. Since subsequent leaching is an oxidation step, the dissolution medium reacts more rapidly with the reduced metal than with the lower oxides, thus giving a certain degree of selectivity. The 100,000-ton/year ilmenite beneficiation plant of Benilite Corporation of America (C15) for the production of a feed for chloride-process titanium dioxide plants employs a partial reduction of iron oxides prior to leaching with hydrochloric acid.

The extraction of alumina from silicates as reported by Iverson and Leitch (I5) involves the melting of the charge at temperatures up to 1700°C and quenching the melt with cold water. The resulting glossy, amorphous material after crushing and grinding is more readily leached with sulfuric acid than the completely or partly crystallized material.

The effect of roasting on the recovery of uranium and vanadium from carnotite ores by carbonate leaching was studied by Halpern *et al* (H8) The extraction of uranium and vanadium from carnotite ores by leaching under pressure has been observed to be dependent on the composition and roasting conditions, such as temperature, atmosphere above the charge during the roasting and cooling steps, and the presence of other chemical reactants. Variation in leaching conditions was found to be of no significance. A systematic examination of the above variables during the roasting process and of the subsequent carbonate pressure leaching of the calcined products to yield high recoveries of uranium and vanadium was made.

The extraction of vanadium was found to increase when the ore was roasted at around 850°C in the presence of a calcium salt which is more acidic in reaction than $CaCO_3$ and does not decompose to CaO at the roasting temperature. The charge, however, needed to be maintained in an acidic condition during the process of roasting in order to prevent the simultaneous reduction in uranium extraction. The optimal results obtained for ores with low lime content were at roasting temperatures of 850°C in the presence of 3–5% $CaSO_4$. Between 90 and 95% U_3O_8 and 70–80% V_2O_5 were extracted from the roasted charge in the subsequent leaching step. Ores high in $CaSO_4$ required no reagent addition prior to roasting and were readily leached. For ores with high lime content, an increase in roasting temperature to effect the solubilization of vanadium resulted in very poor uranium extraction. A uranium recovery of 90–95% and 40–60% vanadium could be attained by roasting between 500 and 600°C which is below the decomposition temperature of $CaCO_3$. Additional advantages of roasting the carnotite ore prior to leaching are the improvement in the settling and filtering characteristics of the ore, the vaporization

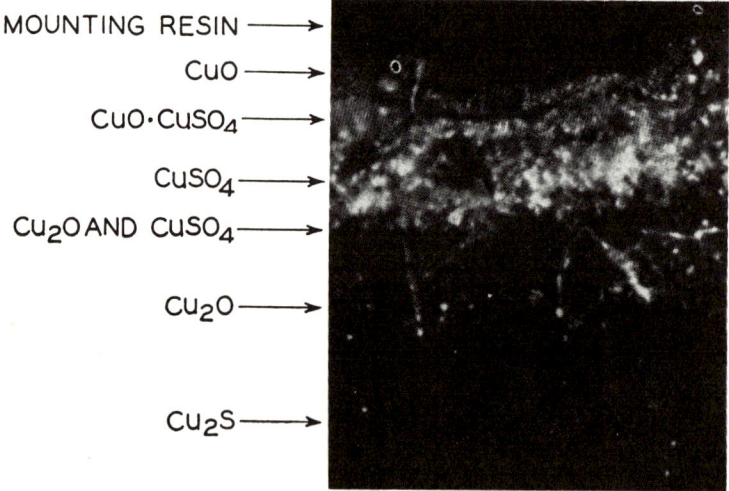

FIG. 1. Micrograph shows outer oxide and sulfate layers formed during the roasting of Cu_2S. Unetched specimen was photographed under polarized light, ×100. Area reduced approximately 10% for reproduction. McCabe and Morgan (M20).

of the ore's carbonaceous matter which would otherwise contaminate and consume excessive leaching reagent, and the appreciable increase in the recovery of vanadium from the ore.

The experience gained from the roasting of sulfide ores has established that the lower the temperature and the higher the pressure of SO_3, the more sulfate will be formed in the product. McCabe and Morgan (M20) have demonstrated that the sulfate does not form at the sulfide surface and that the reaction at the sulfide surface is the same regardless of whether sulfate is formed or not. An examination of a cuprous sulfide cube (made from reagent grade Cu_2S powder) which had been roasted at 600°C in the presence of oxygen gas for 24 hr resulted in the various layers of copper compounds shown in Fig. 1. The compositions of these various layers have been confirmed by Debye–Scherrer X-ray patterns. The mechanism proposed for the sulfate formation during the roasting of cuprous sulfide is based on the assumption that the rate-determining step is the diffusion through the oxide and sulfide layers, rather than the reaction at the sulfide-oxide interface. The following is the sequence of reactions.

(1) The cuprous sulfide, Cu_2S, is oxidized with oxygen which has been transported through the oxide–sulfate layer by either gaseous or solid-state diffusion. The reaction products at the Cu_2S surface are Cu_2O and SO_2.

(2) SO_2 diffuses outwardly through the pores in the Cu_2O until it reaches a place in the Cu_2O where the oxygen and the SO_2 pressure are thermodynamically favorable for the formation of $CuSO_4$. The Cu_2O reacts with the mixture SO_2, SO_3, and oxygen to form $CuSO_4$.

(3) The basic sulfate is formed by the decomposition of $CuSO_4$ with the evolution of SO_3 or SO_2 and O_2 since the SO_3 gradient will decrease toward the oxygen gas phase from the $CuSO_4$ layer.

(4) The basic sulfate will decompose to form CuO for the same reason.

III. Leaching

The factors that determine the choice of solvent for a given ore are dependent on the nature of the mineralization and association of the metal values with the unwanted bulk of the ore. Since in the process of dissolution of the desired metal, other metals are also simultaneously going into solution, the minimization of the unwanted side reactions becomes very important in the final choice of a solvent. It must be remembered that the concentration of the recoverable metal in the ore very seldom exceeds 1% and in most cases is only a fraction of 1%. In the dump leaching of copper waste dumps, an average value of 0.30% is considered good. In the case of uranium, 0.10% or more U_3O_8 is considered ore; and for gold bearing ores, a concentration of 0.015% (5 oz/ton) is economically attractive.

The compositions of ore minerals and other raw materials used as sources of metals are shown in Table I.

A closer look into the principal considerations involved in the development of the leaching process for uranium will help elucidate the above conditions. Gaudin (G1) brought out the following significant features.

Prior to the atomic age uranium was leached in a small-batch-scale operation and was recovered as a by-product of vanadium or radium processing. With the increased demand for uranium, the lower-grade deposits became economically acceptable. The lower concentration of uranium in the ore made large-scale operations a must in order to maintain economic viability. This in turn led to the development and use of the continuous leaching process. Improvements in the analytical detection techniques had to be developed in order that highly accurate process control of the operation could be instituted. In a large-tonnage operation, a fraction of 1% uncertainty in the chemical analysis could mean the difference between profit or loss. The leaching process had to be carried out at ambient temperature and pressure whenever feasible and the choice of the solvent had to be limited to relatively noncorrosive leach liquors.

The two most important sources of uranium are the minerals carnotite, where uranium occurs in the hexavalent oxide or hydrated oxide, and pitchblende, where uranium occurs mostly in the tetravalent state as a compound salt with other metals. It also occurs as a mixed oxide with titanium, thorium, and niobium in the tetravalent form. The tetravalent uranium minerals appear to have been geologically formed in the presence of reducing agents such as hydrocarbon minerals, graphite, native metals, and sulfide minerals, while such association is rarely observed with the hexavalent uranium minerals.

The hexavalent uranium compounds more easily form water soluble complexes than do the tetravalent compounds. In order to carry out successfully the dissolution of the tetravalent uranium minerals (or of the hexavalent uranium minerals admixed with reducing agents), the introduction of an oxidizing agent such as manganese dioxide, ferric ion, chlorine, chlorate, or nitrite ion is required. Some of the more effective oxidizing agents for an acid leach are MnO_2 and $NaNO_3$. Ferric iron has chemical properties similar to those of uranium, thus making the separation step more of a problem. The corrosive properties of chlorine-containing substances limited their usefulness and the more common oxidizing agents although very effective are impractical from an economic consideration.

TABLE I
Principal Ore Minerals of the Metals[a]

Mineral	Composition	Pure mineral	Ore[b]
Aluminum			
Bauxite	Mixture of hydrous aluminum oxides		Major source of aluminum
Antimony			
Stibnite	Sb_2S_3	71.4% Sb	Chief ore of antimony
Arsenic			
Arsenopyrite	FeAsS	46% As	Ore of arsenic
Chromium			
Chromite	$FeCr_2O_4$	68% Cr_2O_3	
Copper			
Chalcopyrite	$CuFeS_2$	34.5% Cu	0.70% Cu for
Chalcocite	Cu_2S	79.8% Cu	sulfides are
Bornite	Cu_5FeS_4	63.3% Cu	considered ore
Tetrahedrite	(Cu, Fe, Zn, Ag)As_4S_{13}		
Enargite	Cu_3AsS_4	48.3%	
Antlerite	$Cu_3(OH)_4SO_4$		
Gold			
Native gold	Au		> 0.3 oz. Au/ton presence of silver
Iron			
Hematite	Fe_2O_3	70% Fe	Most important iron ore
Magnetite	Fe_3O_4	72.4% Fe	Important iron ore
Geothite	$HFeO_2$	62.9% Fe	
Taconite	Silica rich iron ore		Important iron ore
Lead			
Galena	PbS	86.6% Pb	Primary ore of lead, important source of Ag
Cerrussite	$PbCO_3$	83.5% PbO	Important ore of lead
Anglesite	$PbSO_4$	73.6% PbO	Lead ore
Magnesium			
Magnesite	$MgCO_3$	47.8% MgO	Source of magnesia
Sea Water			Primary source of Mg
Manganese			
Pyrolusite	MnO_2	63.2% Mn	> 45% Mn is ore
Manganite	MnO(OH)	62.4% Mn	
Mercury			
Cinnabar	HgS	86.2% Hg	Major source of Hg

continued

TABLE I—continued

Mineral	Composition	Pure mineral	Ore[b]
Molybdenum			
Molybdenite	MoS_2	59.9% Mo	
Nickel			
Pentlandite	$(Fe, Ni)_9S_8$		> 1.2% Ni is ore
Garnierite	$(Ni, Mg)SiO_3 \cdot nH_2O$		> 1.0% Ni is ore
Platinum			
Native platinum	Pt		
Silver			
Native silver	Ag		
Argentite	Ag_2S	87.1% Ag	Important silver ore
Tin			
Cassiterite	SnO_2	78.6%	Principal ore of tin
Titanium			
Ilmenite	$FeTiO_3$	31.6% Ti	Titanium ore
Rutile	TiO_2	60% Ti	
Tungsten			
Wolframite	$(Fe, Mn)WO_4$	76% WO_3	Principal tungsten ore
Scheelite	$CaWO_4$	80.6% WO_3	Presence of Mo, 0.4% WO_3 is ore
Uranium			
Uraninite or Pitchblende	UO_2		Presence of rare-earths > 0.5% U is ore
Carnotite	$K_2(UO_2)_2(VO_4)_2 \cdot 3H_2O$		
Vanadium			
Carnotite	$K_2(UO_2)_2(VO_4)_2 \cdot 3H_2O$		Vanadium ore
Zinc			
Sphalerite	ZnS	67% Zn	Principal zinc ore
Smithsonite	$ZnCO_3$	64.8% ZnO	
Hemimorphite	$Zn_4(Si_2O_7)(OH)_2 \cdot H_2O$	67.5% ZnO	
Willemite	$Zn_2(SiO_4)$	73% ZnO	

[a] From Hurlbut (H33).
[b] Ore grade is dependent on other metal values present.

Inasmuch as the raw material involved in a leaching process is a heterogeneous mixture of a small quantity of the valuable component and a large quantity of the unwanted material, the development of chemical techniques to leach selectively only the desired or more valuable materials will be a major step forward in any hydrometallurgical processing. To some degree, this has already met with some success in the leaching of very-low-grade ores and of beneficiated concentrates.

A. Atmospheric Pressure Leaching

The majority of the ore tonnage treated by hydrometallurgical means is processed at atmospheric pressure because of the obvious economic advantage. In the case of the secondary recovery of additional copper values from mine waste dumps, mined-out ore bodies, and marginal deposits of ores with complex mineralization, the only economic process has to be at ambient pressure.

1. *Dissolution Media*

a. Acid Solution. The final choice of a dissolution medium for any process is dictated by economics. Most materials in the presence of strong acids such as sulfuric, hydrochloric, nitric, or hydrofluoric readily go into solution at ambient temperatures. In some instances a slight increase in temperature increases the effectiveness of the leaching operation. Sulfuric acid is the favored acid solvent for most leaching processes because of its effectiveness in reacting with most of the metals, their sulfides and oxides, its availability and low cost, and its less corrosive properties than either nitric or hydrochloric acid.

Cohen and Ng (C21) described a process of improving the chrome–iron ratio of reduced pelletized chromite concentrate by acid leaching of iron without significant loss of chromium. A 5% sulfuric acid solution was used to quench and leach the fired pellets from 400°C to room temperature.

The dissolution of cuprite (Cu_2O) in sulfuric acid is known to be by an oxidation–reduction couple, which in the absence of oxygen results in the formation of 50% metallic copper and 50% cupric (Cu^{++}) ion. The mechanism of dissolution taking place during the leaching of cuprite in sulfuric acid at various temperatures was reported by Wadsworth and Wadia (W3). Diffusion was eliminated as the slow step in the overall reaction by stirring. Two simultaneous dissolution reactions were observed, both of which were dependent upon the hydrolytic adsorption of H_2SO_4. The first was attributed to the thermal decomposition of a surface site containing adsorbed H_2SO_4. The second was attributed to the reaction between H^+ ion and a surface site containing absorbed H_2SO_4. The activation energies calculated for the two processes are 10,300 cal/mole and 9,935 cal/mole respectively.

Under certain conditions natural acidic mine waters have successfully been used in leaching copper waste dumps (J2, S10). Fresh water, during a few recycling stages through the dump, reacts with the soluble sulfates, nitrates, and halides present in the dump to develop acidic properties. For increased efficiency and control in the process, sulfuric acid is commonly added to maintain a fixed leaching solution pH. After precipitation of the copper from the leach solution by reaction with scrap iron, the barren solution is recycled back to the dump.

A major source of water for leaching is the normally acidic run-off water from the mining operation itself. O'Leary (O2) describes a technique of cleaning up to 900 gal/min of mine and tailings water at the Butte operation of the Anaconda Company. The suspended solids, which at times are in excess of 2% by weight, are equivalent to about 50 tons of mud per day. Since copper is recovered from this stream by precipitation and the barren liquor is used to leach the mine waste dumps, solids, and colloidal materials have to be removed. With a flocculant, a 98% reduction in solid content is attained with a feed stream containing 0.22% solid.

Under certain conditions, there are definite advantages in using hydrochloric, nitric, or other acids to carry out a dissolution step. In their evaluation of proposed processes for the recovery of alumina, Peters *et al.* (P8) cited earlier experimental work which showed that both hydrochloric and sulfuric acid are equally good in extracting alumina from calcined clay (T10). In the separation of the leach liquor from the silica residue by filtration, the chloride solution rapidly separated, while the sulfate solution did not separate easily. In addition to ease of filtration, the hydrochloric acid leach also made the later removal of iron easier. The insolubility of titanium dioxide in hydrochloric acid also eliminated another separation problem. Under this particular situation, hydrochloric acid was the natural choice. As in most large leaching operations, the acid would be recovered and recycled.

An improved process for the recovery of precious metals from silver refinery slimes using nitric acid has been developed by Tougarinoff *et al.* (T13). The nitric acid dissolves palladium, rhodium, indium, and silver, leaving gold and the bulk of the platinum as gold sand. The separation of palladium, rhodium, and indium from silver is accomplished by selective thermal denitration of the nitrates. The nitric acid is recycled to the electrolytic cell as silver nitrate and the balance is recovered from the nitrogen oxides evolved during the denitration. The nitric acid leach gives better separation than the conventional sulfuric acid process, resulting in easier refining of the metals and increasing the recovery of the gold in the electrolytic step.

Baroch *et al.* (B5) used a nitric acid leach in their proposed process for treatment of bastnaesite ore to recover rare earths. The ore contained 0.1% ThO_2, 25–35% calcite, 10% rare earth oxides, 15–20% silica, and 30–40% barite. The rare earth assay of their ore was 50.7% CeO_2, 4.2% Pr_6O_{11}, 11.7% Nd_2O_3, 1.3% Sm_2O_3, and 34.3% La_2O_3. The ore was calcined at 900°C for 1 hr to reduce the $CaCO_3$ to CaO effectively. Leaching the calcined bastnaesite for 8–10 days with 30% nitric acid gave a rare earth recovery of only 45–50%. Increasing the nitric acid concentration to 57% for a retention time of 1 hr dissolved all of the rare earths. The combined rare earth nitrates were recovered by solvent extraction with tributyl phosphate. Approximately 60% of the nitric acid consumption was by the formation of calcium nitrate and other by-products. Reacting this with sulfuric acid regenerated the nitric acid. Over 98% of the rare earths could be covered by this process.

b. Basic Solution. The use of a caustic solution as a leaching reagent is not as common as the use of a sulfuric acid solution because a high concentration of sodium hydroxide is necessary to carry out the dissolution of the silica and other silicates normally present in the ores. Then the spent liquor must be treated with lime in order to precipitate the dissolved silicates and thereby regenerate the sodium hydroxide for recycle.

Two methods of approach are available in upgrading and extracting metal values from their raw materials. The first is by dissolving the desired metal values, followed by separation and concentration stages, and the other is by putting into solution the undesirable components, leaving a residue rich in the desired metals to be used as high-grade raw material in standard metallurgical processes. Alkaline leaching has found application in both approaches.

In an attempt to upgrade the siliceous bauxites and other aluminous materials, such as clay where the alumina usually is in the form of complex silicates, Skow and Conley (S19) considered the removal of the silica by caustic digestion. Prior to leaching, the aluminuous materials were calcined at 970°C, resulting in an optimal removal of silica with little loss in alumina content in the subsequent leaching step. The desilication step is necessary prior to using the raw materials as feed to the standard Bayer process (E3). The pilot plant operation to extract alumina from an anorthosite ore has been reported by St. Clair *et al.* (S24).

Lundquist (L17) proposed a leaching scheme in which silica and some silicate minerals are removed by sodium hydroxide solution from very-low-grade manganese ores and manganese flotation concentrates, leaving a product with an upgraded manganese content. Separation of the residue by filtration was carried out at 60–70°C. The spent leach liquor was regenerated for recycling by reacting it with newly calcined lime. Perkins (P5) reported the pilot plant results of this method and produced a product with low enough silica and alumina content to meet metallurgical-grade specifications.

Extensive work on the upgrading of iron ore containing high concentrations of silica, alumina, calcite, phosphorus, and sulfur by removal of up to 80% of the silica by warm caustic leach has been reported by Herzog and Backer (H28). As much as 40% of the ferrous iron compounds were dissolved, while ores containing only the ferric iron compounds had negligible solubility. The carbonates in the ore were decomposed, the evolved carbon dioxide gas being precipitated as sodium carbonate which dissolves during the washing of the concentrate. The formation of hydrated lime from the calcite made available a binding reagent to help agglomerate the concentrate by cold compression, in addition to increasing the basicity index. Up to 95% of the CaO was left in the concentrate. As much as 90% of the alumina and phosphorus contents were put into solution. The vanadium and sulfide contents were completely dissolved.

A very important result of this work was the transformation of the iron oxides from nonmagnetic to magnetic forms. Besides making possible the production of high-purity iron concentrates by magnetic separation, other possible applications include the recovery of iron from the red mud formed during the production of alumina from bauxite by the standard Bayer process, and the elimination of iron oxides and phosphorus from low-grade manganese ores to produce a high-quality metallurgical grade concentrate. The difficult problem of filtration normally associated with silicates in solution has been bypassed by magnetically removing the iron oxide particles.

Tiemann (T8) studied the dissolution of silica from a siliceous iron ore by sintering the ore with sodium carbonate followed by leaching the sodium silicate with water. The reaction rates were found to be low after sintering for 4 hr at 1450°F. The residual concentrate was analyzed to be 56% iron, corresponding to 88% dissolution of the silica. Partial to complete fusion resulted when the temperature was increased.

In the alkaline leaching of uranium ores with sodium carbonate, the uranium must be present in the hexavalent state. An oxidation step, such as making use of the oxygen in the air at temperatures above 71°C or adding an oxidizing agent such as potassium permanganate, is required when uranium is present in the tetravalent state. Stephens and Macdonald (S25) discussed carbonate leaching methods used by several of the industrial operations in the United States. The advantages of using an alkaline leach are: (1) the corrosion problems are negligible compared to an acid circuit; (2) in most instances, the spent liquor for leaching can be recovered for reuse; (3) where uranium is extracted from limestone and high-lime ores, the high consumption of acid requires that a carbonate leach be used; (4) the other metal impurities are not dissolved, resulting in selective leaching of only uranium and vanadium; (5) the relatively low cost of sodium carbonate; and (6) the negligible disposal problem involved in an alkaline circuit. The disadvantages include: (1) it usually is not as effective as an acid leach; (2) ores high in sulfides or gypsum require high consumption of the reagent and, therefore, are more suitable for an acid leach; (3) the solid–liquid separation (filtration and thickening) is more difficult; and (4) some of the more complex uranium minerals such as euxenite, betafite, samarskite, brannerite, and davidite have negligible solubility in carbonate solution. A prior fusion step usually breaks down the complex mineral bonding to effect satisfactory solubility in the alkaline reagent.

Butler (B39) described the leaching of carnotite ores that contained high-vanadium–low-lime and low-vanadium–asphaltic-type minerals. The high-vanadium–low-lime ore (less than 2%) is roasted with 6–9 wt. % salt at 850°C and is immediately quenched in 3% hot sodium carbonate solution. The calcine is ground to 65 mesh and uranium and vanadium are dissolved by agitation leaching at 92–96°C. 93% of the uranium and 85% of the vanadium are extracted from the ore. The asphaltic type ore is roasted at 550°C and 80–85% of the uranium and 35–40% of the vanadium are extracted. The leach solution concentration was maintained between 7 and 9% sodium carbonate. The pulp density was between 45 and 50% solids.

c. Bacterial Media. The successful application of bacteriological leaching to low-grade ores and mines that have run out of economic grade ores in order to recover metals that normally would be discarded or left in place has extended and increased the tonnage extracted from the dwindling natural resources (M9). The technology of biological leaching (K10) is only now being understood, although the same mechanism of autotrophic bacterial oxidation of the sulfide minerals to their water soluble sulfate form has been going on for centuries. The leaching of copper at Rio Tinto in Spain has been going on since 1725 and the success of the process was for years attributed to something unique, but unknown to that part of the world. It was not until 1962 that the presence of bacteria in their leach waters was determined (T17).

The natural oxidation of the sulfide minerals by oxygen in the air is dependent on many variables such as temperature, the intensity of the light, the pH of the water, the presence of oxidizing agents in the ore, the mineralogy of the ore, and the presence of a catalyst that would increase the reaction rate.

In 1929, Sullivan (S30) found that the oxidation of the sulfide minerals by air alone is very slow, but is accelerated in the presence of dilute sulfuric acid and ferric iron.

The isolation of the species *Thiobacillus thioxidans* from the soil was accomplished by Waksman and Joffe (W4) in 1922. This autotrophic bacterium derives its energy from sulfur, sulfur dioxide, and thiosulfate; its carbon requirement is derived from carbon dioxide. The outstanding characteristics of this microorganism are its ability to withstand very high acid concentration and the ability to grow rapidly at the pH range of 2–3. At about the same time, Rudolfs and Helbronner (R16) in their studies on the extraction of metals from low-grade sulfide ores detected the presence of unidentified sulfur-oxidizing microorganisms which converted iron sulfide or pyrite and zinc sulfides to the sulfate.

The pollution of streams by acidic water and high concentrations of iron near bituminous coal mines where pyrite is also present led to the isolation of a new autotrophic iron bacteria by Colmer *et al.* (C23). This bacterium belonging to the genus Thiobacillus was given the name *Thiobacillus ferroxidans* (T2). It rapidly oxidizes ferrous ions in acid solution, does not grow on sulfur, uses thiosulfate as the sole energy source, and increases the formation of acid from pyrite (T3). Leathen and Braley (L3) isolated *Ferrobacillus ferroxidans* in 1954, a bacterium with the ability to oxidize rapidly ferrous iron to the ferric state at pH 2–4.5 without appreciably oxidizing acid thiosulfate and sulfur (L5, S17). An autotrophic bacteria that uses both ferrous iron and sulfur as energy sources, called *Ferrobacillus sulfoxidans*, was isolated by Kinzel (K6) in 1960.

The Kennecott Copper Corporation received a patent for the leaching process using iron-oxidizing bacteria in 1958 (Z2). The streams near Bingham Canyon, Utah, have been found to contain two strains of autotrophic bacteria, the *Thiobacillus thioxidans* and the *Thiobacillus ferroxidans* (B31–B33). These were found to grow very slowly on thiosulfate media and to obtain their energy by the oxidation of ferrous to ferric ion or sulfur to sulfate (B10). In a survey of copper mine waters in Arizona, Corrick and Sutton (C25) identified the presence of: (1) *Thiobacillus concretivorus*, a sulfur oxidizing bacteria that derives its nitrogen from nitrates and ammonium ion; (2) *Ferrobacillus ferroxidans* which uses iron; and (3) a microorganism that uses both ferrous iron and sulfur, referenced under catalog No. 13728, American Type Culture Collection in Washington, D.C.

Three mechanisms have been proposed by which these bacteria accelerate the leaching process of minerals, specially sulfide minerals. In their patent, Zimmerly et al. (Z2) claim that the iron-oxidizing bacteria help release the metals directly. In the case of iron pyrites, the bacteria oxidize the ferrous iron to the ferric sulfate and at the same time produce an excess of sulfuric acid. Since the precipitation of ferric iron as sulfate in the copper mine waste dump must be minimized, the practice at Kennecott is to add sulfuric acid to the leaching solution to suppress the biological oxidation of ferrous iron.

Other investigators (I4, T3) contend that the bacteria biologically oxidize the ferrous to the ferric state as sulfate, which in turn attacks the metal sulfides ores, resulting in both the dissolution of the metal and the oxidation of the sulfide. In the process the ferric iron reduces back to the ferrous state. An increase in the pH causes the basic ferric sulfate to precipitate (self-buffering action) with corresponding release of sulfuric acid. This has the cumulative effect of increasing the acidity of the mine water.

The possibility that the direct action of the bacteria on the sulfide minerals of copper and nickel results in metal release, and the possibility that the oxidation to ferric sulfate increases the biological dissolution rate have been proposed by Razzell (R6).

More and more minerals are being found amenable to bacteriological leaching. The copper sulfide minerals, such as chalcopyrite (B31–B33, D22, D24), chalcocite (B35), and tetrahedrite (B32, D21) are among the best studied. The iron sulfide (pyrite) (B31, B33, C22, L4) and sulfur (B33, B34, C22, L4) oxidation processes are the best understood. Investigations on the leaching of nickel sulfides (D21, D24, T17), lead sulfide (E4), molybdenum sulfide (molybdenite) (B17, B31, D24), cobalt sulfide (D9), zinc sulfide (D24), and uranium oxide (D24, F2, H13, H14, M1) have been reported in the literature.

In addition to application of bacterial leaching to low-grade ores (C26), and mill tailings (M9), the dissolution of several sulfide minerals in mill products, including pyrite cinders has been reported (D23). Kajic (K1) describes a process that may make economically feasible the dissolution of minerals in the presence of acid-consuming gangue materials. An inorganic nitrate solution at a pH 7–9.5 under anaerobic conditions in the presence of denitrifying bacteria is used to treat the ore. The nitrate is reduced to nitrite and ammonia. This process is ideal for metals that form amine complexes in this pH range, such as copper and zinc. The energy requirement of the bacteria can be provided as nutrient hydrocarbon. In the presence of toxic materials in the leach liquor, a tolerant strain of bacteria will have to be bred.

The only metals presently recovered on a commercial scale from their minerals using bacterial leaching as part of the process are copper (Z2) from mine waste dumps and uranium (M2) from old mine workings and mine waters.

d. Cyanidation. Until recently, the gold and silver industries have been the only metal users of aqueous cyanide solution as a leaching reagent. Mellor (M26) mentioned that the jewelers of the eighteenth century probably took advantage of this solubility of gold by using it in gilding. He also reported that the solubility of gold in aqueous cyanide solution was noted by Scheele in 1783 and by Bagration in 1843. Elsner (E7) in 1846 was the first to discover that oxygen was necessary for the dissolution process. It was not until 1887 that the application of this knowledge to the recovery of gold and silver from their ores was introduced with the issuance of a patent to McArthur *et al.* (M18, M19) for their cyanidation process which replaced the amalgamation process then in use.

The mechanism of the dissolution of gold and silver by cyanide remained in dispute until 1966 when Habashi (H2) presented conclusive evidence that the overall chemical reaction representing the dissolution of gold in cyanide is

$$2\,Au + 4\,KCN + O_2 + 2\,H_2O \rightarrow 2\,KAu(CN)_2 + 2\,KOH + H_2O_2$$

The dissolution of silver in aqueous cyanide solution can be represented by an analogous chemical reaction. A comprehensive review of the theoretical aspects of the cynidation process was made by Habashi (H3) in 1967.

The reactions of cyanide to various minerals are summarized in a report by Hedley and Tabachnik (H21). The conditions for easy extraction of gold from its ore include a high degree of liberation of the gold minerals, a cyanide solution free of impurities, and a sufficient amount of oxygen available in the solution during the reaction. Quartz, silicate minerals, and alkali metal carbonates are relatively insoluble in cyanide solutions. The other metals that are normally associated with most gold and silver ores cause complications in the chemistry of cyanidation. The solubilities of minerals and metals in cyanide solutions are presented in Table II.

The presence of copper, even in concentration below 0.10% affects not only the dissolution of gold but also the precipitation of pure gold from the solution by zinc. The copper minerals increase cyanide consumption by going into solution and forming the copper cyanogen complex. Zinc in the ore reacts in the same way as copper.

Small quantities of nickel have negligible effects on the dissolution step, but have a pronounced effect on the precipitation of gold. In practice, the nickel content of the pregnant solution is kept below the critical

TABLE II
Solubility of Minerals and Metals in Cyanide Solutions[a]

	Mineral		Percent dissolved in 24 hr	Reference
Gold minerals	Calaverite	$AuTe_2$	RS[b]	(J8)
Silver minerals	Argentite	Ag_2S	RS	(L8)
	Cerargyrite	$AgCl$	RS	
	Proustite	Ag_3AsS_3	SS	
	Pyrargyrite	Ag_3SbS_3	SS	
Copper minerals	Azurite	$2\ CuCO_3 \cdot Cu(OH)_2$	94.5	(L8)
	Malachite	$CuCO_3 \cdot Cu(OH)_2$	90.2	
	Chalcocite	Cu_2S	90.2	
	Cuprite	Cu_2O	85.5	
	Bornite	$FeS \cdot 2\ Cu_2S \cdot CuS$	70.0	
	Enargite	$3\ CuS \cdot As_2S_5$	65.8	
	Tetrahedrite	$4\ Cu_2S \cdot Sb_2S_3$	21.9	
	Chrysocolla	$CuSiO_3$	11.8	
	Chalcopyrite	$CuFeS_2$	5.6	
Zinc minerals	Smithsonite	$ZnCO_3$	40.2	(L8)
	Zincite	ZnO	35.2	
	Hydrozincite	$3\ ZnCO_3 \cdot 2\ H_2O$	35.1	
	Franklinite	$(Fe, Mn, Zn)O \cdot (Fe, Mn)_2O_3$	20.2	
	Sphalerite	ZnS	18.4	
	Gelamine	$H_2Zn_2SiO_4$	13.4	
	Willemite	Zn_2SiO_4	13.1	
Iron minerals	Pyrrhotite	FeS	RS	(H19)
	Pyrite	FeS_2	SS	
	Hematite	Fe_2O_3	SS	
	Magnetite	Fe_3O_4	PI	
	Siderite	$FeCO_3$	PI	
Arsenic minerals	Orpiment	As_2S_3	73.0	(H19)
	Realgar	As_2S_2	9.4	
	Arsenopyrite	$FeAsS$	0.9	
Antimony minerals	Stibnite	Sb_2S_3	21.1	(H19)
Lead minerals	Galena	PbS	SHA	(L10)

[a] Habashi (H3).
[b] Key to abbreviations: RS, readily soluble; SS, sparingly soluble; PI, practically insoluble; SHA, soluble at high alkalinity.

concentration by removal of barren cyanide solution high in nickel. Arsenic and antimony minerals dissolve in the cyanide solution to form thioarsenites and thioantimonates. These react with the oxygen in the cyanide solution to form arsenites and antimonites and leave little or no oxygen for the dissolution of gold.

The presence of carbonaceous material in the ore precipitates the gold in cyanide solutions, thus resulting in an increase of gold content in the cyanidation residue. For such complex gold-bearing ores, a roasting operation oxidizes the carbonaceous matter and thereby makes the ore more amenable to cyanidation. However, the alteration of the associated minerals presents new dissolution problems which must be solved.

Since its discovery the cyanidation of high-grade gold and silver ores has been well discussed in approximately twenty-five books. Continuing research on the subject is focused on the recovery of gold and silver from very-low-grade ores (D14, E8), mine waste dumps (H24, R13), and carbonaceous ores (L6, S4).

The use of cyanide solution as a chemical reagent in the improvement of the recovery of copper during the benefication process in flotation is not new (G2, M41, S11). Its use as a solvent in leaching copper-bearing ore fractions is once more being reexamined (D25, H20, L9). Lower and Booth (L16) reported the recovery of copper by a cyanidation–precipitation–regeneration technique. Compared to the cyanidation of gold, the leaching of copper minerals required higher concentration of cyanide and shorter reaction periods, but no oxidation step was necessary. The copper dissolves in cyanide to form cyanide complexes with $[Cu(CN)_3^{-2}]$ being predominant (H18, H21, P4). The precipitation of copper is accomplished by the addition of acid to the copper-bearing pregnant solution; this causes reversal of the dissolution reaction. The cyanide is regenerated by the acidification step, being converted to hydrocyanic acid. This is recovered by air, inert gas, or steam stripping after precipitation.

Rose et al. (R13) described the operation of a 24-ton/day pilot plant for a patented cyanide leaching process that uses a cyclic operation of leaching, copper precipitation by acidification, and regeneration of the cyanide. The copper is recovered from flotation sand tailings containing an average of 4 lb copper/ton. About 90% of the copper was leached from the material with negligible loss of cyanide ions in closed oxygen-free tanks. There is also renewed interest in studying the effect of ammonia in cyanide solution on the leaching of copper-bearing ores (C24, J4, L7, L16, S34).

e. Aqueous Ammonia Solution. Some of the common metals such as copper, zinc, cobalt, and nickel are known to react with aqueous ammonia to form metal amine complex ions. The stability, solubility, decomposition reactions, and other chemical properties of these metal amine complexes are well discussed by Bjerrum (B20). As early as 1917, Kennecott (D20, E2, L2) was leaching by percolation techniques copper carbonate mill tailings assaying 0.80% copper with ammonia–ammonium carbonate solution. Metallic copper, averaging 0.4% in mill tailings at Calumet and Hecla (B12–B15) was being leached with aerated ammonia–ammonium-carbonate solutions.

Caron (C7) received the basic patent for ammonia leaching of nickel ores in 1924 and it was applied on a large scale in the Nicaro operations in 1943. Caron (C8) reported the results of ammonia leaching tests carried out on different nickel and cobalt ores originally formed from weathering of peridotites or similar basic rocks such as iron laterites. The ammonia leaching process essentially consists in reducing nickel or cobalt oxides to metal with H_2, CO, or H_2S at temperatures up to 950°C, dissolving the reduced ore at ambient temperatures with ammonia–ammonium carbonate solution in the presence of oxygen from the air, and recovering relatively pure nickel and cobalt as hydroxide or carbonate and copper as the oxide by a simple boiling operation to recycle the NH_3 and CO_2. The iron is reduced to magnetic (Fe_3O_4) and, being insoluble in ammonia, is readily separated from the solution. A high degree of selectivity during the leaching process is possible since the reaction of ammonia to nickel and cobalt is very specific. Dufour and Hills (D19) have described the pilot plant and development of the ammonia leaching process for the recovery of nickel oxide from lateritic iron ores at Nicaro. The fact that metallic nickel reacts energetically with ammonia and ammonium salts in the presence of oxygen determined the sequence of process steps in the design of the commercial plant. In the reduction of the ore prior to leaching, the conditions were controlled so that iron and magnesium do not become soluble in ammonia. The choice of a mixture of ammonia–ammonium hydroxide solution as the leaching reagent was based on the fact that basic ammonium compounds exert the best solvent action on metallic nickel. Multiple leaching stages provided a means of increasing the nickel concentration in the pregnant liquor without decreasing the nickel recovery.

The extraction of nickel from Cuban lateritic ores on a commercial scale has been described by Baragwanath and Chatelain (B4). The Nicaro nickel project was operated for almost two years during the Second World War and produced 2,600,000 lb/month of nickel oxide. The laterite ores, containing 1.5% nickel, were dried to reduce the moisture content of the ore from 28 to 2.5%, ground to 90% −100 mesh, roasted in a reducing atmosphere, and leached with liquid ammonia. The leach solution was steam distilled to recover ammonia and precipitate the nickel as basic nickel carbonate. The ammonia was regenerated and the nickel carbonate calcined to produce the nickel oxide.

The leaching was carried out in thickeners that were covered with dome-shaped roof to prevent ammonia losses. Compressed air was introduced into the thickeners to dissolve the nickel and oxidize and precipitate the iron in solution. The ammonia saturated air from the thickeners was passed through absorption towers where it was scrubbed with water to recover the ammonia.

The production of lead using ammoniacal ammonium sulfate solutions to dissolved oxidized lead compounds has been reported by Bratt and Pickering (B27). The lead sulfide present in the ores, concentrates, or mattes is converted to lead sulfate, lead monoxide, or a basic lead sulfate by roasting or by oxidation in aqueous suspension. Lead carbonate can be treated by thermal decomposition or acid treatment. Lead silicate can be treated by acid treatment. The pretreated material is leached under ambient conditions, yielding pregnant liquor containing up to 100 gm/liter Pb and a barren residue. The precipitation of lead can be carried out by electrolysis to produce the metal, by removal of free ammonia, or by heating of the supersaturated leach liquor to produce ammonium sulfate, monobasic lead sulfate or a mixture of the two. The remaining ammoniacal ammonium solution is recycled to the leaching stage after removal of other accumulated metals such as copper and zinc.

Panlasigui and Wheelock (P2) investigated the rate of dissolution of copper by oxygen and by the cupric ammine complex as a function of oxygen, copper, ammonia, and ammonium carbonate concentrations, and stirring velocity using the rotating disk method. The dissolution rate was found to be first order with respect to oxygen and the cupric ammine complex and is limited by the transport of oxygen and/or of the cupric amine complex form the bulk of the solution to the surface of the metal. The diffusivity of oxygen in $0.5\ N$ aqueous ammonia is 5.57×10^{-5} cm^2/sec and that of cupric ammine complex is 8.93×10^{-7} cm^2/sec.

f. Other Solvents. Water is the most common solvent and its application to metal recovery has not been overlooked by researchers. The important properties to be considered are the solubility of the compound in question and the effect of the accompanying hydrolysis reaction. Wilson and Sullivan (W14) described the leaching of copper by water after roasting of blast furnace matte, followed by leaching with brine (sodium chloride solution) to recover lead and smelting of the iron left in the residue. A process to recover sulfur from molybdenite by reduction of the flotation concentrate with aluminum powder at 800°C and leaching the product with water to form an impure molybdenum alloy plus aluminum sulfide has been described by Haver et al. (H15). The aluminum sulfide hydrolyzed to yield pure hydrogen sulfide from which 90% of the total sulfur can be recovered by the Claus process (A5). A 95% recovery of the molybdenum as 99.5% pure molybdic acid was obtained.

The use of brine as a solvent in the hydrometallurgical separation of lead from its ores was extensively studied by Lyon and Ralston (L18). Saturated sodium chloride solution and neutral ammonium acetate solutions were found to be good solvents for lead chloride and lead sulfate. Lead oxide and lead carbonate became soluble if the brine was first acidified with either sulfuric or hydrochloric acid. The dissolved lead was recovered electrolytically (S18). Marsden (M13) used this method in a process to recover lead from zinc plant residues. About 80% recovery of the lead was obtained by leaching at ambient temperature and the recovery of lead was increased to 98% when hydrochloric acid was added to the brine.

The possibility of using brine to slurry the ore in the presence of an oxidizer such as chlorine in order to extract metals from the more common sulfide minerals has been studied by Strickland and co-workers (J1, S12, S13). The reactions of acid chlorine solutions with galena (PbS), pyrite (FeS_2), sphalerite (ZnS), chalcocite (Cu_2S), covellite (CuS), chalcopyrite ($CuFeS_2$), bornite (Cu_5FeS_4), pyrrhotite (FeS), and arsenopyrite (FeAsS) were examined with respect to their reaction rates and mechanisms.

Three primary reactions were observed between aqueous chlorine and the base-metal sulfides. Elemental sulfur was produced during the reaction with chalcocite, bornite, and covellite. A rapid oxidation of the pyrrhotite, pyrite, and arsenopyrite to the sulfate form was observed. The formation of sulfur monochloride was indicated with sphalerite, galena (under most conditions), and chalcopyrite. The ratio of sulfur to sulfate was close to what could be expected if the sulfur monochloride hydrolyzed to form sulfur. Thermodynamic considerations indicated sulfate formation as the primary product.

An examination of the kinetics showed that the consumption of chlorine by the sulfide, with the exception of galena, is first order and transport controlled. The overall rate of reaction with galena is both chemical and transport controlled.

The presence of iron and/or sulfur in practically all of the ores used for metal recovery provides a readily available source of leaching reagents in the presence of water. Acid solutions of iron sulfate and sulfuric acid are produced by natural processes and are readily available in the streams near most mining operations, or in the mine waters themselves. The ability of acidic ferric sulfate and sulfuric acid to oxidize most of the metal sulfides and oxides during the process of dissolution is well known. The ferrous sulfate that is produced needs to be reoxidized to the ferric state if recycling of the reagents is to be done for economic reasons. Methods that have been tried include air oxidation of the spent leach liquor, anodic oxidation, many sulfur dioxide adsorption–oxidation processes, bacterial oxidation, and addition of sulfuric acid to the leaching circuit.

A recent report by Haver *et al.* (H16) described the recovery of lead and sulfur from a lead sulfide concentrate using ferric sulfate as the leaching reagent. Elemental sulfur was produced during the oxidation of lead sulfide to lead sulfate by hot ferric sulfate solution. The lead sulfate in the residue was changed to acid-soluble lead carbonate by treatment with ammonium carbonate, and ammonium sulfate was a by-product. Hydrofluosilicic acid dissolved the lead carbonate and an electrolysis step regenerated the H_2SiF_6 and deposited 99.9% pure lead metal. The recovery for lead was about 90% and for sulfur, 67%, half as elemental sulfur and the other half as ammonium sulfate.

Dutrizac *et al.* have studied the reactions of chalcopyrite ($CuFeS_2$) (D27), bornite (Cu_5FeS_4) (D28), and cubanite ($CuFe_2S_3$) (D29) in acidified ferric sulfate solutions. In all three studies sintered disks of chalcopyrite, bornite, and cubanite with a well-defined surface area were used. Both bornite and cubanite are important copper minerals in their own right and also occur in association with one of the most common of the copper sulfide minerals, chalcopyrite. The use of pure synthetic samples is necessary if one is to obtain information regarding mechanisms that are in effect during the dissolution process in a commercial dump leaching operation.

For chalcopyrite, at temperatures between 50 and 94°C and ferric ion concentration from 0.001 to 0.6 M, dissolution in acidified ferric sulfate followed the parabolic law. This indicates that the reaction was diffusion controlled through a constantly thickening layer of reaction product, possibly sulfur on the surface of chalcopyrite. At ferric ion concentrations less than 0.01 M, the rate of leaching was controlled by the inward diffusion of $Fe_2(SO_4)_3$ and at higher ferric sulfate concentrations, the outward diffusion of ferrous sulfate through the sulfur layer controlled the rate. Changes in acid concentrations and in speed of disk rotation did not have any effect on the leaching rate, indicating that the rate of agitation in a commercial operation would have no effect on the rate of leaching.

The dissolution of bornite in acidified ferric sulfate solutions at temperatures between 5 and 94°C occurred in two steps. A nonstoichiometric bornite with up to 25% copper deficiency was first formed and then converted to both chalcopyrite and elemental sulfur which accumulated on the bornite surface. The reaction followed the parabolic law below 35°C and stopped at the nonstoichiometric bornite stage. Above this temperature, it continued through to chalcopyrite, following a linear process. At increased temperature, the reaction rate was diffusion controlled through the liquid boundary layer as indicated by its sensitivity to the speed of rotation of the disk.

Between 45 and 90°C, the reaction of cubanite with acidic ferric sulfate solutions followed linear kinetics, indicating that the rate-controlling step was some reaction occurring on the surface of the cubanite. The dissolution rate increased with ferric ion concentration and decreased with increasing concentration of sulfuric acid and ferrous sulfate. The naturally slow reaction was accelerated with the addition of NaCl or HCl. The addition of salt in a dump leaching operation would be a relatively easy and cheap procedure to attain increased reaction rates.

The leaching of mercury ores and flotation concentrates is an attractive alternative to the recovery of mercury by the direct retorting process especially when the ore is low grade, wet, high in pyrite, and contains arsenic and antimony. In addition to a low recovery (as low as 60%), the health hazards from the vapors and dust must also be minimized if not completely eliminated. Volhard (E1) in 1878 reported the solubility of mercuric sulfide in basic sulfide solutions. This technique was applied in a commercial scale by Thronhill (T5, T6) with the recovery of mercury from amalgamation tailings in 1915. Additional studies on the caustic sulfide treatment of mercury sulfides by Town et al. (S27, T14) showed that the mercury in cinnabar readily dissolves in sodium sulfide solution, is partly dissolved in potassium sulfide, and is insoluble in ammonium sulfide. Addition of sodium or potassium hydroxide to sodium sulfide slightly increased the extraction of mercury and addition to the potassium or ammonium sulfide brought the extraction up to the level of the sodium sulfide. Potassium sulfide was found to selectively extract 90% of the arsenic and about 10% of the mercury. The balance of the mercury was leached by a solution of 3 parts sodium sulfide and 1 part sodium hydroxide. Applications of this process to several different ores have been reported by Butler (B40).

There are several processes (D8, R5, V3) for the recovery of manganese from low-grade ores by leaching an aqueous slurry of the ore with a gas mixture containing nitrogen, oxygen, and sulfur dioxide. The manganese dioxide is reduced, going into solution, while the sulfur dioxide is oxidized to the sulfate together with the formation of an appreciable amount of dithionate and sulfurous acid. The presence of dithionate in the solutions causes operating difficulties in the recovery of manganese (R2, R4, V3). Back *et al.* (B3) found that the dithionate formation could be controlled by properly controlling ore and sulfur dioxide feed rates, the concentration of sulfur dioxide, the agitation rate, temperature, pH, cell design, and by the use of appropriate chemical reagents. Falke (F1) found that in leaching manganiferous slime and limestones with sulfur dioxide, the iron did not dissolve while the manganese content went into solution.

Herring and Ravitz (H27) studied the rate of dissolution of electrolytic manganese dioxide in sulfur dioxide solutions in the absence of oxygen. The rate of reaction was determined as a function of dissolved SO_2 concentration (0.03 to 0.5 M), stirring speed, pH, and temperature. The dissolution of MnO_2 in SO_2 solution followed two reaction paths; the first involved undissociated SO_2 in which a transport process is the rate-controlling step and the second involved the reaction of bisulfite ions and accounted for less than 10% of the reaction. Under conditions where the transport of SO_2 to the MnO_2 surface was the overall rate-limiting step, the activation energy for the dissolution reaction was in good agreement with the activation energy at infinite dilution for the diffusion of SO_2 in water.

Dean (D5–D7) describes the use of ammonium carbonate, obtained by passing CO_2 into a concentrated aqueous solution of ammonia, for extracting manganese from its ores. It is based on the rapid formation of an aqueous manganese–ammonia complex in the presence of a high concentration of ammonia. Manganese carbonate can be precipitated quantitatively from the anionic manganese complex in solution by heating under pressure, at the same time regenerating the ammonium hydroxide for recycle. A large degree of selectivity during the leaching process is possible because most metal oxides are insoluble in the lixiviant or are not precipitated with the manganese carbonate. Other impurities are controlled by the addition of sulfides. Studies on the extraction of manganese using ammoniacal ammonium carbonate solutions from manganiferous materials (T19), steel plant slags (H22, H23), and low-grade ores (W9) have been reported.

In the treatment of complex sulfide concentrates containing copper, nickel, cobalt, lead, and iron, roasting with a moderate excess of sodium chloride at temperatures up to 400°C prior to leaching with hydrochloric acid has been successfully used by Kershner and Hoertel (K4) in the recovery of more than 95% of cobalt, nickel, and copper. The chloridized product is treated with steam at 300°C to make most of the iron insoluble before leaching at a pH of 1.0. The advantages of a salt roast prior to leaching include the low cost of introducing chloride ion in the form of sodium, calcium, or magnesium salt compared to hydrochloric acid or chlorine when there is a need to break down the complex associated minerals; the corrosion problems are fewer; the roasting temperature is no greater than a straight chloridizing process; the roasting can be made open to the atmosphere compared to a closed system with toxic chlorine or hydrogen chloride gas; the salt reacts with iron to form iron chloride which is oxidized to ferric oxide and liberates chlorine gas which takes part in the chloridizing reaction; and in the presence of sulfur the salt forms sulfur chlorides which are also good chloridizing reagents.

A process using a mixture of sulfuric acid and sodium chloride to produce hydrochloric acid as the leaching reagent for oxidized copper ore has been described by Silo (S16). The cupric chloride is reacted with calcium carbonate powder derived from powdered sea shells to precipitate the copper as a pulp. The leaching solution is regenerated by the addition of sulfuric acid to the residual solution from the precipitation process.

The studies of Bhappu et al. (B18, B19) on the extraction of molybdenum from sulfide and oxide ores showed that oxidizing reagents such as hypochlorite, acid chlorate, and manganese dioxide–sulfuric acid solutions can leach molybdenite (MoS_2). The molybdenum oxide ores, ferrimolybdite $[Fe_2(MoO_4)_3 \cdot nH_2O]$, and molybdenum-bearing limonite, are soluble in strong acid and decompose in strong alkali solution. A sodium carbonate leach appears to be the most effective for the oxide ores since no insoluble molybdate compounds are precipitated and the problem of reaction with acid-consuming waste materials is eliminated.

A hot water leach for lateritic ores that have been mixed with sulfuric acid prior to roasting has been developed by Zubryckyj et al. (Z3). The dissolution of the sulfated ore constituents in water was carried out at about 80°C for about 30 min. The extent to which the nickel and cobalt oxide minerals retained the sulfur trioxide during the roasting operation was enhanced by the use of an activating addition agent like magnesium or sodium sulfate, the presence of oxygen under pressure, and the minimization of the gas velocity in the roasting zone. An ore with a high magnesia content using these procedures with all the other variables at optimal conditions resulted in nickel and cobalt extractions of 85 and 90% respectively. A pilot plant campaign of over six months, continuous operation has confirmed the laboratory results.

g. Nonaqueous Solvents. Forward et al. (F12) developed a process for producing a silver-free lead with a purity of 99.99+% from lead sulfide ores. The lead sulfide is selectively oxidized by reaction with sulfuric acid and oxygen under pressure to lead sulfate, the zinc and iron sulfide remaining unreacted. The lead sulfate and lead oxide are then dissolved in aqueous solutions of alkylene amines at room temperature with the formation of soluble lead amine complexes. The lead is precipitated from the pregnant solution by passing CO_2 through the solution at room temperature. The lead carbonate is fired to produce high-purity lead and CO_2. The amine solution is regenerated by precipitation of $CaSO_4 \cdot 2H_2O$ with CaO. The diethylene triamine put into solution as much as 650 gm/liter of lead. The other metals that form amine complexes such as Cu, Ni, Co, Zn, and Cd are not precipitated by CO_2.

The use of ethylene glycol as the solvent for leaching high-grade scheelite concentrates to produce high-purity tungstic acid from which high-purity tungsten oxide is obtained has been described by Forward and Vizsolyi (F13). The process consists of treating the $CaWO_4$ with sulfuric acid at 100°C to produce a solid mixture of tungstic acid (H_2WO_4) and $CaSO_4 \cdot 2H_2O$, together with SiO_2 and other insoluble impurities. This solid mixture is reacted with recycled ethylene glycol solution containing ethylene chlorohydrin at 100–120°C. The pregnant solution containing from 80–150 gm/liter WO_3 is hydrolyzed by addition of 10–15% water as condensate and introducing enough HCl or chlorohydrin to provide the equivalent of 2–3% HCl in the mixture. Maintaining this concentration up to 72 hr at 70°C precipitates the crystalline H_2WO_4. Calcination of the high-purity tungstic acid produces a 99.90% WO_3, with 0.3% Ca, 0.04% Fe, 0.02% Mg, and less than 0.001% of all other metal, including Mo. This process is much simpler than the conventional techniques now in use (R3).

The direct leaching of uranium ores with organic solvents was an attempt to get around the high acid consumption required in processing some of the ores. Magner and Bailes (M8) described the results of their experiments using different organic solvents on several types of uranium ores. In a proposed process based on laboratory data, the ground ore on a moving belt would pass under a spray of solvents such as a solution of alkyl phosphoric acid in an organic diluent. The solvent, acidified with sulfuric acid, would percolate through the bed of ore and dissolve the uranium. The uranium-loaded solvent would be collected, stripped of its uranium, reacidified with sulfuric acid, and recycled. The entrained solvent in the ore could be displaced by washing with fresh diluent and the diluent removed from the ore residue by evaporation, recovered by condensation, and reused. The process variables to be considered are the type of solvent, quantity of acid addition, type of diluent, phase ratio of solvent to ore, and reaction time during leaching. Satisfactory recoveries have been obtained with carnotite ores, but not with uraninite ores.

2. Methods of Leaching

a. Heap Leaching. The method of heap leaching was first used in Rio Tinto, Spain (T1). This technique has been successfully applied to certain types of ores, such as the highly porous copper oxide ores that are not amenable to upgrading by the presently available methods of mineral beneficiation (H32, I3, M27, S10, S14, S31). The ore as mined is piled in prepared drainage pads and the readily soluble copper oxide is leached with sulfuric acid solution. The leaching process takes months to complete. The leach heaps are deposited either on existing topography when favorable or the drainage pad is prepared by cementing the soil and sealing with dilute tar to minimize the loss of pregnant solutions. The commercial operations have heaps from 100,000 to 500,000 tons of leach material, the dimension of the largest piece of material being no more than 2 ft. The relatively low capital expenditures required for a heap leaching operation make it ideal for small ore bodies such as the Bluebird mine of Ranchers Exploration in Miami, Arizona (M27, S10) which cannot otherwise be economically developed.

The primary drawbacks of this method are the low copper recoveries and the relatively high acid consumptions. Johnson and Bhappu (J7) examined the factors that influenced these shortcomings with the purpose of establishing a basis of correlation between laboratory and field testing. These studies covered laboratory and field heap leaching of several types of copper oxide and oxide–sulfide ores with acid and acid–ferric sulfate solutions. The results indicated that higher recoveries and lower acid consumptions can be obtained by crushing the ore, maintaining a high degree of ore–solution contact, controlling and optimizing the leaching time, and circulating a relatively large volume of low-strength solutions through relatively thin heaps.

b. Tank or Vat Leaching. In tank or vat leaching, the raw material is prepared by crushing and grinding to optimal size before being bedded into large tanks where the solvent is allowed to percolate through the mass of the material in a matter of days. This technique is used in large-tonnage operations to recover copper from nonporous oxidized copper ores or mixed oxide–sulfide ores containing at least 0.5% acid soluble copper (M17, S10, V1). The increased cost due to crushing, grinding, and screening can be offset by the negligible loss of pregnant liquor, higher copper recovery in a few days, and a higher concentration of copper in the solution (as high as 30 gm/liter), making possible the precipitation of high-purity metallic copper (such as by electrolysis) instead of lower-grade cement copper obtained by precipitation with scrap iron. Appreciable tonnages of gold and silver ores (E9, M36) have been treated in this way, in addition to small tonnages of uranium and vanadium.

Tank or vat leaching has traditionally been carried out as a batch process. The eight concrete vats used at the Weed Heights operation of the Anaconda Company (H36) are 120-ft wide, 139-ft long, and 19-ft deep. This operation handles 8000 tons of ore per week. A batch process gives the advantages of operational flexibility and versatility, less dependence on tight process control, and better adjustments to slow reaction which will not decrease extraction of metal from the ore. The disadvantages are the ineffective control of undesirable side reactions, the inability to make optimal use of heat since the heat load in a batch process is not constant, and the long period of time needed to train operators.

Detailed studies on percolation as a leaching process for different minerals, especially the copper minerals, have been reported by Sullivan (S32, S33), and McKinney and Rampacek (M23).

The various factors that influence percolation leaching were examined by Seidel (S6). The percolation flow rate through the ore bed is very dependent on the permeability or texture of the bed. It decreases with increase in air or gas bubbles present in the bed and can be increased by increasing the temperature of the leaching solution. The size distribution of the particle, the porosity, temperature, leaching solution concentration, and percolation rates influence the mass-transfer rate.

Lodding (L15) described an apparatus that may find applications in continuous percolation leaching. The dewatered feed material, together with the solvent, is fed into a large insulated silo which has provisions for the continuous withdrawal of the leached material at the bottom. A water spray pointing at an angle of 45° is used to activate the smooth and uniform withdrawal of material, containing as much as 70% solid for a sand-size material.

In addition to being continuous, the advantages of this design include: the negligible loss of heat, especially if leaching is done at elevated temperatures; a higher concentration of solvent since dilution is only due to the moisture content of the ore; a minimal power requirement since the ore bed moves by gravity; and long treatment time whenever needed.

Donald (D13) presented a comprehensive review of the theory and practice of percolation leaching as applied to a number of materials including metallic ores. The well-documented section on metallic ores includes the practices as early as the eighteenth century. Percolation leaching in tanks, heap leaching, and in-place solution mining of ores from worked out mines up to the present century are reviewed.

c. Dump Leaching. In the large-scale open-pit method of mining, a large portion of the tonnage moved consists of below-grade materials which must be removed in order to reach the ore. The waste materials are moved by trucks or train and then deposited in nonmineralized areas of the mine, taking advantage of existing topography. The selection of the dump site is made on the basis of having an impermeable bottom and being able to direct the downward flow of pregnant solution to a common collection point by taking advantage of the canyons and valleys. When this type of location is not available, special preparation of the surface is made by grading; compacting; adding sized hard material such as slags and coating the compacted slag with asphalt primer, a layer of asphalt and asphaltic sealant, a 12-in. minimum layer of fine dump material, and a 5–6-ft layer of coarse dump material. A very high copper recovery has been obtained by the preparation of such an impervious pad.

The increasing application of dump leaching can be attributed to the large tonnage of below-grade material accumulating every day, the relatively low capital outlay required to put a leaching and precipitation plant on stream, the small labor force required to operate the plant, and the simplicity of the operation itself.

The leaching and precipitation of copper at Rio Tinto in Spain this century is described by Taylor and Whelan (T1); and an extensive report on the present-day technology of copper leaching practices in the western United States is discussed by Sheffer and Evans (S10). Modified practices have been successfully used throughout the world both in low- (S10) and high-rainfall areas (B7).

The introduction of leach solutions into the dump can be made by means of spraying, flooding, or by injecting into the dump through buried vertical pipes. The decision as to which method is most desirable is dependent on several variables such as climatic conditions, height of the dump, surface area of the dump, mineralogy, the size of the operation, the size of the materials, and most important, the experience of the operating personnel (C2, M10, R11, W18).

The complex nature of the leaching of copper sulfide waste dumps was recently the subject of a very comprehensive theoretical and practical analysis by Bhappu *et al.* (B16). The important factors and variables that could be used in the improvement of leaching operations and the understanding of some of the mechanisms taking place during the process were covered both from actual operating experience and relevant simulating laboratory experiments. The results from the simulating laboratory experiments effectively demonstrated the influence of chemical, physical, and biological factors on the actual processes taking place in the dumps. By no means were all the problems solved, but sufficient information on the mechanisms and controlling variables in the various stages of leaching has been developed to make possible the initiation of mathematical modeling of the process.

The quantity of leach solution needed for application each wetting cycle can be calculated if the permeability or void space in the dump is known. At the initial leach cycle, the volume of the leach solution used is minimized to that required to dissolve the copper soluble salts on the surface of the particles in order to obtain a concentrated pregnant liquor. It is estimated that the leach solution needed to fill the pores of the broken ore mass requires a mositure content of 12–15%, which is approximately the same volume of solution needed to saturate the broken ore. This much solution is estimated to remain on the surface after washing. The flooding of the dump or the presence of an excess of leaching solution is unnecessary and not desirable. A period of inactivity to allow the dump to dry must then be calculated based on experimentally determined reaction rates between the solvent and minerals involved, the mass-transfer and diffusion rates, and other parameters which may be found controlling for a given dump. After the drying cycle, the soluble salts on the surface of the particles are removed by a wetting cycle. A continuous cycling of the wet and dry periods will optimize the extraction of copper from the waste dumps.

The necessity of good contact between the copper-bearing materials and the leach solution as it percolates down the waste dump has led to the use of radiation logging. Howard (H31) describes the use of natural γ-ray logging, neutron–neutron or moisture logging, and γ–γ or bulk density logging. The major advantage of radiation logging is the ability of γ rays and neutrons to penetrate the steel casings used in drilled holes. The porosity, saturation, and percent moisture of the material in the dump can be calculated from calibrated bulk density and moisture-content logs, knowing the specific gravity of the material. The unequal distribution of leach solution in the dumps has been shown to be caused by the presence of a layer or clay material, precipitated iron forming an impermeable layer, and the greater compaction of material delivered by trucks than by other means.

d. Solution Mining. The dissolution of the valuable component of the ore at its original place of formation and the subsequent recovery of the pregnant solution at the surface is referred to as solution mining, *in situ*, or in-place leaching.

The best known and most successfully used solution mining technique is the Frasch process for the recovery of elemental sulfur. In the last decade, the focus of attention has mainly been on trona (B9, C1), potash (C14, D3, 03), and uranium (D2, S15) in addition to copper. The technology developed in the oilfields, such as the multiwell pattern (D26) and the hydraulic fracturing techniques (C19), has been adopted with some degree of success to solution mining. Some of the advantages of solution mining for soluble minerals are the use of drilled holes for entry of leach solution and exit of the minerals in solution rather than the excavation and movement of a large tonnage of overburden materials to reach it; the usefulness of the technique above or below sea level, regardless of depth; and its applicability to thin bedded deposits which are normally uneconomical by conventional mining.

The first commercial application of this technique was in the recovery of copper in ore that could not be economically removed by mining (A6, C13, D13, T4, V1). Weed (W8) describes the leaching in place which began in the 1920s at Cananea in Mexico. The various areas being leached, in addition to the waste dumps, were the mined-out and caved underground stopes. The leach solution is sprayed at the top of the caved-out areas and is allowed to percolate down to the bottom where it is collected behind specially built bulkheads. This pregnant solution is pumped to the surface for precipitation of the copper with iron.

The recovery of U_3O_8 by leaching in place has been described by MacGregor (M1). A predetermined cycle of fast, high-pressure washing and rewashing of stopes and the intermittent spraying of low-grade heaps with fresh water and recirculated acid water has been found most effective. The bacteria present in the mine waters were found to develop slowly a tolerance for U_3O_8 as indicated by the steady increase in U_3O_8 concentration of the pregnant solution. The presence of thorium and rare earth metals in the pregnant solution has also been confirmed.

Sievert et al. (S15) examined the current state of the art for the solution mining of uranium. In a large-scale operation, the problems of preventing the loss of solvent from the mining area and the recovery of the pregnant liquor are the major considerations. The complexity of the ore–solution–fluid flow interaction in the reservoir fractures, fissures, and heterogeneities can probably be solved by the creation of hydrodynamic barriers. The experience and knowledge developed on miscible slug flooding, waterflooding, mathematical fluid-flow model, etc., by the petroleum industry are applicable to solution mining.

Davis (D2) reported the results of a field test that involved the isolation of a uranium ore body by surrounding it with a curtain of grout material. The test site was located in a highly permeable, dry sandstone at a relatively shallow depth, with modest uranium content. The technique of filling only the natural porosity with grout material was chosen because the ore deposit was above the water table and had good natural permeability. The flow paths between the injection wells were preheated with hot air before the grouting material, in this case refined oils, was pumped through the pores. The oil appeared to follow closely the path of the heated air between the two injection wells and appeared to stay in place at least a week. The use of the hot oil grouting technique appears to be very promising and its extension to permeable ore bodies beneath the water table is a possibility.

The results of a field test on a thin-bedded potash deposit have been reported by Davis and Shock (D3). Kalium Chemicals, Ltd., in Saskatchewan, Canada (M29), a pioneer in the solution mining of potash since 1965, have been joined by Texas Gulf Sulphur in their Cane Creek property in Utah (C14).

B. Elevated Pressure Leaching

One of the most important developments in the field of hydrometallurgy has been the application of elevated pressures and temperatures to complex sulfide and oxide ores (B21, F8, G8, M5, M6). The pressure-leaching of bauxite ores by the Bayer process (E3) is probably the first successful commercial application of this technique. The bauxite ore is leached with sodium hydroxide solution with a specific gravity of 1.36–1.4 at 160–170°C for $1\frac{1}{2}$–2 hr under a working pressure of 100 psig. The alumina is produced by calcining the aluminum hydrate precipitated from the leach liquor.

The incorporation of both increased pressures and temperatures during the dissolution process has opened a new approach to solving hydrometallurgical problems that could not be handled at atmospheric pressure. Some of the advantages of using this techniques are: (1) the reduction of reagent consumption by generation of the required reagents *in situ*, in the presence of water and oxidizing agent; (2) rapid reaction rates, easily attained, resulting in a shorter reaction period which in turn increases the capacity of the plant; (3) a favorable shift in the thermodynamics of a given reaction brought about by increasing the pressure; (4) the lack of need to use concentrated solvents in most cases because of favorable kinetics; (5) the possibility of a more complete extraction of the metal values from the ore; (6) the possibility of a high degree of selective leaching, resulting in negligible concentration of impurities in the leach liquor; (7) the ability to maintain the concentration of gaseous or highly volatile reagents in the aqueous solution by increasing their partial pressure;

(8) the elimination of the need for raw material pretreatment such as roasting as in the case of the Sherritt Gordon process; (9) the possibility of also recovering the valuable minor metals; (10) the production of high-purity powdered metals; and (11) the small number of steps required in the processing exemplified by the pressure leaching, solution purification, and precipitation in the Sherritt Gordon operations (B25).

1. *Dissolution Media*

   a. Acid Solution. Some of the low-grade mineral deposits containing one or more metal values may justify the development of new separation and concentration processes on the basis of having more than one marketable product. Iwasaki and Carlson (I6) developed a process for a manganiferous iron deposit which was too low in iron to be considered as ore and not high enough in manganese to be upgraded and marketed as concentrate. The iron in this low-grade ore was reduced to metallic iron by reaction with solid carbon fines at about 1900°F. A magnetic separation step recovered the metallic iron in the magnetic product and the manganese in the nonmagnetic tailing. The 2–8% manganese present was converted to manganous oxide during the reduction stage and upgraded to 20 or 30% during the magnetic separation for iron. The formation of manganese silicates during the reduction tied up a good fraction of the manganese which was insoluble in dilute sulfuric acid. It also resulted in the dissolution of a large amount of soluble silica. The dissolution of iron and silica was suppressed by leaching with sulfuric acid at elevated temperatures under oxygen pressure in an autoclave. The manganese silicates formed dissolved under pressure. The iron in the leach solution was eliminated by aeration of the solution above a pH of 5.5. Manganese metal was recovered by electrolysis.

   The extraction of cobalt from arsenical concentrates consisting of autooxidation acid leaching under pressure, separation, purification, hydrogen reduction of ammoniacal leach solution, and removal of sulfur and granulation of the metal was described by Mitchell (M37). The final product contained 95.6% cobalt, 3.90% nickel, and 0.03% arsenic compared to the feed concentrate with an assay of 17.5% cobalt, 1% nickel, and 24% arsenic.

   Forward and Halpern (F7) described extraction of uranium by generating sulfuric acid directly from the sulfide minerals present in the ore at elevated temperatures and pressures. The optimal conditions found for most of the ores studied are a grind of 50–65% −200 mesh, a pulp density of 65% solids, leaching temperature of 130°C, and an oxygen partial pressure of 10 psi to give 90–95% uranium dissolution in 4–6 hr. This technique is of particular advantage with ores that have very high concentration of sulfides which would otherwise consume an excessive amount of solvents and with ores that are refractory in that they need increased pressure to complete the reaction.

The treatment of uranium ores with acid-containing refractory uranium minerals or a complex uranium–rare earth–iron–titanate such as davidite, is best accomplished under elevated pressure and temperature in the presence of oxygen. Gray (G9) obtained approximately 90% dissolution of U_3O_8 from davidite containing 0.7% U_3O_8 by leaching the 150 mesh ore with dilute acid at 150°C and about 100 psig. The acid consumption was only a third of that required at atmospheric pressure and the titanium and iron concentration in solution were effectively reduced.

The leaching of uraninite ore, containing iron and copper sulfide and a siliceous slaty gangue, by *in situ* generation of sulfuric acid during the reaction was also tried. The ore was pulped at 50% solids with water in the presence of 3% MnO_2 as an oxidant. In the absence of oxygen, 29% U_3O_8 was dissolved, this being attributed to the presence of sulfate in the ore. A 94% dissolution of the U_3O_8 was obtained by leaching at 140 psia oxygen partial pressure and 140°C for 2 hr. Negligible amounts of iron and aluminum went into the solution. The required reduction in particle size was about 60%-200 mesh compared to 80%-200 mesh for atmospheric leaching. The optimum economic condition found for this process was leaching at an oxygen partial pressure of 40–45 psi at a temperature of 180–200°C. The use of a continuous bench scale reactor with a capacity of 2 kg ore/hr resulted in a 95% uranium dissolution at 180°C and 40 psi oxygen partial pressure with a retention time of only 50 min.

Most of the common metals such as copper, zinc, lead, molybdenum, nickel, and cobalt are produced from their sulfide minerals. These ores are usually upgraded by standard mineral benefication techniques. The present beneficiation techniques are in most instances close to their upper limit of efficiency. This is probably best demonstrated in the flotation of copper sulfide ores, which are concentrated from about 0.70% to as high as 30% copper. These concentrates are the raw material feed for the presently used pyrometallurgical operations.

In recent years, attention has been focused on the preparation of a concentrated metal solution that can be used directly as a feed to metal winning operations such as electrolysis, reduction with hydrogen under pressure, ion exchange or solvent extraction combined with electrolysis, or by one or more of the other available techniques.

The direct leaching of metal sulfide minerals by dilute sulfuric acid under an oxygen partial pressure at elevated pressures and temperatures has been successfully applied to zinc sulfide concentrates (F11, M7, R12, V4), lead sulfide (M7, V8), natural and artificial pyrrhotite concentrates (D15, D16), copper sulfide concentrate (V5, V10, W6), and nickel and cobalt concentrate (M15). The basic reaction involves the conversion of the solid metal sulfide into concentrated metal sulfate solution and under the controlled conditions of oxygen overpressure produces pure elemental sulfur and a residue, a major portion of which is iron. The main selling point of this type process is the fact that pure elemental sulfur is produced as a by-product instead of sulfuric acid or sulfur dioxide waste gases. An example of this type of approach is the direct acid pressure leaching of chalcopyrite concentrates, as described by Vizsolyi *et al.* (V10). The important chemical reactions in this process are (1) the oxidation of the sulfide sulfur to elemental sulfur and the formation of cupric and ferrous sulfate with the added sulfuric acid and (2) the oxidation of the ferrous to ferric sulfate, the hydrolysis of the iron and regeneration of sulfuric acid. To be useful as feed for further metal winning operations, a leach solution should contain at least 60 gm/liter of copper and no more than 5 gm/liter of iron. It took nearly 12 hr at an oxygen partial pressure of 30 psi to produce a leach solution containing 60 gm/liter copper compared to 3 hr at 500 psi. A grind of at least 95%-325 mesh was necessary to obtain the highest rate of oxidation. A 25–50% percent excess of copper sulfide concentrate was required to prevent formation of excess sulfuric acid due to hydrolysis of the ferric sulfate. The excess concentrate can be recovered from the leach residue and recycled back to the autoclave. The temperature for optimum leaching rate under the above conditions was found to be between 230 and 245°F, in order to have a reaction time of 2.5 hr. Copper recoveries as high as 98% and pure elemental sulfur recoveries of 85% were attained.

The present conventional step after the beneficiation of ores is to smelt the concentrate to a matte, during which the majority of the iron and sulfur in the concentrate is removed. Pearce *et al.* (P3) have developed a process of refining mixed-base-metal mattes of nickel, copper, cobalt, and iron by acid or ammonia pressure leaching followed by hydrogen reduction to powder metal. For mattes that will be refined by hydrometallurgical processes, the iron and sulfur content must be decreased to a minimum consistent with good recovery of the base metals since both constituents will otherwise require additional oxygen or air for oxidation during the processing. The preferred leaching step is influenced by the analysis of the matte, with those containing over 3% cobalt more amenable to acid pressure leaching and those low in cobalt being treated by ammonia leaching.

Laboratory and continuous pilot plant studies were made with matte having the following composition: 32% Ni, 3.5% Cu, 16% Co, 13% Fe, and 30% S. The matte was ground to 99.5%-65 mesh and 84%-200 mesh and washed with 1% sulfuric acid at room temperature and atmospheric pressure to remove the calcium present in the matte as calcium sulfate. The leaching was carried out at air pressures between 95 and 100 psig at 250°F for 8–10 hr, with further increase in temperature resulting in shorter leaching time, but with no change in final metal extraction. Too little or too much sulfuric acid reduced the extraction, the optimum addition being 155 lb H_2SO_4/ton matte (1.37 lb H_2SO_4/100 gal leach solution). The leach solution analyzed 57–60 gm/liter Ni, 28–30 gm/liter Co, 6–7 gm/liter Cu, and 0.5–1 gm/liter Fe, corresponding to a recovery of 98.5% of the nickel, 98% of the cobalt, 90% of the copper, and 3% of the iron in the matte.

A possible alternative process to the pyrometallurgical method of recovering tin from cassiterite concentrates (SnO_2) was described by Vizsolyi and Forward (V7). It consisted of the reduction of cassiterite with hydrogen to produce impure tin, the dissolution in sulfuric acid at 20°C and 5–10 psi overpressure of oxygen, separation of the insoluble residue by filtration, hydrolysis of the tin-bearing solution at 90–110°C under an oxygen pressure of about 10 psi, with the formation of pure SnO_2 and the regeneration of the sulfuric acid. Pure tin metal with an analysis of 99.9% was obtained by the reduction of SnO_2 with hydrogen or carbonaceous reducing agents. The above purity was attained so long as the cassiterite concentrate did not have more than 1% each of iron, titanium, tantalum, columbium, lead, tungsten, etc., as impurity.

For rapid and effective dissolution of tin in sulfuric acid, the leach solution must contain at least 3 moles of SO_4 per mole of Sn, and at the completion of hydrolysis the resulting solution must have a ratio of 1.5/1 to 2.5/1 of Sn(IV) to Sn(II) and the H_2SO_4 concentration must be more than 100 g/liter.

Gerlach and Pawlek (G7) investigated the dissolution reactions of speiss, a complex smelter by-product which contains appreciable quantities of metals such as Cu and Pb, sulfides such as Cu_2S, FeS, and PbS, arsenides such as Fe_2As, FeAs, Cu_3As, NiAs, $Ni_{11}As_8$, and Ni_5As_2 and, in minor amounts, Ag, Au, Sb, and Sn. About 78% of the Cu and As and around 90% of the Ni and Co were extracted with sulfuric acid at a leaching temperature of 80°C and an oxygen partial pressure of 10 atm. Leaching with ferrous sulfate and with pyrite and pyrrohotite as a source of iron and sulfuric acid at the same temperature and oxygen overpressure resulted in lower extraction of the above metals.

Mackay and Wadsworth (M3) reported the dissolution kinetics of pure UO_2 in sulfuric acid under oxygen pressure of as high as 900 psi and 270°C. It was observed that the rate of dissolution follows a linear rate with time, that by proper agitation the reaction rate may be limited to a surface reaction alone, that the dissolution rate is directly proportional to the partial pressure of oxygen above the solution up to 900 psi, and that a single UO_2 surface site adsorbs a water molecule to form 2 hydroxyls and act as a weak acid.

b. Alkaline Solution. The leaching of uranium ores by alkaline solution has not met with much favor in industry because of the inability of this process to oxidize the tetravalent uranium to the soluble hexavalent state. Forward and Halpern (F6) have developed a process of treating uranium ores by carbonate leaching at elevated pressures and temperatures and of the direct recovery of high-grade uranium oxide form the basic leach solutions by reduction with hydrogen in the presence of a catalyst. Uranium and vanadium are selectively dissolved by a carbonate leach, the solution containing negligible amounts of other metals, and are recovered as oxides by precipitation under pressure. The successful application of this technique to the primary ores of uranium, such as pitchblende, is carried out at elevated temperatures under a positive partial pressure of oxygen or compressed air. A final U_3O_8 extraction of better than 92% is obtained in all such cases. Where the ore contains an appreciable amount of carbonaceous matter, a prior roasting of the ore at 400–600°C eliminates side reactions, which consume oxygen and carbonate solution, and at the same time improves the filtering and settling characteristics of the ore.

Forward and Halpern (F6) found that the variables which influenced the leaching rates and recoveries of uranium included the particle size, the concentration of the carbonate and bicarbonate leach solutions, temperature, and oxygen pressure. In the case of pitchblende, the observed rate of leaching suggested that the rate is directly proportional to the exposed surface area (F6, P7); for carnotite ore, grinding to the natural grain size of the mineral was found to be satisfactory. A solution containing 40–50 gm/liter Na_2CO_3 and 10–20 gm/liter $NaHCO_3$ gave optimal leaching rates and recoveries. The pulp density could be varied from 25–60% solid without appreciable change in extraction. The optimal leaching temperature for pitchblende ores was around 100°C; and above 120°C, a large amount of silica went into solution. For carnotite ores, 60–70°C was found to be sufficient.

For pitchblende ores, the rate of leaching increased in direct proportion to the square root of the oxygen pressure which has been observed to be the rate-determining variable. The oxygen pressure was varied from 0.5–4.0 atm with an oxygen partial pressure of 1 atm at 100°C resulting in rapid oxidation and leaching. Increased extraction under oxygen pressure was observed with carnotite ores, with some samples being leached even in the absence of oxygen. The advantages of using oxygen in compressed air as an oxidizing agent include its low cost and the fact that the only product of its reaction is water.

The removal of silica from a siliceous iron ore, such as the taconites found in Minnesota and Wisconsin, has been studied by Tiemann (T7, T9). Caustic concentrations from 25–500 gm/liter were used to digest the ore in a bomb at temperatures from 312 to 408°F. The leaching pressures in the bomb correspond closely to the equilibrium vapor pressures of the sodium hydroxide solutions used. A residual concentrate containing around 65% iron was obtained with −200 mesh material in 60 min of contact time. The high rate of dissolution of the silica was attributed to its occurrence in the form of microcrystalline (chalcedonic) varieties with high specific surface. The dissolution rate of pure quartz is directly proportional to the surface area and an average rate of 17×10^{-10} gm moles/cm^2 sec was obtained for a 100 gm/liter NaOH solution at 312°F for the −400 mesh fraction.

Stone and Tiemann (S28) have reported the specific rates of silica extraction from taconite ores as a function of time, temperature, and sodium hydroxide concentration. The microcrystalline varieties of quartz associated with the taconites are easily disintegrated by sodium hydroxide under pressure. The rate of silica dissolution has an activation energy of 15,000–19,000 cal/mole.

Stone et al. (S29) developed by a mathematical analysis the functional relationship between the rate of extraction of silica from pure quartz in sodium hydroxide solution and time, temperature, sodium hydroxide concentration, and particle size. With the use of response surface methodology, a comprehensive picture of this dissolution process was obtained from a few well-chosen experiments. The fractional extraction of silica can be expressed by a second-order equation. The effect of quartz particle size and temperature are predicted to be about equal and greater than the influence of sodium hydroxide concentration and reaction time. The reaction rate is controlled by the surface area of the quartz. An increase in sodium hydroxide concentration increases the activation energy for the reactions and is found to be independent of quartz size.

Maslenitsky and Perlov (M16) developed an autoclave soda process for treating low-grade scheelite concentrates and later applied it to high-grade scheelite and wolframite. The leaching of scheelite, of low-grade concentrates containing 40–65% WO_3, of tungsten concentrates and of middlings, with 1–24% WO_3 at temperatures from 200 to 225°C resulted in tungsten recoveries of 94–99%. The autoclave–soda process gives high recovery, dissolves very little of the gangue material, and is suitable for most tungsten-containing material. The decomposition and leaching process is carried out in one vessel at pressures close to the equilibrium vapor pressures of the soda solution. The interaction between scheelite and sodium carbonate is dependent on the calcium carbonate formed during the reaction as a coating on the mineral surface. The initial high rate of reaction decreases as the thickness of the coating increases. The additions of steel balls to the pulp causes new surfaces to be continuously produced, giving rise to increased tungsten recovery. For high-grade scheelite concentrate, two-stage leaching gives 99.5% recovery. The conversion of Na_2CO_3 to $NaHCO_3$ is suppressed by the addition of lime.

In their kinetic study of the leaching of molybdenite by KOH under an oxygen overpressure, Dresher *et al.* (D18) found that a linear mechanism controlled the leaching process in the temperature range 100 to 175°C and in the pressure range 0–700 psia of oxygen. The leaching rate is controlled both by the oxygen overpressure and KOH concentration. In the case where rhenium is present in appreciable quantity in the ore, the pregnant solution will contain potassium perrhenate and potassium molybdate, from which both rhenium and molybdenum can be separated and recovered by ion exchange or solvent extraction. The ease of dissolution of molybdenite and the noncorrosive conditions of the leaching process made its application to the production of ferromolybdenum and molybdenum chemicals very promising.

c. Aqueous Ammonia Solution. The chemistry of the aqueous ammonia pressure leaching process for the extraction of copper, nickel, cobalt, and sulfur from the high-grade nickel concentrates produced by Sherritt Gordon in Canada has been reported by Forward and Mackiw (F5, F9). The sulfide flotation concentrate averages 12–16% Ni, 1–2% Cu, 0.2–0.5% Co, 33–40% Fe, 28–34% S, 8–20% insoluble materials, and less than 0.02 oz/ton of precious metals. It was found that to successfully carry out the dissolution of the sulfide concentrate in an aqueous ammonia solution in the presence of oxygen, the following conditions must be maintained: (1) a sufficient amount of oxygen must be present in solution to complete the reaction with the sulfides; (2) the acid formed by the oxidation of the sulfides must be completely neutralized by supplying a sufficient amount of NH_3; (3) additional ammonia must be present to form the higher amines of copper, nickel, and cobalt, but the excess ammonia should not be so high that the solubility of the amines is diminished; (4) sufficient anions such as SO_4^{--}, etc. must be present; and (5) the reaction temperature must be such that the dissolution rate is high without causing the precipitation of basic salts or insoluble amine complex.

The reaction of a strong solution of aqueous ammonia with the sulfide concentrate in a strongly agitated pressure vessel at a temperature between 160 and 190°F under an oxygen partial pressure of about 10 psi, either as pure oxygen or as compressed air, fulfills the optimal conditions for the above requirements. The iron present in the concentrate is oxidized to hydrated ferric oxide which, together with the silicates is insoluble in aqueous ammonia. The copper, nickel, and cobalt form their amines, while the sulfides are oxidized to sulfates, thiosulfates, and polythionates.

The temperature and NH_3 concentrations in the leach liquor have the most influence on the rate of leaching, followed by oxygen partial pressure and amount of agitation. A pregnant solution with sufficient thiosulfate and polythionate content must be produced to react with the copper present in the subsequent boiling stage. It must also be regulated to produce an iron oxide residue with very little absorbed nickel.

Forward *et al.* (F10) modified the Caron ammonia process for the recovery of copper and nickel from a nickel sulfide concentrate, ore, and matte. The concentrate is roasted to remove the bulk of the sulfur, and the calcine is reduced with hydrogen at around 900°F to convert most of the nickel and copper into the metallic state while most of the iron is kept as Fe_3O_4. The presence of reduced iron inhibits the solubility of nickel in the leaching solution. This is completely overcome by digesting a 30% pulp made up of the reduced calcine and ammonia leaching solution for about an hour at 100°F in the presence of oxygen. The leaching of the reduced digested calcine is carried out with ammonia–ammonium carbonate solution in a 5% pulp for 2 hr at 80°F and an oxygen pressure between 1 and 2 atm. With this process, the nickel and copper extraction for a 17% Ni and 1.7% Cu concentrate were 95 and 80%, respectively. The copper is completely precipitated from the agitated leach solution by reduced nickel at 150°F for half an hour in the absence of oxygen. Although this process gives good metal extractions, the large number of operations involved and the anticipated plant control problems do not make this very attractive on a large scale.

Iwasaki *et al.* (17) reported a segregation process in which nickel is recovered from iron laterites and oxidized nickel ores. Halide salts such as sodium or calcium chloride, and a solid reductant such as metallurgical coke were mixed with the ore prior to roasting. The amount of halide salts was varied from 5 to 16% and the coke containing 85% fixed carbon from 3 to 5%. The optimum roasting temperature was around 900°C where the metallization of the nickel content of the calcine was the highest. Leaching of the roasted product was carried out with equimolar NH_3–$(NH_4)_2CO_3$ solution in an autoclave at 70°C and an oxygen pressure of 10 atm. A leaching extraction of 92% was obtained for ores containing 1.1% nickel.

The application of ammonia pressure leaching as a method of refining nickel–cobalt matte has been studied (B11, P3) and is considered an economically attractive process. Nickel–copper mattes with low cobalt content have been found to be readily leached with ammonia (P3). A high-nickel matte with a composition of 77% Ni, 0.1% Cu, 1.8% Co, 0.7% Fe, and 20% S is amenable to both acid and ammonia leach, but the sulfur deficiency can be made up more economically by circulation of ammonium sulfate solution within the process than by providing the sulfur through addition of sulfuric acid.

A matte with a higher copper content is more commonly encountered in practice because of the natural association of copper and nickel in most of the sulfide ores. A nickel–copper matte with 54% Ni, 12.5% Cu, 0.5% Co, 6.8% Fe, and 22.4% S was leached in a series of laboratory experiments followed by a continuous pilot plant campaign of over three months (P3). The matte was ground to 99.5% -100 mesh without giving problems in the filtration of the residue. The flow sheet followed in this test together with analytical data from the process streams, are shown in Fig. 2. The extractions for nickel, cobalt, and copper were 98.5, 95, and 98.5% respectively.

Vizsolyi et al. (V9) proposed a scheme to produce Pb from galena concentrates without producing elemental sulfur via aqueous oxidation of the lead sulfide in ammonia solutions at temperatures below 100°C and in the presence of 20 psi of oxygen. A 3-to-1 molar ratio of ammonium sulfate to galena is maintained in the presence of associated iron in the concentrate to assure rapid reaction rate. Lead sulfate and ammonium sulfate are produced and the lead can be recovered following a procedure similar to the amine leaching process proposed by the same authors (F12).

Stanczyk and Rampacek (S23) described a method of leaching copper sulfide minerals and flotation concentrates with ammoniacal solutions under an oxygen overpressure at elevated temperatures and pressures. The utilization of sulfur in the sulfide ore during the leaching reaction could make the roasting step prior to leaching unnecessary. The extraction of copper from bornite, chalcocite, and covellite improved with addition of ammonium sulfate to the leach solution and, together with chalcopyrite, decreased the amount of free ammonia required. The chalcocite was found

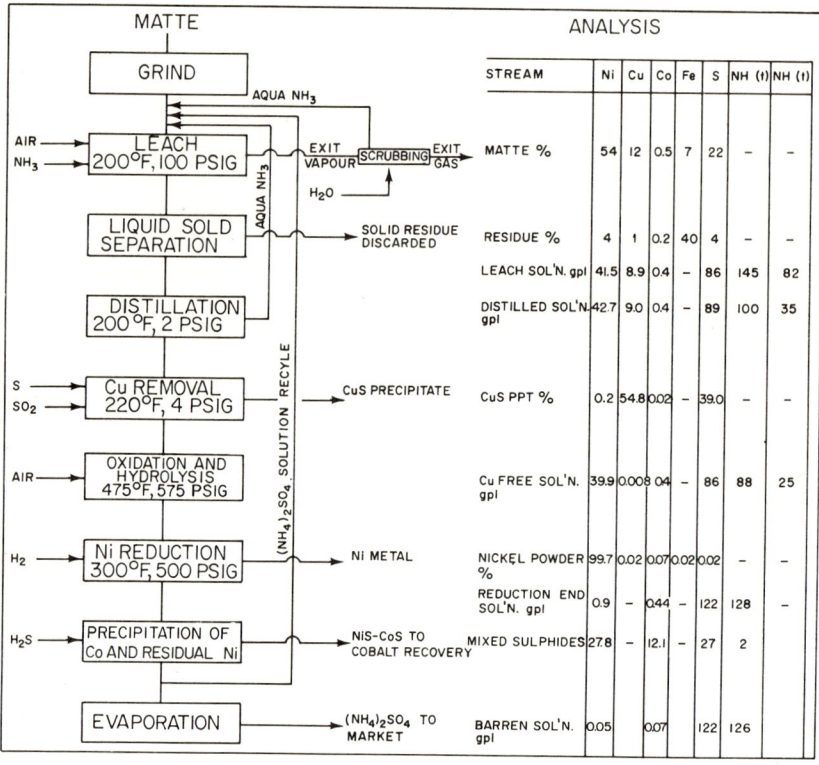

FIG. 2. Ammonia leach and preferential hydrogen reduction for a nickel-copper matte. Pearce et al. (P3).

to convert to covellite as one step during the dissolution reaction. Iron in the copper ore formed hydrated iron oxide on the mineral surface. Almost complete dissolution of copper from bornite, chalcocite, and covellite was obtained at 75°C under an oxygen partial pressure of 100 psig for 60 min, and 95% copper extraction from chalcopyrite was obtained under the same conditions. The leaching of zinc sulfide with ammoniacal solution under an oxygen overpressure at elevated temperature has also been reported (S22).

d. Other Solvents. Seraphim and Samis (S8) studied the oxidation of galena in an almost neutral solution of ammonium acetate under an oxygen overpressure. There was no sulfate and sulfur in the leach solution, the lead going into solution and the liberated sulfur remaining as a film of elemental sulfur on the mineral surface. It was found that the reaction rate is independent of the oxygen and ammonium acetate concentrations and is dependent on the diffusion of lead atoms, lead ions, or lead sulfide molecules from the mineral to the aqueous phase through a film of molten sulfur. A parabolic oxidation rate was obtained experimentally, consistent with diffusion-controlled reactions.

Johnson (J6) described a pressure oxidation process for the generation and regeneration of ferric sulfate and sulfuric acid solutions from pyrites and spent iron sulfate leach liquors. Solutions of ferric and ferrous sulfates and/or pyrites are converted to sulfuric acid and/or ferric sulfate solutions. The spent leach liquor containing ferrous sulfate is mixed with pyrite at high pulp densities in the presence of oxygen at elevated pressures and temperatures to generate and regenerate acid–ferric sulfate. The sulfuric acid solution produced by this reaction contains up to 12% H_2SO_4 and up to 6% ferric iron.

The sulfide minerals are at present the major source of the base metals. Associated with most of the sulfide ores are the minerals pyrite and pyrrhotite. If the hydrometallurgical processing of ores becomes the predominant method of metal extraction, the recovery of elemental sulfur as a by-product is a very promising possibility. The formation of elemental sulfur has been observed by many investigators as a reaction product of sulfide minerals under certain experimental conditions.

Downes and Bruce (D15, D16) reported the direct recovery of elemental sulfur from pyrrhotite and indirect recovery by thermal decomposition of pyrite to artificial pyrrhotite at around 700°C. An aqueous suspension of finely ground pyrrhotite is oxidized in an autoclave under an oxygen pressure of about 30 psig. No outside heating was necessary since the reaction to produce elemental sulfur and iron oxide is exothermic, raising the temperature to between 110 and 125°C. The melting point of monoclinic sulfur is around 120°C and when the reaction temperature exceeds this value, the sulfur formed during the reaction agglomerates into globules which form nodules on cooling. Separation of the sulfur from the iron oxide is readily accomplished by screening. The iron oxide formed is relatively free of other metals, making it a desirable raw material for the blast furnace. If the sulfur content is high, this can be sintered to produce a low-sulfur feed. The dissolution of the sulfides is by the action of acid ferric sulfate formed during the reaction.

McKay and Halpern (M21) studied the kinetics of pyrite oxidation in aqueous suspension under an oxygen pressure of up to 4 atm and temperatures between 100 and 130°C. The oxidation of ferrous sulfate was also studied since this appears to be an important reaction in the overall oxidation of pyrite. This reaction was established to be homogeneous and independent of the rate of solution of oxygen. The rate of oxidation of $FeSO_4$ also increased in the presence of $CuSO_4$.

A heterogeneous process on the pyrite surface involving an O_2 molecule was found to control the oxidation of pyrite. High acidities and low temperatures favor the formation of elemental sulfur and at low acidities and high temperatures, the formation of sulfuric acid is favored. Available data indicate that the oxidation of the pyrite is via the formation of $FeSO_4$ and S.

A process to recover nickel from iron laterites and the high silicate nickel oxidized ores has been proposed by Seidel and Fitzhugh (S7). The ore after separation from the barren material is mixed with elemental sulfur in a conventional ball mill to liberate to natural grain size the nickel minerals. The pulp, which consists of entirely -200 mesh material at pH 7, is transferred to an autoclave and the reaction is carried out at 230–240°C and 380–480 psig. The elemental sulfur reacts with the oxide to form the iron and nickel sulfides and the balance is oxidized to form sulfuric acid, which in turn reacts with the acid soluble constitutents such as the nickel–magnesium silicates. An oxidation step is carried out in another autoclave at 200°C at an air pressure of about 400 psig. The nickel sulfide is oxidized to soluble nickel sulfate and the iron sulfides are oxidized to sulfuric acid and ferric sulfate. Hydrolysis of the ferric sulfate results in the formation of ferric oxide and additional sulfuric acid, which in turn reacts with the balance of the nickel–magnesium silicates resulting in nickel going into solution.

The nickel in solution in the slurry is completely precipitated without liquids–solids separation with metallic powered iron at about 150°C under a pressure of 150 psig. The precipitated nickel contains occluded basic ferric sulfates which are decomposed by calcining at 950°C to produce a mixture of metallic nickel, metallic iron, and iron oxides. Melting of this mixture with a slag is calculated to yield a ferro–nickel containing more than 55% nickel.

2. *Pressure Leaching Equipment*

 a. Laboratory Units. In carrying out heterogeneous reactions in solid–liquid–gas systems at elevated pressures and temperatures, the choice of equipment is limited to either a bomb or an autoclave unit. Both types are available commercially in varying sizes and capabilities. The design and sophistication of the equipment depends on what type of information is desired from the experiment. The simplest bomb unit can be made of a closed stainless-steel tube with a pressure gauge and a thermometer cell which is brought to temperature by external heating with a gas burner. Bombs are usually small vessels with capacities around 200 ml and usually do not have mechanical stirring arrangements. Autoclaves vary in size from experimental to commercial units and have stirring arrangements as part of the design.

Lilge and Siebert (L14) described multiple-bomb leaching equipment that incorporated the desirable features of both the autoclave and the bomb. The basic unit in the assemblage consisted of 6 stainless-steel 100-ml bombs, a gas compression and measuring system, a furnace housing the bombs, a pressure–temperature recording system, an agitation system for the bomb, a water cooling system for protection of pressure transducers, and a furnace temperature control system.

For experimental work designed to obtain fundamental information such as mechanisms and kinetics of reactions, rather than the feasibility of a reaction, a high-pressure–high-temperature reactor meeting a number of special conditions is necessary. Some of the special conditions are: (1) the accurate regulation of the pressure and temperature of the reaction, (2) a provision for extracting liquid samples from the reactor during the course of the reaction without stopping the experiment, and (3) the elimination of time-consuming adjustments and repairs by improved mechanical design.

Dresher *et al.* (D17) described an elevated pressure–temperature unit incorporating these special conditions for laboratory work with heterogeneous liquid–solid systems for pressure up to 1000 psi. The provision for sample removal during the course of a reaction makes possible the accumulation of a complete set of data from a single experiment. The outstanding features of this reactor are: (1) the solution temperature in the autoclave is constantly measured, recorded, and controlled; (2) the gas overpressure of the system is constantly regulated and recorded; (3) the stirring power mechanism, such as the agitator-drive motor, reduction gears, and tachometers are isolated from the autoclave in a separate, pressurized chamber; (4) the sample can be lowered into the reaction zone after the desired thermal equilibrium has been attained; (5) samples can be removed during the run without stopping the operation of the unit; and (6) the parts in contact with the solution or its vapors are all made of 316 stainless steel. A schematic diagram of this autoclave is shown in Fig. 3.

b. Commercial Models. Commercial pressure leaching equipment is designed according to specific conditions dictated by the chemical reaction itself and the physical requirements for handling large tonnage of raw material. Mitchell (M37) described a horizontal pressure vessel 40 ft in

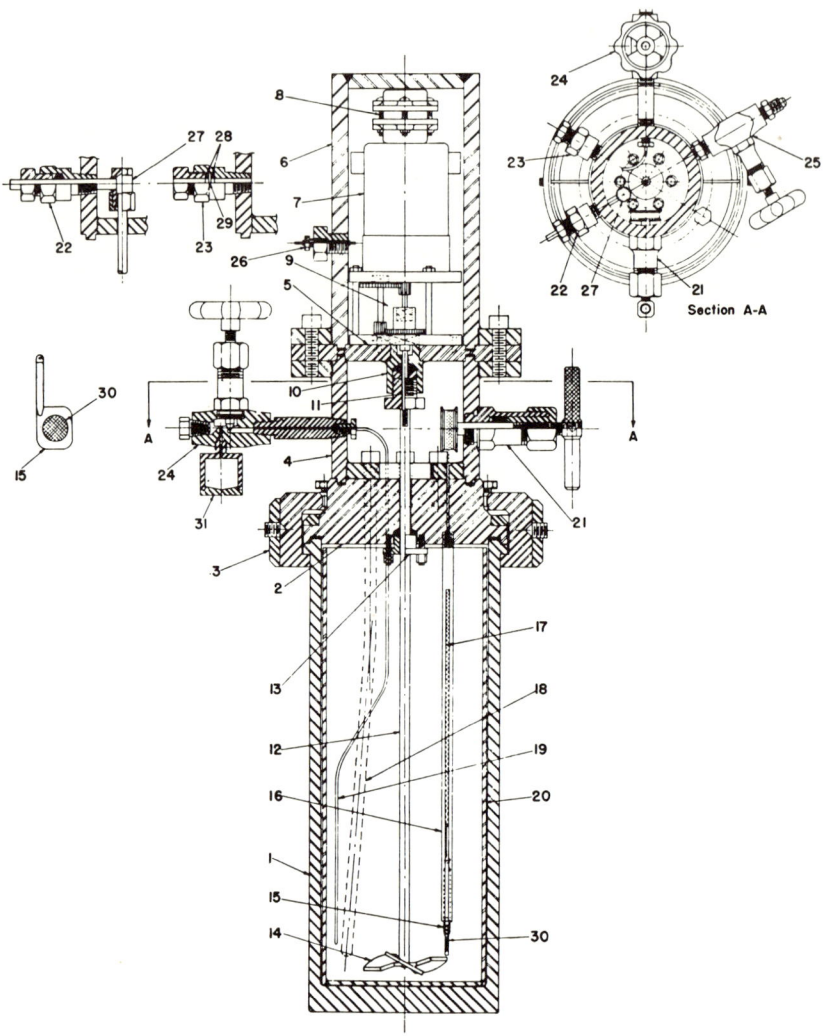

FIG. 3. High-pressure–high-temperature autoclave. Dresher *et al.* (D17). (1) Autoclave reaction compartment, (2) autoclave head, (3) closure rings, (4) intermediate compartment, (5) pressure diaphragm, (6) motor compartment, (7) agitator–drive motor, (8) tachometer, (9) reduction gear train, (10) packing gland, (11) connecting shaft, (12) drive shaft, (13) bearing, (14) agitator, (15) sample holder, (16) sample holder guide, (17) silver chain, (18) thermocouple well, (19) sampling tube, (20) corrosion resistant liner, (21) sample lowering winch, (22) thermocouple well gland, (23) safety blowout assembly, (24) sampling valve, (25) gas inlet valve, (26) electrical lead gland, (27) thermocouple well elbow, (28) pressure disks, (29) silver blowout disk, (30) sample disk, (31) sampling bomb.

length and 6 ft in diameter that was used in the acid pressure leaching of arsenical concentrates to recover cobalt. Carbon steel was used for the shell with a $\frac{3}{16}$-in. 316 stainless steel cladding, 12-lb lead and acid brick lining. The autoclave was divided into six equal compartments by acid brick walls with overflow weirs, which allowed the transfer of the oxidized slurry from one compartment to the other. An agitation mechanism in each compartment provided mixing and aeration and provisions were also made for sparging air into each compartment. Considerable corrosion of the stainless-steel clad was observed when the acid brick lining was replaced after three years of service.

The dissolution of uranium ores in sodium bicarbonate leaching solution under pressure has been described by Mancantelli and Woodward (M11). The autoclaves are horizontal tanks 25 ft long and 8 ft in diameter, insulated with 3 in. of fiber glass. There are two end turbomixers, with 42-in. impellers operated at 62 rev/min to keep the pulp in suspension and a center mixer with a hood ring and a 27-in. impeller operated at 140 rev/min to thoroughly mix air with the pulp. The autoclaves are operated at 80 psig and 230°F with a 20% headroom to allow for the circulation of 1000 ft^3/min of air, which is three to five times the theoretical oxygen requirement. The slurry flows by gravity from one autoclave to another by mounting two parallel banks of autoclaves on an 8° slope. Corrosion problems, while appreciable, are not severe in nature.

Nashner (N5) described the commercial scale leaching of the Ni–Cu–Co concentrate at the Sherritt Gordon refinery in Canada. It is carried out in four compartment autoclaves 45-ft long and 11 ft in diameter with 316 stainless-steel cladding. This is shown in Fig. 4. Agitation is provided by

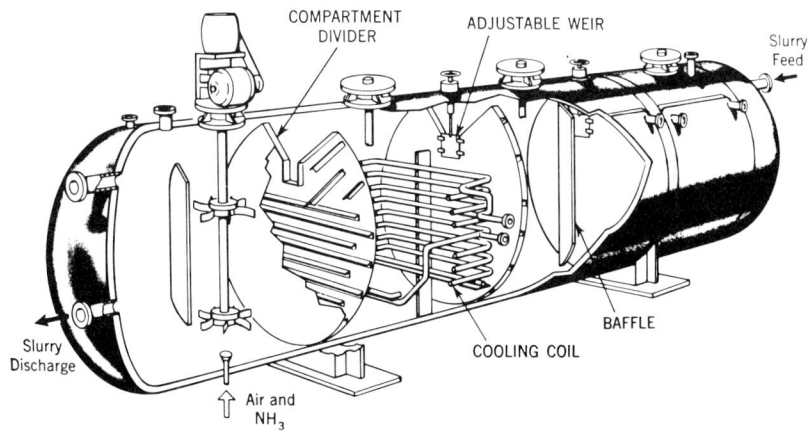

FIG. 4. Cutaway view of Sherritt Gordon pressure autoclave. Boldt and Queneau (B25).

impellers 36 in. in diameter run between 80 and 100 rpm. The autoclaves are operated at 170–180°F and 100–110 psig total pressure. The agitators are equipped with mechanical seals and a water-loaded internal seal which allows for maintenance work on the drive and mechanical seal without releasing the pressure in the autoclave. The retention time in the autoclave as well as in its various compartments is controlled by externally operated level control gates. The retention time is dependent on the amount of thiosulfate and thionate sulfur required in the subsequent copper separation stage. The reaction temperature is automatically controlled by means of cooling water flow. A level-control mechanism on the final cells in each train provides for automatic discharge from the autoclaves. The design considerations of this autoclave have been discussed by O'Kane (O1).

C. Leaching at Reduced Pressure

The dissolution of solids under pressures of less than 1 atm and elevated temperatures has not attracted the interest of most investigators in the field of hydrometallurgy. The technology needed to carry out such a process is readily available and the conditions necessary to carry out the desired reactions are more readily attained than those at elevated pressures. This undeveloped area of research should offer the innovative researcher another avenue of approach in the development of new hydrometallurgical processes.

1. *Dissolution Media*

Acid Solvent. In the dissolution of sulfide ores, a large amount of the iron associated with the ore together with the sulfide sulfur must be separated from the desired metals. For economic reasons, the iron must also be separated from the sulfur. Bjorling and Kolta (B22) studied the extraction of the metals from sulfide concentrates and other materials by reaction with nitric acid at pressures below 1 atm. The reaction of most metal sulfides with a dilute solution of a strong acid can be represented by the reaction

$$\text{MeS} + \tfrac{1}{2}\text{O}_2 + 2\,\text{H}^+ \rightarrow \text{Me}^{++} + \text{S} + \text{H}_2\text{O}$$

Under the same condition, the iron sulfide is oxidized and hydrolyzed, the reaction being

$$2\,\text{FeS} + \tfrac{3}{2}\text{O}_2 + 3\,\text{H}_2\text{O} \rightarrow 2\,\text{Fe(OH)}_3 + 2\,\text{S}$$

The oxidation of the metal sulfides in the presence of free oxygen is very slow. Increasing the pressures and temperatures does not accelerate the oxidation of the metals, but instead causes the formation of sulfates or other salts containing sulfur and oxygen, which are not desirable sulfur by-product forms. The addition of dilute nitric acid has been found to accelerate the oxidation reaction of the lower metal sulfides according to the equations

$$3\,\text{MeS} + 2\,\text{HNO}_3 + 6\,\text{H}^+ \rightarrow 3\,\text{Me}^{++} + 3\,\text{S} + 2\,\text{NO} + 4\,\text{H}_2\text{O}$$

$$2\,\text{NO} + \tfrac{3}{2}\,\text{O}_2 + \text{H}_2\text{O} \rightarrow 2\,\text{HNO}_3$$

The chemical reaction of iron under this condition is represented by the equation

$$\text{FeS} + \text{HNO}_3 + \text{H}_2\text{O} \rightarrow \text{Fe(OH)}_3 + \text{S} + \text{NO}$$

In a leaching reaction where elemental sulfur is one of the by-products, it becomes important to carry out the reaction at temperatures below the melting point of monoclinic sulfur (119°C). Liquid sulfur will prevent the reactions of the other constituents in the aqueous phase.

The economic advantage of leaching at low pressure compared to elevated pressure is obvious. Consideration of the required chemical conditions suggests that for optimum leaching of sulfides, a dilute solution of a strong acid should be used to extract the metal salts, and to produce sulfur in elemental form, the leaching temperature should be under 119°C, a moderate pressure of oxygen should be used, and a promoter should be added to accelerate the reaction with oxygen. Leaching with nitric acid satisfies all of the above requirements.

The decomposition of the lower sulfides of the heavy metals and the recovery of the metal as soluble salts and of sulfur in the elemental form have been demonstrated for pyrite, pyrrhotite, chalcopyrite, sphalerite, galena, molybdenite, and associated metals such as nickel and cobalt. Pyrite and chalcopyrite are higher sulfides and to be amenable to this treatment have to be thermally decomposed at 600–650°C prior to leaching. The reactions with nitric acid are exothermic, and are carried out below 1 atm and at around 100°C. In addition to the sulfides, this technique has been applied successfully to the extraction of nonferrous metals from partly oxidized sulfide ores, fayalite slags, copper scrap, and other intermediate products, such as residue from electrolytic zinc plats.

IV. Separation and Concentration Processes

The leaching process produces a solution rich in the primary metal value or values but with an appreciable amount of undesirable impurities.

There are many methods of separating and concentrating the elements in a given solution as exemplified by the techniques used in analytical chemistry. Oxidation, hydrolysis, and precipitation are among the techniques used in large-scale operation.

In industrial hydrometallurgical systems, the two most widely used unit operations for separating and concentrating process liquors are ion exchange (C16, C17, E12, H25, H26) and solvent extraction (B29, M14, R9, Z1). The first significant applications in hydrometallurgy of both techniques were in the nuclear raw material processing programs. Both techniques are applicable to dilute and concentrated solutions.

The ion-exchange process involves the reversible interchange of ions between a solid and a liquid. Ion-exchange resins can be characterized as water-insoluble cross-linked polyelectrolytes of high molecular weights with ionic functional groups. The solid, which may be either natural or synthetic, inorganic or organic, is classified as either a cationic or an anionic exchanger. The active chemical groups in most cationic exchangers are sulfonic, carboxylic, or phosphoric groups, while the active groups in the anion exchangers are the amino groups in weakly basic resins, quarternary ammonium groups in strong base resins, and for intermediate strength resins, a mixture of the amino and quaternary ammonium groups. Many different types of resin are available which have been found useful for specific metal systems (K7, K8, M28).

The mass transfer in solvent extraction takes place when a water-immiscible organic phase is intimately mixed with an aqueous phase from which one or more constitutents are transferred to the organic phase. Stripping of the loaded organic phase with an aqueous solution concentrates the desired components, while at the same time regenerating the organic solvent for recycle.

Solvent extraction has found wider acceptance in hydrometallurgical processing than ion exchange because of the advantages inherent in a hydraulic system. The fast mass-transfer rates result in a shorter contact time and a high concentration factor. The main disadvantage of solvent extraction is the loss of organic phase through entrainment and solubility in the barren aqueous phase.

A. Resin Ion Exchange

The general characteristics of ion-exchange resin have been presented by Preuss and Kunin (P12) with special emphasis on those used for uranium recovery. In the selection of ion-exchange resins for use in any process, a study must be made of their general chemical and physical properties such as exchange capacity, hydration, particle size, density, hydraulic characteristics, rates of exchange, selectivity, and life expectancy.

The exchange capacity of a resin is the number of fixed ion sites on the resin; this is determined by measuring the number of mobile ions that are adsorbed and desorbed from a given weight or volume of resin. The total exchange capacities of resins used in uranium recovery processes ranged from 3 to 5 milliequivalents/gm and from 1 to 1.8 milliequivalents/ml of wet settled resin. All ion exchange resins are hydrophilic and the degree of hydration is a function of the exchange capacity, the counter ion as well as the nature of its functional group, and the type of cross-linked polymer base. From 40 to 60% moisture is present in resins used in uranium recovery.

The resin beads used in most columnar operations range in size from 0.3 to 0.9 mm in diameter, which is a compromise based on the effect of ion-exchange rates, capacities, and hydraulic characteristics. The especially made resins used in resin-in-pulp operations range in size from 0.8 to 1.6 mm in diameter. The apparent density of a resin is defined as that weight of backwashed and settled wet resin per cubic foot, which for resins used in the uranium industry is about 38–45 lb/ft^3. In column operations, the attrition losses due to swelling and contraction of resin, abrasion of resin–resin surfaces, and abrasion of resin–equipment surfaces are negligible. In resin-in-pulp operations, an appreciable amount of attrition loss is encountered.

Satisfactory column operation is dependent on temperature, particle size, density, and compressibility. A minimum of pressure drop, of the order of 1 lb/in.^{2}/ft^3 resin depth is obtained with clean resin beds. The presence of slimes and precipitates, which causes channeling, increases the pressure drop. A backwashing step, involving the upward flow of fluid to expand the resin particles so that the fine, extraneous material in the ion-exchange columns can be removed is used to prevent channeling.

The rate-determining step in an ion-exchange process is the diffusion of the adsorbable ions into the resin matrix. Retention times of 2–10 min are used in the uranium industry to attain full equilibrium. The metal ion to be recovered must almost completely occupy the resin functional sites to attain a very high degree of selectivity. Resins should be useful for at least two years if clean clarified leach liquors containing no poisonous ions are used. A drastic reduction in the usefulness of the resins is observed in the presence of such ions. Everest *et al.* (E11) studied in detail the deleterious effects of thiocyanates, polythionates and sulfur, cobalt cyanides, silica, and molybdenum compounds on the ion-exchange recovery process of uranium from sulfate solutions.

Dasher et al. (D1) discussed certain engineering aspects involved in the first application of ion exchange in hydrometallurgy, the extraction of uranium. This ion-exchange operation included clarified solution–resin contact in fixed-bed closed tanks and in columns. The resin during a cycle comes in contact with: (1) partially depleted solution for complete removal of all its uranium content, (2) pregnant leach liquor until the resin is saturated, (3) wash water to remove the pregnant solution, (4) backwash water (upward flow) to remove slimes and precipitates so that the pressure drop in the bed will be low, (5) eluting solution to remove the adsorbed uranium from the resin beads, and (6) a rinse to remove the eluting solution.

A relatively high loading of the ion-exchange resins at a fast rate with a high degree of selectivity for the metal value and a rapid elution with minimum chemical makeup and volume are desirable for an economical ion-exchange operation. The optimal economic condition for a given ore can be determined by considering the general trend of the effect of column size, retention time, ionic concentration, elution and recovery of the valuable metal, presence of poisonous ions, the leaching method used, and the total cycle which includes the adsorption, desorption, and the washings after each step.

1. *Bedded Resins*

   a. Fixed Bed. Powell and Spedding (P11) reviewed the use of chelating agents in the separation of rare earths by ion exchange and described the successful pilot plant operations of a fixed bed ion exchange column technique to economically produce large quantities of high-purity rare earths. Ethylenediamine-N, N, N', N'-tetraacetic acid (EDTA) was used as the eluant and copper as the retaining ion. The retaining ion causes the redeposition of the rare earths on the resin bed and transports the chelating agent off the column in soluble form. To prepare the leach solution, finely pulverized xenotime sand is digested in 93% sulfuric acid preheated to 190°C. The temperature of the reaction rises to 240–250°C and is maintained for 6–8 hr. The digested mass is flushed into a dilution tank where it is leached with water. After filtration the combined batches of digested pulp are digested repeatedly to obtain feed solution for the ion-exchange columns. The ion-exchange resin used was 40–50 mesh Amberlite IR-120 (8% crossed-link sulfonated styrene-divinylbenzene copolymer).

   The rare earths are adsorbed in three major bands: the heavy-rare-earth cut consisting of lutetium, ytterbium, thulium, erbium, and about half of the holmium; the light-rare-earth cut consisting of lanthanum, cerium, praseodymium, neodynmium, promethium, samarium, europium, and gadolinium; and the dysprosium–holmium cut which also contains the yttrium and terbium. The individual rare earths are separated from each group by passage of the solution through a series of different columns.

The ion-exchange plant consists of three stages. The primary stage includes 12 columns 30 in. in diameter and 10 ft in height and three auxilliary columns 30 in. in diameter and 4 ft in height. The secondary stage consists of 100 Pyrex columns 6 in. in diameter and 5 ft high. The third stage of the plant includes 32 4-in. diameter Pyrex columns 5 ft tall. All the columns have flexible connections and are arranged in banks to be used as required.

In connection with the above separation of rare earth mixtures by ion exchange, Powell and Spedding (P10) showed that a simple countercurrent separation theory can be used to predict the minimum number of displacements of an adsorbed band necessary to separate the components of binary mixtures. It was also shown how the theory can be applied to more complex systems. When the ion-exchange columns are arranged in series, the bulk of the rare earths in a band 100 ft or more long can be separated in a single pass. The longer the absorbed band, the greater is the percentage of pure material obtainable at a given flow rate. An increase in the flow rate does not increase appreciably the fraction of material that must be recycled. Arrangement of the columns in rows makes possible the elution of one band behind another with only small gaps between bands.

Frisch and McGarvey (F16) derived equations based on equilibrium concepts for use as a source of important process information in the design of ion-exchange processes. The derived equations are useful as guides to maximum regenerated capacity and minimum exhaustion leakage. In cases where precise kinetic data are not available for determining the degree of approach to equilibrium, the qualitative information on the rate behavior of ion-exchange resins is found to be useful. The highly dissociated resins such as the sulfonic or quaternary-ammonium types usually react rapidly with common low-molecular-weight ions. The weakly dissociated acidic or basic resins have been found to react more slowly, except possibly in neutralization reactions. The high-molecular-weight ions are usually exchanged at slower rates. The prediction of ion-exchange performance is best approached by a combination of kinetic information such as given above with the equilibrium relationships.

The kinetic relationships describing an ion-exchange process are usually based on a mass balance, a rate equation, an equilibrium isotherm, and a set of boundary conditions. Moison and O'Hern (M38) developed a generalized correlation applicable to the design of both batch and continuous countercurrent packed-bed ion-exchange equipment. Vermeulen and Hiester (V6) considered on a fundamental theoretical basis the effectiveness of ion exchange as a separation method. The purity and recovery of product streams in fixed-bed operations are determined by the time-dependent solution concentration levels leaving the bed. The prediction of column performance over a wide range of variables for a given system can be made by consideration of equilibrium, diffusional mechanisms, and stoichiometry.

b. Movable Bed. The development and performance of a modified ore dressing jig as a continuous ion-exchange contactor has been evaluated by McNeill et al. (M24). A multistage countercurrent operation consisting of a number of swinging sieve contactors arranged in tower form has been used to increase ion-exchange efficiency by improved resin–solution contact. The semifluidized beds of resin cascade from plate to plate down the tower through staggered downcomers while the liquor flows up the tower, pulsated by the diaphragm jig. All particles have a uniform residence time on each plate because of the low turbulence in the pulsated and semifluidized bed. For a large-scale multistage operation, a series of standard diaphragm jigs connected end-to-end was proposed.

Arden et al. (A7) used the jigged-bed ion exchange technique for the extraction of uranium from acid-leached pulps. This method treats desanded pulp containing 20–40% by weight of solids with the +300 mesh particles removed by classification in single hydrocyclones. The volume of the slime pulp is 1.2–2.0 times the volume of the leach solution compared to 2–3 times if conventional filtration is used.

The desanded leached pulp is pumped upward through a bed of 10- to 20-mesh anion-exchange resins held against the underside of a 30-mesh screen across the top of the column. A pneumatic system provides a 30-mesh screen across the top of the column. A pneumatic system provides a gentle pulsation to the flow of pulp; this prevents the blockage of the bed and the screen. This nonturbulent semifluidized bed helps retain the multistage efficiency of a conventional ion exchange bed. The uranium-loaded resin, which is denser than the eluted resins segregates to the bottom of the bed coming into contact with the incoming pulp while the outgoing barren pulp comes into contact with the lighter eluted resin. The eluted resin fed at the top spreads over the exposed surface of the bed during each downward stroke of the pulsation, ensuring an even distribution of the resin to the bed. The uranium-loaded resin is withdrawn uniformly from the base of the bed in a stream of turbulent pulp. A 30-mesh vibrating screen separates the resin from the pulp stream. The loaded resin is washed and eluted in a separate column, with the eluant solution flowing downward through a bed of resins moving upward by pulsation applied to the column. Columns up to 4 ft in diameter have been applied with success to uranium and other solid–liquid adsorption systems.

A contactor developed by Higgins and Roberts (H29) makes use of a small-diameter column that utilizes a solution downflow and an upward movement of the resin by application of hydraulic impulses to the base of the resin bed. In operation, a period of several seconds is used to force the resin an incremental distance up the column while the solution flow is stopped. A predetermined interval of several minutes is then used to pass the solution through the stationary resin bed. An extremely large number of cycles consisting of alternate countercurrent flow of resin and solution is necessary to effect a satisfactory separation. Sodium and lithium separations have been successfully carried out in this device.

Arehart et al. (A8) described the operation of a continuous countercurrent ion-exchange column with the resin bed moving downward as a dense bed. From the top, the column has a hydraulic ram, loading, loading wash, stripping, and stripping wash sections with appropriate numbers of inlets and outlets. The resin withdrawn at the bottom is returned to the top of the column. The hydraulic ram causes the resin to move downward like a piston due to the frictional drag of the liquid. This force is transmitted all the way to the bottom of the column effectively preventing the fluidization of the resin. This column has been operated at severalfold the fluidizing velocity with no indications of fluidization. The successful operation of this type of column is dependent on keeping the hydraulic balance by preventing fluidization and cocurrent flow of liquid streams and resin.

The observation that acid leach solutions from copper dump leaching operations contain from 2 to 15 ppm U_3O_8 led to the development of a process to recover specification-grade uranium concentrate by a combination of ion exchange resin and solvent extraction with a liquid anion exchanger. The Bingham Canyon mine of Kennecott has a rate of solution flow of 28,000 gal/min from waste ore dumps with about 13 ppm U_3O_8. George et al. (G5) evaluated a system of multiple-compartment countercurrent ion-exchange columns using a $-16 + 20$ mesh strong base anion exchange resin. Each column consisted of eight 4-ft flanged sections 14 in. in diameter, with a short conical bottom section and a 24-in. diameter top section fitted with a peripheral overflow launder. A sampling port in each section permits withdrawal of resins and solution. The flow rate of the feed solution (introduced at the base of the column) is maintained at a rate that fully fluidizes the resin charge and keeps the overflow of the depleted solution constant without the resin being lost with it. A unique feature of this column is the presence of orifice plates between each of the flanged sections. Six inclined tubes, which are located just above each orifice plate, impart a swirling motion to the resin thereby increasing resin–solution contact. The orific plates give the effect of using several columns instead of a long one-stage unit. The overall rate of ion exchange is increased over that in a single stage unit.

The inlet solution valve is closed at regular intervals for about 30 sec to allow a volume of the loaded resin to be withdrawn from the bottom compartment through the resin outlet valve and at the same time an equivalent volume of the resin is moved through the orifice holes from each compartment into its succeeding compartment. An equivalent amount of eluted resin is added at the top section before resumption of the solution flow.

The regeneration or elution of uranium was carried out in a countercurrent column, with the loaded resin being fed at the top and the eluted resin continuously withdrawn by gravity through a drop hole in the base of the column. The eluting solution was 1.5 M H_2SO_4. The weighted average assay of approximately 1200 gal of pregnant eluate produced during the test was, in grams per liter, 3.02 U_3O_8, 0.68 Fe, 0.18 Si, and 100 H_2SO_4 with copper and aluminum varying in concentration from 0.5 ot 1.5. About 180 gal of pregnant eluate analyzing 2.42 gm U_3O_8/liter was concentrated by solvent extraction using a kerosene solution containing 5 vol. % of a tertiary alkyl amine and 2% isodecanol. Stripping the kerosene solution with an aqueous solution of 1 M $(NH_4)_2SO_4$ at 50°C and a pH of 4.3 ± 0.1 gave a solution with 34.7 gm U_3O_8/liter. Neutralization of the aqueous solution with NH_3 precipitated the uranium. Calcination of the precipitate at 675°C for 2 hr produced U_3O_8 with an analysis of 98.4%, meeting all specifications required by the nuclear power industry and the AEC.

2. *Resin-in-Pulp*

a. *Intermittent.* The development of the resin-in-pulp techniques for the recovery of uranium directly from slurries was spurred by the difficulties encountered in the filtration and decantation of the leached pregnant solution from the barren residue. The direct adsorption of uranium sulfate complexes on anion-exchange resins and their subsequent separation from the pulp by a screening operation is of definite advantage when treating ores from different mining operations. This approach is analogous to the cyanide–charcoal process in which gold is adsorbed from the cyanide pulp by activated charcoal.

Hollis and McArthur (H30) described the process development of the pilot plant stage of the resin-in-pulp technique. Coarse bead resins, predominantly 10–20 mesh, were developed together with a mechanical method of conducting the basic adsorption and elution steps. A basket containing the resins was moved up and down in a flowing stream of uranium-bearing pulp to contract and expand the bed of beads in order to increase contact with the pulp and at the same time to cause a mild agitation which improves the circulation and prevents settling of the solids inside and outside the basket. For maximum screening efficiency, the pulp was desanded, leaving the slimes with the pregnant leach liquor. This process gives better than 99% recovery of the dissolved uranium resulting in a final product averaging 70% U_3O_8.

An example of a full-scale resin-in-pulp uranium operation handling 1800 tons/day of uranium ore has been described by Painter and Izzo (Pl). A sand–slime separation of the sulfuric acid leach pulp is made to produce a pulp that will flow freely through the ion-exchange units without settling and immobilizing the ion-exchange resin beads. All of the $+325$ mesh material is removed and the specific gravity of the pulp is adjusted to about 1.06 by addition of water. A highly efficient desanding operation minimizes the addition of water resulting in increased throughput, produces a higher concentration of uranium in the liquor which improves the resin loading giving an overall increase in recoveries, and increases the ion-exchange recovery for a given tonnage by longer contact time between the pregnant liquor and the resins. The elution of the resin is carried out in the same type of equipment by reacting with nitric acid solution at pH 1.2.

Hussey (H35) presented the results of the application of ion-exchange resins in the recovery of gold and silver from a clayey, slimy ore which is not amenable to treatment by the conventional cyanidation process. The ore was leached with cyanide solution followed by countercurrent adsorption of gold and silver from the leach pulp by the anion-exchange resin Amberlite IR-4B. About 78% of the gold and 51% of the silver was leached from the ore. The overall recoveries of gold and silver in the ore by countercurrent adsorption and elution were 74 and 40% respectively. A leaching time of 48 hr was required. The resin was agitated with the pulp for 30 min. Sodium hydroxide was used as the eluting solution.

b. *Continuous.* A multistage countercurrent resin-in-pulp process without desanding of the leached uranium pulp has been described by Read (R7). The pilot plant consisted of 12-stage vertical columns in series into which the resin and the pulp were fed at the top and air was pumped in at the bottom and near the top to provide agitation. The specific gravity of the pulp was as high as 1.5. The sink–float principle was used in the separation of the resin from the pulp. A homogeneous mixture of resin and pulp was found necessary to attain a maximum rate of ion exchange. Losses of the ion-exchange resins were attributed to entrainment and attrition.

Carman (C6) described the successful development and application to a uranium resin-in-pulp process of a continuous countercurrent ion-exchange pilot plant. This new technique is based on the observation that the resins at the correct level of air agitation float in close proximity to the surface of the pulp. So long as the resin beads are able to move about gently but freely in the surface layer, a satisfactory rate of ion exchange is possible. Under this condition, the mechanical damage to the resin due to attrition is negligible.

The factors considered as essential to this floating resin technique are: (1) the pulp must circulate freely through the resin layer to obtain a high rate of ion exchange without damaging the resin beads, (2) the resin must be completely floated at all times so that it does not get into regions of intense agitation, (3) the transfer of resin from stage to stage must be

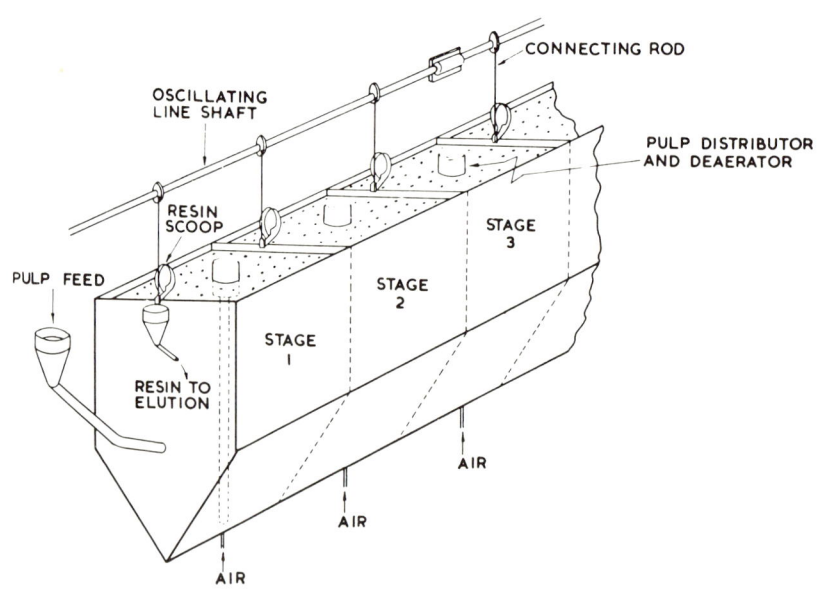

FIG. 5. Pilot plant, floating resin process. Carman (C6).

equal to the rate of new feed, (4) the optimum quantity of resin must be maintained in each stage by controlling the ratio of pulp volume to surface area, and (5) the specific gravity of pulp should be between 1.4 and 1.5. Air agitation of the pulp gave the best results. The equipment used for the continuous air agitation in the floating resin process is shown in Fig. 5. No entrainment problems were encountered in this equipment and all 14 stages were used for the adsorption steps. The pulp was moved continually from stage to stage by gravity in a fashion similar to the flow of pulp through a bank of flotation cells. The resins were transferred by mechanical or hydraulic devices. The resins used were all greater than 28 mesh.

B. Solvent Extraction

Perry (P6) defines solvent extraction operations as those in which the separation of mixtures of different substances is accomplished by treatment with a selective liquid solvent with which one of the components of the mixture must be immiscible so that at least two phases are formed over the entire range of operating conditions. The solvent extraction process, also referred to as liquid–liquid extraction, consists of mixing the two immiscible phases to effect the transfer of the desired ions or complexes into the solvent phase and of separating the loaded solvent phase from the barren phase. In hydrometallurgical operations, the solvent is usually an organic liquid and the other phase is an aqueous solution of metal ions or complexes. The metal ions or complexes in the organic phase are recovered by stripping the loaded organic phase with an aqueous solution which at the same time regenerates the organic extractant for recycle. The highly selective organic extractant is usually dissolved in a carrier or a diluent such as kerosene to obtain maximum contact with the aqueous phase and to minimize the chemical inventory in the circuit. In some systems, it is necessary to add an additive to ensure that intermediate products of reactions do not precipitate or form a third phase. This process involves the transfer of metal values from the dilute aqueous leach liquor into a relatively small volume of organic extractant. The stripping operation further concentrates the metal values in an even smaller volume of aqueous solution.

The application of solvent extraction to metallurgical systems is classified into two groups based on the nature of the extractable species involved. The first group involves an ion-pair transfer mechanism and the second group is the liquid ion-exchange mechanism, which is essentially the interchange of anions or cations between the aqueous and organic phase.

1. *Solvent Extraction System*

    a. Ion-Pair Transfer. The ion-pair transfer mechanism involves the interaction of an electrically neutral hydrated metal complex or configuration of closely associated cations and anions reacting as a single unit with the organic phase. The reaction of hydrated metal-bearing ion pairs with organic solvents such as alcohols, ethers, ketones, and esters causes dehydration of the ion pair and results in the formation of a neutral molecular species that is preferentially soluble in the organic phase. The rate of transfer is promoted for most metal systems by the use of concentrated solutions of salts and acids, and by the addition of electrolytes containing the anion of the extractable species with a nonextractable cation.

    Bautista and Hard (B8) made a comparative study of the extractability of several of the first-transition metals from thiocyanate solutions using methyl isobutyl ketone as the organic solvent. The transition metals readily extracted were scandium (III), iron (III), and cobalt (II) while chromium (II) and manganese (II) were not. The principal extractable species were found to be the neutral scandium and iron trithiocyanate complexes, while the extractable cobalt complex was the negatively charged tetrathiocyanate radial $Co(SCN)_4^{--}$. The distribution ratio for scandium, iron, and cobalt decreased with increase in metal ion concentration but increased with increasing ionic strength of the solutions.

    Carlson and Nielsen (C3) described the pilot and full-scale plant separation of an ore containing more than 30% combined columbium and tantalum oxide using a sulfuric-hydrofluoric acid leach and methyl isobutyl ketone (MIBK) as solvent in pulsed columns. The -200 mesh columbite–tantalite ore was digested with 70% HF until the combined $(Ca + Ta)_2O_5$ in the leach liquid reached 3 lb/gal at which time it was diluted to $15N$ free acid and clarified by filtration. This solution was contacted countercurrently in the pulsed column where Ta and Cb were extracted by MIBK. Columbium was stripped from the organic with demineralized water which diluted the free acid in the solvent, making possible the transfer of all the Cb; the Ta-loaded solvent was then stripped with demineralized water causing the transfer of Ta to the aqueous phase. The oxides were then precipitated with 28% ammonium hydroxide solution. Conversion to the respective oxides was by calcination of the precipitates.

    In another plant (C18) the separation of columbium from tantalum is carried out by the digestion of the -100 mesh ore with anhydrous HF and deionized water for 10 hr at 175°F in Haveg-lined wooden tanks. The pregnant solution is contacted with MIBK in a cascade of mixers and settlers. The solutions from the Ta and Cb effluent streams are precipitated with ammonia to form their respective oxyfluorides. Reduction to metal is by heating to 2000°C at 1μm − mercury pressure.

Stickney (S26) described the operation of a plant that was used to produce low-hafnium zirconium oxide (40 ppm) and low-zirconium hafnium oxide (less than 2%) from an impure zirconium tetrachloride feed. The aqueous acid–chloride solution containing Zr and Hf as thiocyanate complexes was contacted countercurrently with a mixture of MIBK and thiocyanic acid in a column. The Hf and about 30% of the Zr was selectively extracted in the organic. The aqueous zirconyl chloride solution which was essentially Hf free was treated to recover Zr. The Zr in the organic stream was stripped with HCl solution and the aqueous solution recycled to the extractor. The organic phase from the stripping was scrubbed with H_2SO_4 solution to remove the Hf. The thiocyanate was removed from the scrubbed organic with 28% ammonium hydroxide. The cleaned organic was used as solvent feed for the thiocyanate recovery column.

b. Liquid Ion Exchange. Liquid ion-exchange processes involve the interchange of ions between two immiscible phases, a light organic phase and a heavy aqueous phase. The chemical extractants used in liquid ion exchange can be classified as cationic or anionic exchangers. Examples of cationic exchangers are the acidic organic extractants such as the organophosphorus compounds and carboxylic acids. The basic organonitrogen compounds such as the amines are anionic extractants. Some metals such as uranium and molybdenum exist in the form of anion and cation in equilibrium in their acid solution. It is possible then to extract either metal by the use of an anionic or cationic exchanger. For most metals, however, the choice of the extractant is dictated by the species in which the metal is present in solution.

Fundamental studies have been reported using the cationic liquid ion exchanger di(2-ethylhexyl) phosphoric acid in the extraction of uranium from wet-process phosphoric acid (H34), yttrium from nitric acid solution (H11), nickel and zinc from a waste phsophate solution (P9), samarium, neodymium, and cerium from their chloride solutions (I2), aluminum, cobalt, chromium, copper, iron, nickel, molybdenum, selenium, thorium, titanium, yttrium, and zinc (L11), and in the formation of iron and rare earth di(2-ethylhexyl) phosphoric acid polymers (H12). Other cationic liquid ion exchangers that have been used include naphthenic acid, an inexpensive carboxylic acid to separate copper from nickel (F4), di-alkyl phosphate to recover vanadium from carnotite type uranium ores (M42), and tributyl phosphate to separate rare earths (B24).

Casto *et al.* (C10) described the use of a pilot plant scale multistage, box-type mixer-settler extractor in separating lanthanum from a mixture of rare earth chlorides of monazite origin using di(2-ethylhexyl) phosphoric acid (HDEHP) in Amsco odorless mineral spirits for the organic solvent and dilute HCl for the aqueous scrub solvent. The HDEHP concentration was 0.5 M, well below the regions of gel and emulsion formation, the HCl concentration was 0.1 M and the rare earth chloride feed was 2.6 M. The use of a 20-stage extractor produced a 97% lanthanum raffinate product with a 60% lanthanum recovery from a feed containing 45.6% lanthanum, 10.0% cerium, 30.5% neodymium, 3.6% samarium, 8.0% praseodymium, and 2.4% gadolinium.

Black et al. (B23) described the design and operation of a custom uranium mill using dodecyl phosphoric acid (DDPA) as the liquid ion exchanger. The solvent system was a 3% solution of DDPA in kerosene. The design criteria were based on pilot plant studies that confirmed laboratory findings. Sulfuric acid with sodium chlorate as an oxidizing agent was used in the leaching step. Sodium hydrosulfide was used to precipitate the heavy metals and to reduce ferric iron before clarification of the solution. The clear overflow from the solid–liquid separation containing 1.25 gm/liter U_3O_8 went to the extraction units. The extractors were 20-ft diameter acid resistant tanks with cone bottoms equipped with a gas-dispersion agitator to provide the necessary phase contact. The stripping was carried out with 10 N HCl in five stages, each stage consisting of separate mixing and settling tanks. The uranium rich hydrochloric acid, containing between 50 and 100 gm/liter of U_3O_8, was evaporated to recover about 90% of the available HCl as 10 N acid. The concentrated uranyl chloride solution from the evaporator, containing about 800 gm/liter of U_3O_8, was diluted with water and the uranium was precipitated as ammonium diuranate. Calcination of the precipitate to U_3O_8 gave a final product purity approaching 90%.

The organic amine extractants are the most commonly used anion exchangers. Secondary amines have been used to recover uranium from leach liquors (G10); secondary and tertiary amines to recover molybdenum from uranium mill circuits (L13); a primary amine, diethylenetriamine-penta-acetic acid (DTPA) to extract cerium group lanthanides (B6); tri-N-butylamine-3-methyl-2-butanone to separate yttrium and rare earth nitrates (G13); tricaprylyl amine (Alamine 336) and methyltrioctyl-ammonium salt (Aliquat 336) to recover vanadium from acidic solutions (A3); and Aliquat 336 to extract vanadium from slightly acidic or alkaline leach liquor (S36).

The removal of trace amounts of chromium, which degrades the whiteness of titanium dioxide pigments, from sulfuric acid leach solution by liquid–liquid extraction with 5 vol. % Alamine 336 in xylene has been reported by Bonsack (B26). The chromium (III) in the sulfuric acid leach liquor of ilmenite ore was oxidized to the extractable anionic chromium (IV) species. The chromium content of the extracted ore solution containing 110 gm/liter TiO_2 was reduced to less than 1 ppm chromium.

Agers et al. (A4) described a process for the recovery, purification, and concentration of copper values from acidic dump leach liquors employing an extractant, LIX-64, in kerosene diluent. At pH 1.4–3.0, this reagent selectively extracts copper from sulfate solutions, with very slight extraction of iron and other impurities present in the leach liquor. A 10 vol. % LIX-64 in kerosene can be loaded in excess of 2.5 gm/liter of copper from a typical dump leach liquor. This loaded organic phase is stripped with a solution containing 200 gm/liter H_2SO_4 with 8 gm/liter recycled copper to produce a pregnant strip solution averaging 90 gm/liter copper in three stages. One stage stripping is sufficient to yield about 30 gm/liter copper to produce excellent quality electrolytic copper analyzing 99+ Cu. The buildup of ferric iron in the strip solution can be controlled in a closed system without bleeding by installing a scrub stage which uses a dilute solution of ammonium bifluoride between the extraction and stripping sections.

Another extractant, LIX-63, was found to be specially suited to recovering copper from ammoniacal leach liquors. Swanson and Agers (S35) presented the results of using 5% LIX-63 kerosene with feed liquor averaging 11 gm/liter Cu. About 99% extraction of copper was obtained. Stripping with a solution containing about 200 gm/liter H_2SO_4 gave a pregnant strip solution of approximately 50 gm/liter Cu.

Agers and DeMent (A2) presented the results of using LIX-64 in a large pilot plant operation at the Bagdad Copper Company in Arizona. For a year's operation, an average of 93% copper recovery was obtained from a dilute acid solution containing 0.88 gm/liter copper and 2.6 gm/liter ferric iron. The pilot plant program confirmed all the design parameters previously determined in the laboratory. Pilot plant design criteria are available for scale-up to commercial plants.

The improved capabilities of LIX 64 N reagent for recovery, purification, and concentration of copper values from acidic leach liquors were described by DeMent and Merigold (D10). The leaching of copper sulfide flotation concentrates with subsequent recovery of copper by LIX 64 N was shown to be technically feasible. This extractant loads and strips faster, is more effective in extracting copper from a lower pH solution, has considerably less secondary entrainment, has better iron rejection, and may be used at levels up to 30 vol. % in kerosene without aqueous entrainment. Use of the reagent in operating pilot and commercial plants is also discussed.

2. Solvent-in-Pulp Extraction

The solvent-in-pulp extraction process is a good alternative to regular solvent extraction methods where the liquid–solid separation after leaching must be bypassed because of filtering difficulties. In so doing, it is possible to eliminate soluble solvent losses, increase the loading of metal in the solvent, reduce the stripping reagent consumption, and produce a high-purity precipitate. The extraction of metals directly from leached ore slurries by solvent extraction requires especially developed equipment to carry out the separation in the presence of the solid. The relatively low-tonnage throughput and a large organic circuit inventory are some of the disadvantages of this technique.

The early experiments on solvent extraction directly from leached pulp were beset with problems such as losses of solvent in the aqueous phase and the formation of emulsions. The use of mixer-settler, pump mixer, and internal mixer-settler type contactors on a laboratory scale (G11) has demonstrated the feasibility of uranium extraction from desanded slurries with 5–15% solids and from high-density slurries with 48–60% percent solids. The deemulsification rate of a synthetic slurry as a function of the temperature of the system and the pH of the slurries (T12) and the effect of extractant entrainment in the aqueous effluent on solvent extraction of uranium from slurries containing more than 40% solids (E6) have been studied.

Ritcey et al. (R10) described the pilot plant development of a process to recover uranium from acid leach slurries. Scrubbing, stripping, and acid equilibration was carried out in a 10-in. diameter pulsed plate column and three mixer-settlers. A 0.1 M tertiary amine (General Mills Alamine 336) was used as the extractant in a kerosene diluent containing isodecanol as an emulsion suppressor and third phase inhibitor. The slurry feeds, containing up to 35% solids, were processed at a throughput rate of 13.5 tons/day of dry feed. The solvent losses in this process are low, from 0.7 to 0.10 lb/ton dry feed; and the uranium recovery was 98%. The application of the same technique to the recovery of pure uranium and copper and lower grades of cobalt and nickel salts from acid residues of the gravity concentration of uranium ores has also been reported (J5).

North and Wells (N7) described the design and operation of a rotary-film contactor for the solvent extraction of metals directly from leached ore slurries. It consists of an enclosed rotating assembly of closely spaced disks centrally mounted on a common horizontal shaft. The transfer of the metal to the solvent layer is effected by the rotating disks where a film of slurry is formed on contact with the leached ore at the bottom of the vessel. This contactor gave excellent uranium recovery from sulfuric acid leached ores and concentrates using di(2-ethylhexyl) phosphoric acid or trinonylamine as solvents in kerosene diluents. Solvent losses by entrainment were found to depend on the retention time of the slurry, size, and speed of rotation of the disks and on the number of disks in the contactor.

3. *Liquid–Liquid Contactors*

a. Stagewise Extractors. The most commonly used stagewise liquid–liquid contactor in hydrometallurgy is the mixer–settler. The aqueous phase and the organic phase, flowing countercurrently to each other, come into contact with each other in the mixer section and because of their difference in density are separated or disengaged in a separate settling chamber. These units get to be large, but are relatively inexpensive and easy to maintain. Shaw and Long (S9) discussed in some detail the principles, operation, and commercial performance data on uranium solvent extraction equipment with special emphasis on mixer–settler, pump–mixer, and internal mixer–settler equipment. The processing of uranium and its associated metals has received the most extensive development (B30, L12, T15, W12) because of their importance in the nuclear industry.

A continuous laboratory scale multistage box-type mixer-settler extractor has been designed, constructed, and operated by Rahn and Smutz (R1). The horizontal mixer–settler was made in 10-stage units with adjacent stages having common walls. A central drive shaft with timing belts and geared pulleys for each stage maintains the same speed for the pump–mixer in all stages. The light phase existing as a thin layer on the surface of the heavy phase enters the pump–mixer chamber through the adjustable intake tube where both phases are simultaneously mixed and pumped through the discharge tube into the settling chamber. The separated phases flow through the overflow and underflow ports of the settler into the next stage. The flow in each stage is co-current but is countercurrent throughout the extractor.

The stagewise interface control was based on the hydraulic pressure balance used for the first time in a mixer–settler by Whatley (W10). The raffinate product left the extractor through a jack leg to utilize the hydraulic pressure balance in controlling the interface in the first settling chamber. The pump–mixer was designed to take advantage of the energy dissipated in spinning the liquid in order to lift or pump the fluid from stage to stage. Increasing the impeller speed improved the mixing efficiency.

Casto *et al.* (C11) considered the dynamics of this mixer-settler by introducing intentional upsets of varying degrees after steady state operation had been reached. The system 0.5 M HDEHP–0.1 M HCl–2.7 M rare earth chloride in Amsco mineral spirit diluent with base operating conditions chosen to produce an approximately 88 mole percent lanthanum raffinate product was studied to test the effect of upsets on the raffinate product's total rare earth concentration. Step changes of up to 100% in the various flow rates and interface positions were introduced. The $RECl_3$ flow rate was found to be most sensitive, followed by the HCl, the HDEHP, and the settler interface positions.

The design and scale-up of mixer–settlers for the recovery of uranium from sulfuric acid leach solutions using di(2-ethylhexyl) phosphoric acid with tributyl phosphate as the solvent in kerosene have been reported by Ryon et al. (R18). A study of the unit operation variables affecting the rate of extraction in the mixer and the phase separation in the settler was made the basis for equipment design. The results obtained from the three sizes of geometrically similar units tested, together with the available chemical information from labroatory sized multistage equipment were used as the basis of full-scale plant design without benefit of a pilot plant. For this process, the rate of uranium extraction in the mixer was found to be proportional to the cube root of the power input for both batch and continuous flow. The scale-up of the mixer can be done by geometric similitude at constant power input per unit mixer volume. A dispersion band at the interface whose thickness is a measure of approach to flooding condition characterizes the flow capacity of the settlers. The basis for settler scale-up is the flow rate per unit cross-sectional area. The scale-up for both mixer and settler have been confirmed by actual plant data.

Ryon and Lowrie (R19) reported the basis of designing mixer–settlers for the solvent extraction of uranium from acid sulfate solutions using long-chain amines dissolved in a hydrocarbon diluent. A large mass-transfer rate was observed for both uranium extraction from the acid sulfate solutions and the stripping from solvents by sodium chloride, ammonium nitrate, or sodium carbonate. The dispersion band which formed at the phase interface limited the flow capacity of the settler. The basis for settler scale-up is the constant flow rate of dispersed phase per unit settler area and constant dispersion band thickness. The formation of silica-stabilized emulsions can be minimized by maintaining dispersions that have the solvent as the continuous phase.

b. Differential Extractors. The differential extractors such as the packed, pulsed or rotating-disk columns and the centrifugal type extractor provide continuous countercurrent contact between the aqueous and organic phase. The vertical columns take advantage of the gravitational force for flow and the centrifuge type move the liquids in a very short time radially to the axis of revolution. The simplest columns without mechanical agitation give a very low mass transfer rate resulting in the need for a very tall column. Hanson (H9, H10) and Treybal (T16) have reviewed the different available industrial equipment for liquid–liquid extraction.

Dykstra et al. (D30) described the operation of a 5-in. diameter mixer-agitated type extraction column used to recover uranium from uranyl nitrate solutions containing from 2 to 40 gm/liter of uranium with a maximum impurity content of 40 gm/liter tributyl phosphate. The extrac-extraction column was designed for mass transfer of uranium from the continuous nitric acid waste solution to the organic phase with a normal solvent-to-acid feed ratio of 1 to 1. The 20-ft column had 48 vertically stacked stages and settling zones at the top and bottom of the glass column. The column was operated with an aqueous continuous phase. The stripping column, identical in design with the extractor, brought about the mass transfer of uranyl nitrate from the organic to the continuous water phase with a solvent–water feed ratio of 1 to 1. The best operating condition for the column was found to occur 10–20 rpm below flood conditions. The best height equivalent to a theoretical stage was at a feed flow of 20 gal/hr. The extraction column was found to be more effective than the stripping column.

Sege and Woodfield (S5) evaluated the general performance characteristics of sieve-plate, liquid–liquid extraction columns whose contents are pulsed through the plates. For the same extraction requirements, a pulsed column can be used instead of a taller packed column. In the pulse column, the values of the height of a transfer unit are less dependent on superficial volume velocities. The primary disadvantage of a pulse column is the cost and maintenance requirement of the pulse generator. In addition to the specific properties of the liquid–liquid system under consideration, the performance in a pulse column is affected by the (1) pulse amplitude and frequency; (2) solution flow rates and ratio; (3) choice of continuous phase and plate wetting characteristics; (4) plate geometry including hole size, plate spacing, percentage of perforated area, and plate-to-wall clearance; (5) concentration of the diffusing component; and (6) diameter of the column.

Uranyl nitrate was extracted with tributyl phosphate in a 3-in. diameter pulse column with a perforated-plate section height of approximately 9 ft. Plate-free end sections 3–4 in. in diameter and up to 6 ft in length were incorporated in the design to give several minutes of holdup time for phase separation. The pulse was applied by means of reciprocating stainless-steel bellows or a reciprocating piston.

Three different types of phase dispersion behavior were observed in the pulse column as functions of throughput rate and the pulse frequency at any given pulse amplitude. At low throughput rates and frequencies, a mixer–settler type operation characterized by the separation of the light and heavy phases into discrete, clear layers in the interplate space during the quiescent portions of the pulse cycle was observed. Under this condition, the pulse column operation was very stable but relatively inefficient compared with operations of the same column under emulsion type conditions. At higher throughput rates and frequencies, the formation of a uniform dispersion of small drops about $\frac{1}{16}$-in. in daimeter or smaller was observed. This emulsion-type operation gave the best efficiency because of the intimate contact between the phases. The third type of operation is the unstable performance that occurs at still higher throughput rates and frequencies. Complete flooding eventually results at throughput rates and frequencies beyond the unstable region.

A rotary extractor with a rotating inner cylinder was used by Davis and Weber (D4) in processing radioactive solutions to minimize residence time and thereby to prevent the degradation of the organic solvent. The chemical system used to recover uranium included 5 M HNO_3 and 2.5 vol. % tributyl phosphate in Ultracene. The rotor speed had to be well above the critical speed for vortex formation before blending of the dispersed phase could take place. Efficient mass transfer took place after the occurrence of blending. When the organic phase was dispersed, the value for height equivalent to a theoretical stage depended on the motor speed and was indpendent of flow rate while for the cases where the aqueous phase was dispersed, it increased with increase flow rate at low rotor speeds. Residence time was from 5 to 10 sec.

The centrifugal extractor is essentially a horizontally rotating drum with balanced mechanical seals for introduction of feed and solvent and removal of raffinate and extract through passageways in the rigid shaft. The centrifugal force allows the separation of systems with low density differences and the low-holdup and high-throughput capacity of the extractor results in a very short contact time. It occupies a very small volume compared to other extractors and an increase in capacity is readily handled by addition of a similar unit. The light phase enters near the periphery and is moved inward by the outmoving heavy phase. Both phases pass countercurrently through a series of mixing and calming zones which are created by a series of perforated concentric cylinders. The flow of liquids through the Podbielniak centrifugal contactor is shown in Fig. 6. The potential of using this equipment to recover metals has been examined (C12).

Todd (T11) discussed the features of a compact flexible centrifugal

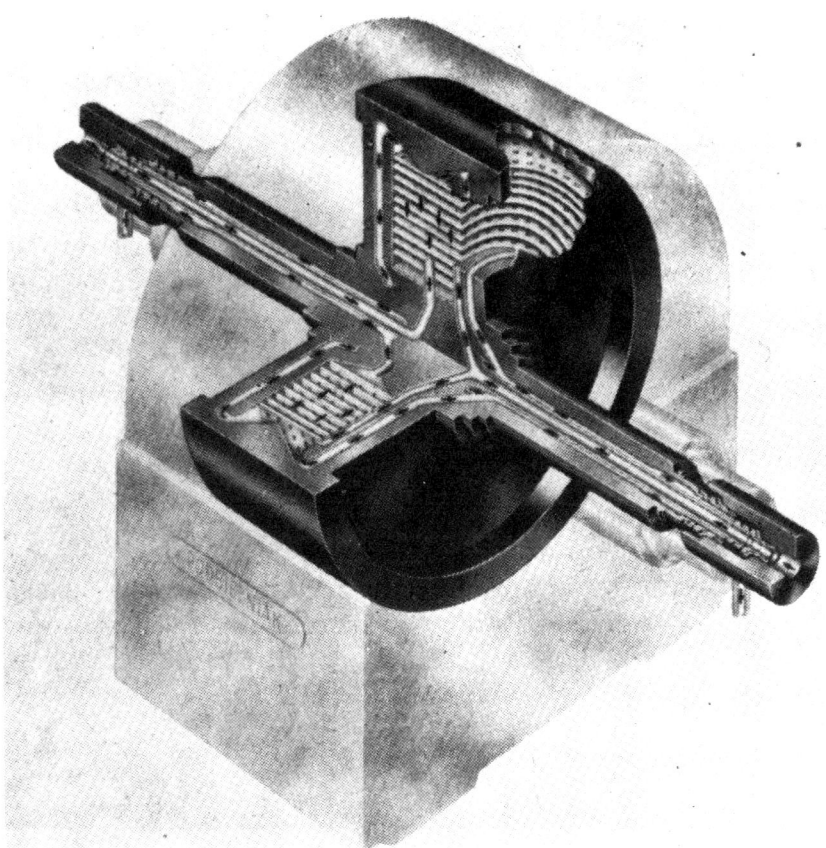

FIG. 6. Flow of liquids through Podbielniak centrifugal contactor. Hanson (H9).

extractor that has provisions for changing feed location, seal arrangements, and internal elements and also has access to the interior of the machine without dismantling or removing the rotor from its base. This extractor provides a large amount of surface for mass transfer by making a fine dispersion, a reduction in backmixing by appropriate baffling to maximize the gradient for mass transfer, and repeated coalescence and redispersion to overcome stagnation within each phase. By handling a large throughput in an equipment of small size and cost, this extractor is economically attractive. The maximum capacity for a given centrifugal extractor is proportional to the speed of rotation.

The operating characteristics of a Podbielniak centrifugal extractor with a combined-stream capacity of 450 cm^3/min at 5000 rpm and a rotor holdup of 529 cm^3 has been reported by Jacobsen and Beyer (J3). The effects of density difference, rotor speed, light-liquid-out pressure, flow rate, holdup, and number of stages on the operational performance were studied. An experimental technique for holdup determination was described and an equation for predicting flooding limits has been verified. The relationship among the number of stages, light-liquid-out pressure, and flow rate ratio was developed and its application to other systems using similar extractors can be made to estimate optimum operating conditions.

Kelsall (K2) reported that attempts to make use of the centrifugal force in a cyclone to separate dispersions of immiscible liquids have resulted in the production of only one pure product; the other is a mixture of both phases. The use of multistage cyclones with suitable recirculation of the intermediate products was found to be more promising. Kindig and Hazen (K5), using a double cyclone system and a diluent that is heavier than kerosene, were successful in discharging pure organic phase from the bottom and aqueous phase from the top of the cyclone. Ten volume percent LIX-64 was dissolved in the common commercial dry-cleaning liquid perchloroethylene (tetrachloroethylene), specific gravity 1.62.

This development could eventually replace the settlers now used for the separation of the organic phase from the aqueous phase after the mixing step. The high cost of centrifugal contactors makes their use for the recovery of large-tonnage, low-cost metal very uneconomical.

V. Metal Reduction from Aqueous Solutions

The reduction and precipitation of metals from their purified solutions is typically carried out industrially through "cementation," by the introduction of nonmetallic reducing agents such as hydrogen gas or SO_2, or by cathodic reduction during electrolysis. Cementation is the process of precipitating a metal from its solution by bringing the solution in contact with another metal that stands higher than the first in the electromotive series. Since for any reducing agent to be effective in a spontaneous reaction the oxidation–reduction potential of the couple involving it must exceed that of the couple involving the substance to be reduced, the only essential distinction between the first two processes lies in the physical state in which the ultimate reducing agent (electron-donor) is introduced. In electrolysis the oxidation–reduction reaction is driven by an external potential so that the electron donor need not be involved in a couple with a potential exceeding that involving the substance to be reduced; the

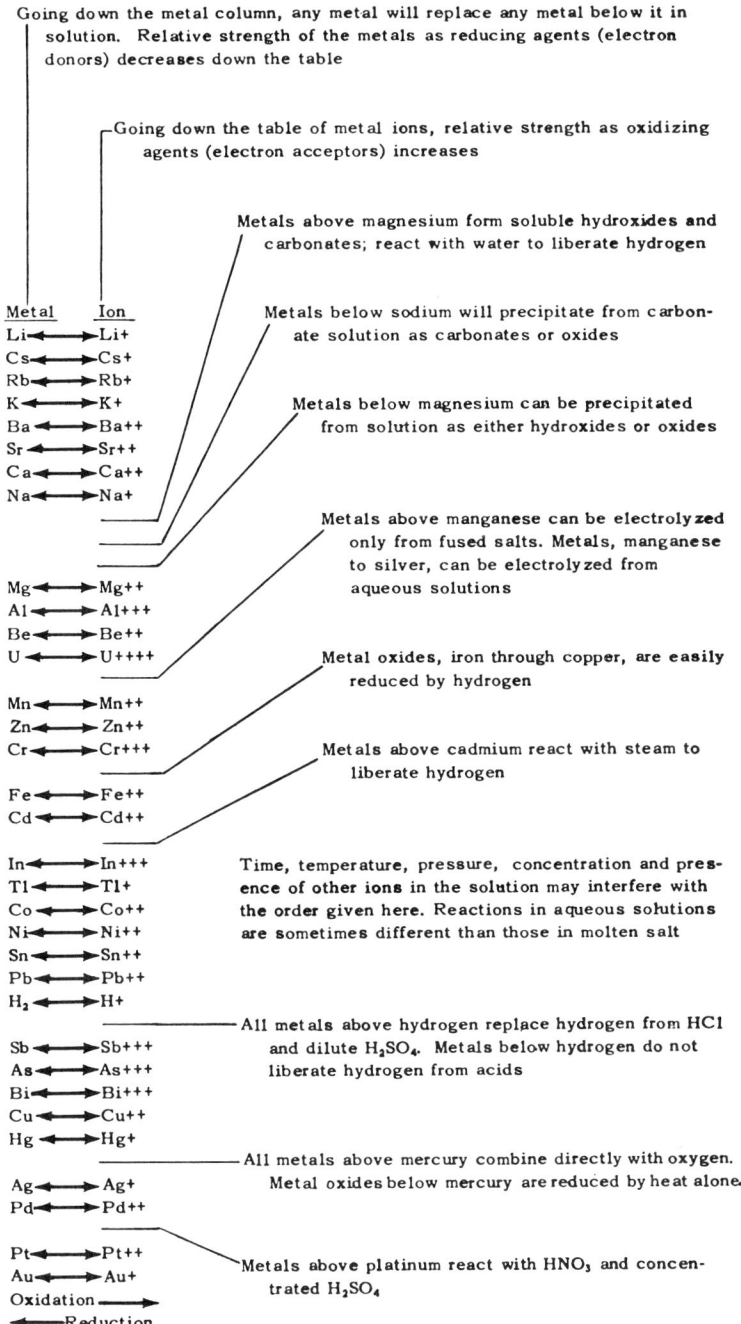

FIG. 7. Electromotive series of metals. Franklin (F15). Copyright *Engineering and Mining Journal*, April, 1958, McGraw-Hill Inc.

donor is a substance initially in the solution and is anodically oxidized. As shown in Fig. 7 a given metal will displace from solution all the metals below it in the series.

A. Displacement Reactions

The copper producers who are leaching their oxide ores, waste dumps, and old mine workings are the largest users of the displacement-reaction technique. The replacement of copper in solution by iron was known as early as 1500 (L1, W16). By 1600 (N4), this method was being used to recover copper at Rio Tinto in Spain. Most of the copper leaching operations produce a leach liquor averaging no more than 2 gm/liter copper in solution. Unless the leach liquor is concentrated by some other means such as ion exchange or solvent extraction, the recovery of metallic copper by electrolysis or by reduction with a chemical reagent is not economically feasible. Another well-established process is the precipitation of gold by zinc powder from cyanide solutions. Many other cementation processes are in use especially in the refining steps of the more rare metals, such as the precipitation of cadmium by zinc prior to the electrolysis of acidic zinc sulfate solution and the cementation of palladium and platinum with zinc dust in the electrolytic refining of gold.

1. *Theory*

The three principal reactions in the precipitation of copper from sulfate solutions by metallic iron have been established by Wartman and Roberson (W7) to be the following:

$$CuSO_4 + Fe \rightarrow Cu + FeSO_4 \qquad (1)$$

$$Fe_2(SO_4)_3 + Fe \rightarrow 3\,FeSO_4 \qquad (2)$$

$$H_2SO_4 + Fe \rightarrow FeSO_4 + H_2 \qquad (3)$$

The first reaction represents the stoichiometric relation in which 0.88 lb of iron precipitates 1 lb of copper from solution. The iron consumption in practice is around 2 lb/lb of copper produced. The second reaction is due to the presence of trivalent iron in solution, the higher its concentration, the more iron is needed to precipitate the copper. The third reaction is due to the addition of fresh sulfuric acid to the recycle solution to maintain it at about pH 2.5. They also observed that the first and second reactions have about the same rate while the third is slower than either (1) or (2). Changes in physical conditions during the precipitation did not affect the rate for (1) and (2), but reducing the contact time during the precipitation significantly reduced reaction (3).

Monninger (M39) reported that to produce a cement copper of high purity and of a size that is relatively coarse, the following conditions must be used: (1) pH control must be initiated to balance hydrolysis and iron consumption due to acid; (2) the iron precipitant must be light and clean with a high surface area per unit volume; (3) a uniform flow of solution at high velocities in relation to the iron must be maintained; and (4) the mechanical design of precipitators must feature adequate circulation of solution with emphasis on uniform and high-velocity flows.

Nadkarni *et al.* (N1–N3) studied the rate of reduction of copper by iron from aqueous solutions, taking into account such parameters as the copper and hydrogen ion concentrations, geometric factors, flow rate, and temperature. An increase in stirring rate accelerated the precipitation of copper by iron reaching a constant value at higher stirring speeds. The precipitates formed at low speeds were spongy in character, becoming fine powders at high speed, and at still higher stirring speeds approaching a colloidal size. The iron impurities caused the precipitated copper to adhere strongly to the iron, causing the reaction to slow down due to a decrease in the available reaction interface area. The precipitation process was found to be first order. The presence of oxygen in solution resulted in excess iron consumption, while lower iron consumption was observed under a hydrogen atmosphere.

The kinetics of the early stages of cementation was studied by von Hahn and Ingraham (V12) using a palladous perchlorate–copper system. This system was chosen for study to take advantage of the simple ion-for-ion reaction and of the absence of side reactions such as the reduction of hydrogen ions and the anion effects in perchlorate solutions. Two stages in the deposition were observed; the first was controlled by the diffusion of Pd(II) ions to the copper surface and/or by chemical reaction at the surface; the second stage was controlled by the diffusion of copper ions from the copper surface, through the deposit, and out to the main bulk of the solution. The extension of this study to silver cementation on copper has been reported by von Hahn and Ingraham (V11). The cementation rate was found to be rapid in acid media with the deposits as finely divided, loosely adhering powders and slower in cyanide solutions with the deposits dense and adherent. Ingraham and Kerby (I1) studied the cementation of cadmium on zinc in buffered sulfate solutions in the presence of impurities as functions of ionic concentration, disk rotation speed, and temperature.

A discussion of the chemical behavior of metals on the basis of their fundamental electronic properties and how they affect metal processing techniques has been presented by Franklin (F15). The characterization of all three methods of metal reduction from aqueous solutions in terms of the electrochemical nature of the associated reactions and the aspects of electrode reactions of importance to all heterogenous reduction systems has been presented by Wadsworth (W1). Nolfi et al. (N6) have presented analytical expressions to represent the exact phenomenological analysis of precipitate growth. These equations include the effects of diffusion controlled migration, interface controlled migration, and mixed control in a single development.

2. Practice

Huttl (H36) described the modified operation of a cementation launder to improve the copper recovery from an oxide ore leaching operation. The leached liquor, containing about 20 gm/liter Cu is precipitated with iron in three stages to provide new surface area and thus ensure complete stripping of the solution. To minimize the channeling of solution through the iron mass in the launders, the solution is introduced into gutters in the floor of the launders and forced to percolate upward through the iron at high velocity. The barren solution contains 0.01 gm/liter Cu and iron consumption is down to $1\frac{1}{4}$ pound per pound of copper produced.

The development of a precipitation cone-type recovery system for use with scrap iron to produce more granular copper precipitate of higher purity at a lower iron consumption than previously possible with other types of precipitators has been described by Spedden, et al. (S20). The operation of this cone precipitator is based on a high-velocity rapid throughput of copper-bearing solutions and good contact between the solution and the clean iron precipitant. This unit is compact and readily automated. It has low iron consumption and is self-cleaning of copper precipitates. The conclusions derived from the kinetic data developed by Wadsworth and co-workers (N1–N3) were successfully translated to a workable full size precipitator. The copper-bearing solutions are pumped into the bottom of the inverted cone through six pressure manifolds that create a vortex and the iron is introduced at the top. The continuous flow of solution from the bottom moves the precipitated copper upward where it overflows into the holding tank. Cement copper produced in this equipment analyzes 90–95% Cu, 0.1–0.2% Fe, 0.1–0.2% silica, 0.1–0.2% alumina, and the balance primarily oxygen. Operation of two cones in series has also given good results.

The use of particulate iron precipitant such as sponge iron gives a relatively faster copper precipitation rate than scrap iron of varying sizes. Back (B1, B2) reported the development of an inverted cone type pre-

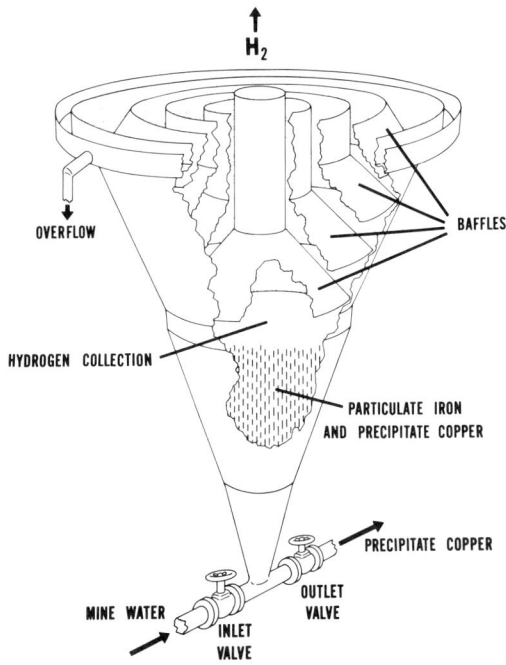

FIG. 8. Kennecott's precipitation cone showing relationship of baffles to distribution. Back (B2).

cipitator, shown in Fig. 8, that dynamically suspends a bed of precipitant at reasonable flow rates resulting in a rapid precipitation rate with iron utilization close to the stoichiometric value. Baffles are provided at the top to channel the hydrogen gas evolved during the reaction which is used as a measure of the extent of the precipitation reactions. This is used as an indicator for process control. The most satisfactory operating procedure is the addition of measured batches of precipitant, with continuous discharge of barren solution at the overflow and intermittant discharge of precitates from the apex of the cone. Under proper operating conditions better than 99% of the metallic iron was utilized to produce metallic copper. Particulate iron up to 10 mesh has been successfully used, although −35 mesh has given the best results.

Fitzhugh and Seidel (F3) found that nickel could be precipitated almost quantitatively from acidic solutions by powdered iron when the reaction was carried out in a closed vessel at around 150°C for 60 min. The cementation product obtained from synthetic liquors analyzed up to 50% Ni and from clarified leach liquor between 25 and 35% Ni. The precipitated nickel was found to be magnetic. Elemental sulfur in the leach slurries needed to be reduced to less than 0.5% before a magnetic nickel precipitate could be produced. Nickel recoveries in excess of 98% were obtained. Chemical reactions analogous to the cementation of copper with metallic iron were proposed.

B. ELECTROLYSIS

The application of an electromotive force to an ionic solution to bring about some desired chemical reaction is taken advantage of in the reduction of metal ions from their aqueous solutions. The passage of a current through a solution of electrolyte causes the movement of the ions toward the electrodes and brings about a gain or loss of electrons. The electrode reactions involved are the deposition of the metal at the cathode and, usually, the liberation of a gas at the anode. Faraday's law of electrolysis governs the extent of the chemical reaction and the quantity of electricity required to bring it about (D11). The techniques involved in industrial electrochemistry are well developed and are covered in some detail in many texts (D12, M12). Two of the most commonly used electrolytic methods are the electrowinning and the electrorefining techniques.

1. *Electrowinning*

The electrolytic reduction of metals ions from purified leach liquors using an inert anode electrode is known as electrowinning. This method requires higher voltages in the electrolytic cells than electrorefining. Electrowinning is applied commercially to copper, zinc, antimony, cobalt, manganese, chromium, iron, gallium, and silver. Figure 9 shows the electrowinning cell used by International Nickel Company.

Walden et al. (W5) made a study of the electrolytic refining of blister copper at current densities of up to 500 A/m² in specially constructed channel cells. Operating difficulties at high current densities are due to Cu^{++} impoverishment of the cathode film and the Cu^{++} enrichment of the anode film. By increasing the relative motion of the electrolyte and the electrodes, the diffusion distance is shortened and the concentration gradient is increased. A channel-shaped electrolysis cell was found to work well at semiindustrial scale in refining anode copper with high-impurity contents at current densities much higher than presently used in the industry.

An examination of the effect of using ultrasonic energy to improve the

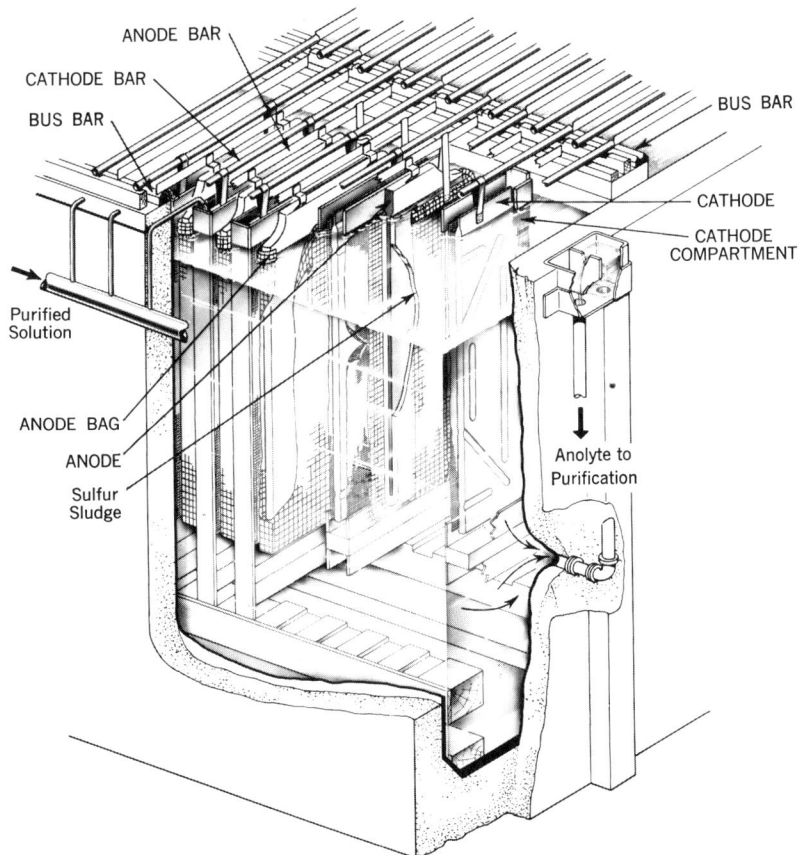

FIG. 9. Cutaway view of Inco's matte anode electrowinning tank. Boldt and Queneau (B25).

electrolytic deposition of copper and zinc at frequencies of 20, 26, 38, and 400 kHz and at accoustical intensities of 0.25–1.5 W/cm² has been reported by Kenahan and Schlain (K3).

Well-consolidated and adherent copper deposits were obtained from acid copper sulfate solution at 50°C and current densities up to 300 A/ft² using ultrasonics. Without ultrasonics, the copper deposits normally become more spongy and less adherent above 30 A/ft². As much as 3-V decrease in cell voltages were observed at the higher current densities. The lower frequencies were more effective than 400 kHz.

Large increases in cathode and anode current efficiencies and large decreases in cell voltages over a wide range of current densities were obtained in the electrodeposition of copper from cyanide electroplating solution at 35°C. Similar results were obtained in the electrodeposition of zinc from a cyanide electroplating bath at 40 and 50°C. The electrolytic deposition of zinc from an acid zinc electroplating bath at 35°C increased the anode efficiency up to 12% at the lower current densities. The cathode current efficiency increased by 34% at a current density of 100 A/ft^2 in the electrolytic recovery of zinc from a zinc sulfate solution at 35°C when ultrasonics was used.

2. *Electrorefining*

In electrorefining, the metal to be refined is used as the anode which dissolves in the electrolyte and is deposited as electrolytic-grade metal at the cathode. The impurities present in the anode remain on it, fall off to the bottom of the cell as slime, or go into solution but are prevented from moving toward the cathode by precipitation with some chemical reagent such as another metal added to the electrolyte. The buildup of metallic impurities that are dissolved but not deposited at the cathode is reduced by circulation of fresh electrolyte through the cells. Electrorefining techniques are used in producing gold, silver, copper, nickel, cobalt, lead, tin, antimony, bismuth, indium, and mercury.

The conventional smelting and refining process for copper, nickel, lead, and zinc sulfide ores result in the evolution of a large amount of sulfur dioxide gas which in most instances is released to the atmosphere. There have been attempts to recover part of the sulfur as sulfuric acid or in some form (H1, H4, H5) at various stages of the pyrometallurgical processing.

Habashi and Torres-Acuna (H6) described a process that involved the direct recovery of copper and elemental sulfur by anodic dissolution of copper (I) sulfide (white metal). The white metal produced from the copper matte analyzed 77.85% Cu and 19.20% S. A closed vessel was used to prevent the evaporation of the copper (II) sulfate acidified with sulfuric acid electrolyte. The electrolysis was carried out at room temperature. The sulfur originally present in the white metal went into the slime as pure elemental sulfur. The elimination of the converting and poling steps in the conventional copper smelting and refining process can be realized by use of this technique.

Spence and Cook (S21) described the commercial operation of the Thompson Refinery of International Nickel Company in Manitoba where the nickel sulfide matte is melted and cast into anodes containing 76% Ni, 20% S, and the balance made up mostly of copper, cobalt, and iron. The direct casting of the sulfide matte into anodes eliminated two pyrometallurgical operations involving the sintering of the matte and the reduction melting of the oxide. Copper, iron, arsenic, and lead impurities are eliminated during the process of electrolyte purification to produce electrolytic grade nickel. Other by-products are cobalt oxide and elemental sulfur.

Caron (C9) developed an electrolytic method for separating cobalt and nickel from alloys of the two metals. The ammoniacal electrolyte composition is chosen to favor the plating of cobalt from nickel. The electrolyte is passed through a number of cells in series, with continuous deposition of cobalt at the cathode and the discharge of a cobalt-free high-nickel electrolyte. Distillation of this solution after the addition of $Ca(OH)_2$ or NaOH converts the nickel to the hydroxide and the regenerated ammonia is returned to the process.

C. Chemical Reduction

The addition of a chemical reducing agent to precipitate metals from their salt solutions has been considered for many years as an alternative to displacement reactions and electrolysis. The developments in the field of powder metallurgy in recent years have increased interest in this reduction method since the final product obtained is high-purity metal powder. The ease with which these high-purity powders can be converted into readily handled forms such as briquettes has improved their marketability to other metallurgical users.

Solids as well as gaseous reducing agents have been successfully used to precipitate metals from their purified aqueous salt solutions. The high reaction rate displayed by gaseous reducing agents, even with solutions of low-metal contents, and the possibility of selective precipitation of several metals from the same solution when the difference between their reduction potentials is sufficiently large are some of the major advantages of this technique. Nickel, cobalt, and copper are the primary metals presently produced in large tonnage by gaseous reduction.

1. *Gaseous Reducing Agent*

a. Hydrogen Reduction. The most commonly used gaseous reducing agent for the precipitation of metals from their salt solutions in present-day hydrometallurgical processing is hydrogen. From a process engineering standpoint, its use decreases the weight and volume of the gas in the process streams, making possible the design of compact mass-transfer equipment. The fact that the reaction products from the reduction of the purified aqueous salts are either hydrogen or hydroxyl ions practically eliminates the problem of contamination. In addition, hydrogen gas is available at a relatively low price in commercial quantities.

Schaufelberger (S2) reviewed the early work on the gaseous precipitation of metals from their salt solutions and reported the work done by Chemical Construction Corporation (now part of Sherritt Gordon Mines, Ltd.) on metal reduction with hydrogen. The precipitation of a metal from purified leach liquor was shown to be controlled by solution compositions, pH, maintenance of proper metal ion concentration through complex formation and hydrolysis, and sufficient agitation to keep the metal in suspension and the hydrogen gas uniformly distributed throughout the solution. The temperature and pressures were maintained above equilibrium values to obtain favorable process kinetics. The reduction of metal from aqueous solution by hydrogen involved nucleation, growth, and agglomeration. It was found necessary to induce nucleation by addition of seed crystals or some other catalyst. The chemistry and thermodynamics of the reduction of copper, nickel, cobalt, and cadmium from their solutions by hydrogen under pressure are discussed by this author. The selective reduction of copper, nickel, and cobalt separately from aqueous solution by hydrogen gas has been described by Schaufelberger and Roy (S3).

Evans (E10) recently reviewed the process technology and chemistry of gaseous redutcion of metals from solution with special emphasis on the present-day commercial operations using hydrogen gas. The many process improvements developed at Sherritt Gordon such as the discovery of new catalysts and the conditions for their use to initiate metal reduction, the prevention of plastering of the powder on the walls of the reduction vessel, the control of the particle size distribution of the copper powder, and the production of low apparent density, high surface area nickel powders were discussed.

b. Others. Other gaseous reducing agents such as carbon monoxide and sulfur dioxide have been used successfully to reduce metals from their solutions. Meddings and Mackiw (M25), Burkin (B37), and Evans (E10) have reviewed the reaction chemistry, thermodynamics, kinetics and catalysis of gaseous reducing agents in the precipitation of metals from their salt solution. The use of SO_2 may lead to contamination with sulfur; and CO, besides being more expensive than hydrogen, reacts with metals such as nickel to form the toxic nickel tetracarbonyl which requires a few more processing steps before metallic nickel can be produced (B25).

2. *Solid Reductant*

a. Inorganic. The production of coatings of some metals on solids suspended in the solution by inorganic or organic chemical reducing agents such as hypophosphite or hydrazine is generally known as electroless plating. A catalytically active surface at which a heterogeneous reaction between the reducing agent and metal ion can occur is necessary in order to produce metal as a coherent layer on the solid. The initiation and maintenance of catalytic activity on the solid surface to be coated is an important step for success of the process. A solution of the metal to be plated and a reducing agent initially added at a relatively high concentration or continuously added during the plating reaction makes up the electroless plating bath. Proper solution temperature and pH control are necessary for optimal results. A complexing agent is usually introduced into the solution to prevent the formation of metal hydroxides and insoluble salts. Brenner (B28), Saubestre (S1), and Burkin (B37) have recently reviewed the technique of electroless plating.

b. Organic. The gold and the palladium and platinum from the discarded electrolyte formed during the electrorefining of gold is normally recovered by precipitating the gold with sulfur dioxide and then mixing the filtrate with zinc dust to precipitate a palladium–platinum concentrate. Elkin and Bennett (E5) developed a process using oxalic acid to precipitate the gold after adjustment of the pH of the discarded electrolyte to 6 with sodium hydroxide. The gold-free filtrate is heated with live steam and powdered sodium formate is slowly added with agitation to prevent the effervescing bath from foaming over. A palladium–platinum concentrate with a typical analysis of 87.5% palladium, 8.0% platinum, and 0.06% gold is produced, which is a marked improvement over product from the older process.

VI. Hydrometallurgical Operations

Hydrometallurgical processing has in the last thirty years found increasing application in the production of many different metals from various raw materials. The copper industry in the United States has maintained the lead in the adaptation of this method to their operations as shown by their production of 15% of the annual output of metallic copper for 1969. The advent of nuclear power helped spur the scientific and technological development needed for the production of uranium oxide and other high-purity metals such as beryllium, vanadium, columbium, hafnium, tantalum, and molybdenum. The chemical processing of nickel and cobalt, as exemplified by the ammonia process at Sherritt Gordon and the acid leaching at Moa Bay and Nicaro, has contributed to the growing maturity and acceptance of hydrometallurgy in the recovery of metals from their ores and other secondary sources. Selected commercial processes are taken to illustrate the broad range of processing capabilities that are necessary in setting up a successful process flow sheet when the raw materials are derived from a large number of different sources to produce a few end-products of simple composition. Such is the case in the mineral and metallurgical industry.

A. *Copper*

The largest tonnage of metal produced by hydrometallurgical processing is copper. This is mostly produced by acid leaching of copper waste dumps, in-place leaching of worked-out mines and of newly mined oxidized ores. The application of newly developed hydrometallurgical process technology can readily improve its share of the total copper production in the United States from the present 15%. The recovery of copper from copper sulfide concentrates and scrap material is a promising area of application. The process flow sheets for the recovery of high-purity copper powder from cement copper, copper oxide ore, and scrap materials by different routes follows.

1. *Feed Materials*

   a. *Cement Copper.* The Arizona Chemcopper Company (Y1, Y2) produces 25 tons of high-purity copper powder per day by leaching cement copper with aerated sulfuric acid and reducing the dissolved copper with hydrogen gas under pressure. The flowsheet is shown in Fig. 10.

   The feed to the refinery averages 82% Cu (dry basis), 2.6% Fe, 0.4% Pb, 0.07% Sn, 0.5% insoluble in acid (dry basis), and 25% moisture. Fifteen-ton lots of the cement copper are slurried to 50% by weight with regenerated acid recycled from the reduction operation and makeup acid to give a copper concentration of the slurry averaging 90 gm/liter. Leaching is carried out in an agitated tank at 180°F while air is introduced to oxidize the metal. Adjustments in the solution pH control the iron buildup in the circuit. The pregnant solution is clarified by pressure filtration using filter aids and by continuoually circulating the content of a 90,000-gal-capacity tank through a second pressure filter for further clarification. To reduce the dissolved copper, two 3600-gal batches of the clarified solution are treated with hydrogen in two mechanically agitated autoclaves operated at 300°F and a total pressure of 425 psig with the partial pressure of hydrogen maintained at 350 psig. Around 2400 lb copper is precipitated from each 3600 gal of solution. Regeneration of the sulfuric acid takes

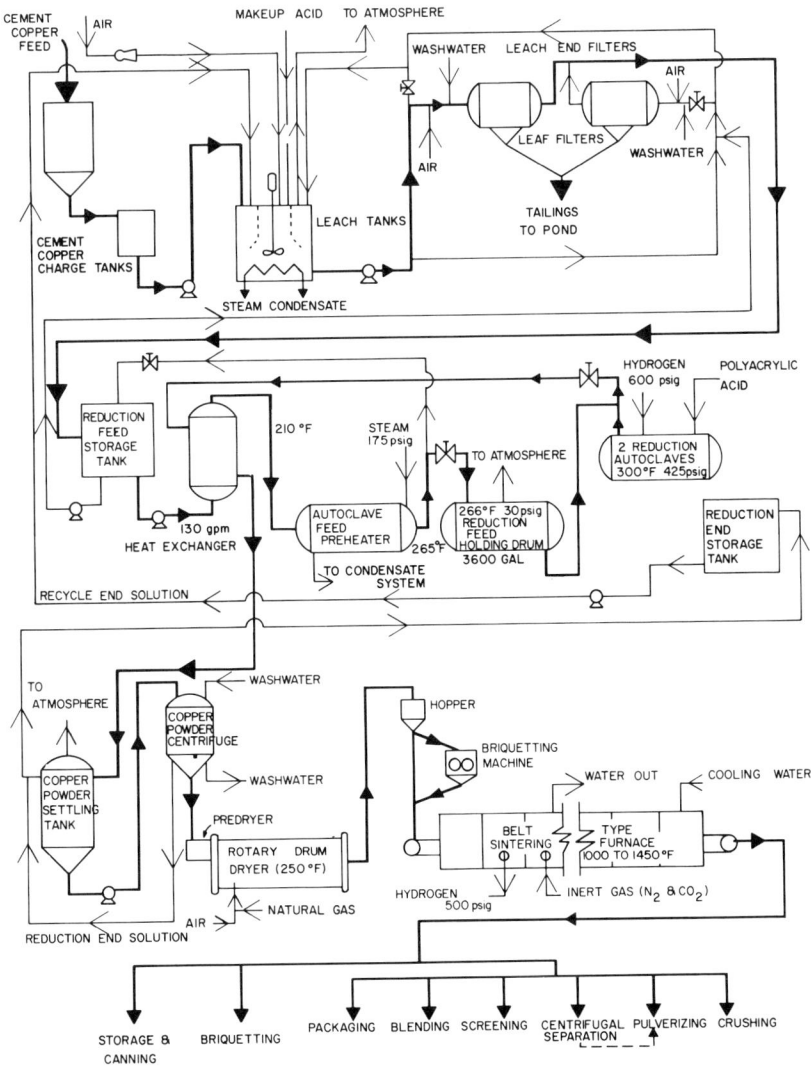

FIG. 10. Arizona Chemcopper Company, process flow sheet. Yurko (Y1).

place during the precipitation of copper powder. To minimize plating and to control particle size distribution requires the addition of 0.008 lb polyacrylic acid/lb precipitated copper powder each reduction cycle.

The copper powder is separated by decanting the reduction end solution in the settling tank. The copper powder slurry is centrifuged and washed.

The high-purity powder contains about 15 wt. % water and 0.05–0.1% carbon from the decomposition of polyacrylic acid used in the reduction autoclaves. Drying the copper powder in the presence of 2–3 wt. % oxygen is necessary to accelerate the removal of carbon during the sintering under an atmosphere of hydrogen.

b. Oxide Ores. The PMC-Powdered Metals Corporation (R8) has developed the Harlan process for recovering copper from small marginal orebodies in which a complete cycle from ore to 99.9% copper powder takes only 4 hr. The entire plant can be dismantled and moved to a new ore body by truck. A plant is now in operation at Aguila, Arizona with a design capacity of 2.5×10^6 lb/yr copper powder from oxide ores. The $-\frac{1}{4}$-in. crushed ore is leached with spent solution from the electrowinning step containing 8–9 wt. % sulfuric acid (93–95% H_2SO_4) in demineralized water. The copper deposition in the electrolytic cells is promoted by the addition of 1 part by weight of alumina. A 1:1 ratio of ore to leaching solution is maintained. Up to 75% of the copper is leached in about 1.5 hr by cycling of the leaching solution through the ore in the glass-fiber leaching tanks. The enriched solution above the ore level is drained off to the pregnant solution glass-fiber settling tanks and the liquid below the ore level is drained off to the slime settling tanks. Demineralized wash water is flushed through the leaching tanks and is added to the pregnant solution settling tanks. The soluble copper recovery at this point is increased to 90%. The copper recovery is further increased to 95% by passing the tailings through a washer and then to a pond where it is leached for 20 days by spraying wash water on it. The barren residue is then moved to a permanent dump. The washwater is added to the leaching tanks. The slime and pregnant solutions are allowed to settle for 48 hr and are filtered to avoid contamination in the cells. A chelating agent is added to the clear solution.

The electrolytic cells use a parallel-electrode arrangement with a lead–antimony alloy of specific composition (88–91% Pb, 9–12% Sb) for anodes, and 99% nickel sheet for cathodes. A three-cell facility has a capacity of 1.25×10^6 lb copper/yr. The temperature during the electrolysis is not allowed to exceed 140°F, by making use of the heat generated to preheat the entering electrolyte, whose average temperature is about 59°F. The initial cathode current density is about 125 A/ft², gradually rising to 250 A/ft². The copper, instead of adhering to the cathodes, falls to the bottom of the cells. The spent electrolytic solution, containing 15 gm/liter Cu, is drained from the cell and used to leach the next charge of copper-rich ore. The copper pulp from the cell (1–25 μm) is collected and washed with hot, deaerated water to remove gas and oxidized particles. The wash

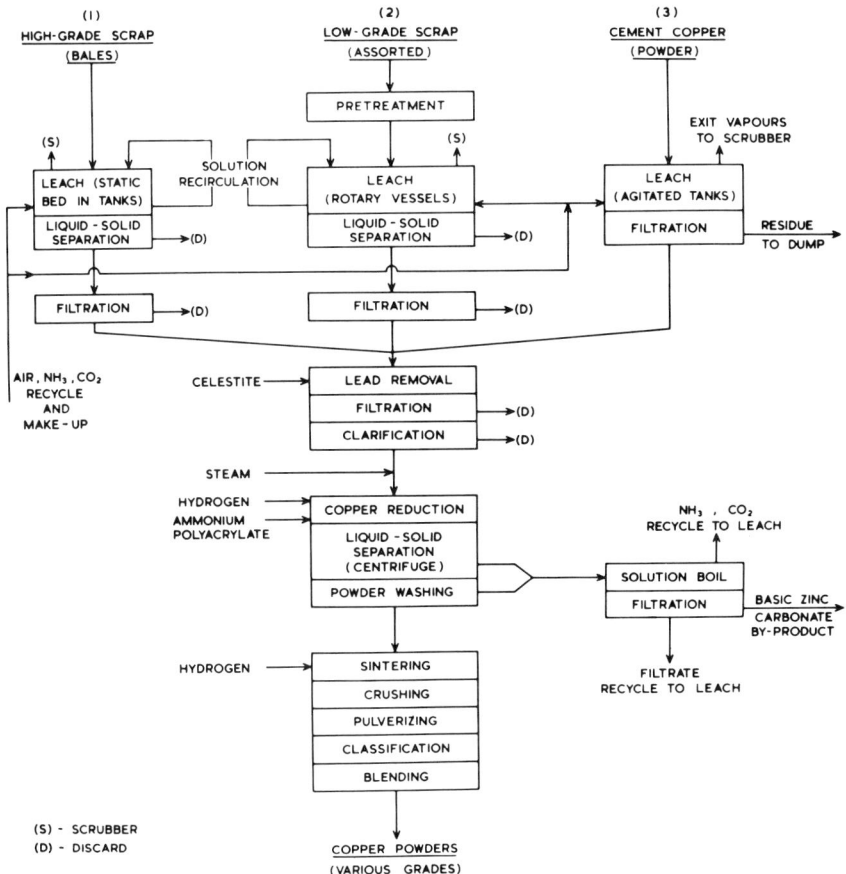

Fig. 11. Universal Minerals and Metals Inc., simplified flow diagram for production of copper powder from scrap, ammonium carbonate process. Evans (E10).

water is filtered to remove entrained solids and combined with demineralized water for reuse as washwater. The washed pulp is dried and is sold as powder or can be processed into ingots, bars, or sheets.

c. Scrap Metal. The Universal Minerals and Metals, Inc. (H17, R17, T18, W11) in Kansas City, Missouri is producing 5–6 tons/day of copper powder from various copper scrap materials by leaching with ammonium carbonate at atmospheric pressure and at 120–140°F followed by the reduction of the copper from the purified leach liquor with hydrogen. The process flowsheet is shown in Fig. 11. The high-grade baled copper scrap is batched-leached by recirculating ammoniacal ammonium carbonate solution through the bed of scrap or by leaching low-grade scrap in a rotating vessel until the copper concentration in the solution reaches 140–160 gm/liter Cu. When cement copper is used as the raw material, agitated leach tanks are used. The clarified pregnant leach solution, containing two-thirds of the copper as cuprous ammines and one-third as cupric ammines, is treated with strontium sulfate in the form of ground natural celestite to remove the small quantity of soluble lead and tin. Two stages of filtration follow.

Reduction of the copper in solution is carried out in an agitated horizontal autoclave at 400°F and a hydrogen pressure of 900 psig. Plastering on the walls of the autoclave is prevented and an acceleration of the reduction rate is accomplished by the addition of ammonium polyacrylate. The slurry is centrifuged and washed. Surface oxidation of the powder during the drying step is allowed, so as to effectively remove the organic contaminant by reaction with the copper oxide during the sintering stage in hydrogen at 1100°F to 1300°F. The sintered cake is broken up; air classification of the cake fragments produces various standard grades of copper powder.

Ammonia and carbon dioxide are regenerated by boiling part of the reduction end solutions. This also maintains the water balance. The basic zinc carbonate in solutions is precipitated during the boiling step as a by-product. The remaining portion of the reduction end solution is used for leaching. As much as 60 gm/liter of Zn and 40 gm/liter of SO_4 have been circulated in the closed circuit without affecting the processing. The average copper powder, as reduced, is 97.5% copper and the sintered product averages 98.9% Cu. The reduction end solution contains 1.5 gm Cu/liter.

B. Molybdenum

The metal values associated with other minerals but not amenable to separation and concentration by the standard beneficiation techniques such as flotation can usually be recovered by chemical methods. The Climax Molybdenum process to recover the "molybdenum oxide" values associated with the sulfide ore and the Union Carbide process for the separation of molybdenum from tungsten in a floatation concentrate are in this category.

1. *Flotation Product*

a. Molybdenum Oxide. The Climax Molybdenum Company (M30, W17) developed a hydrometallurgical process to recover molybdenum oxide values, in association with sulfide ore, from the oxide ore itself or from oxide molybdenum tailings from previous operations. The tailings from the sulfide molybdenum flotation, containing 0.12–0.14% "oxidized molybdenum," are upgraded in three stages of cycloning to 0.25–0.35% molybdenum and to 50% solids. This feed to the leaching stage is preheated with the hot acid slurry tailings from the adsorption circuit using a heat exchanger. A sulfur dioxide–sulfuric acid leach at atmospheric pressure for 12 hr extracts 95% of the molybdenum from the concentrate in the form of "molybdenum blue." The unreacted SO_2 is removed from the pulp by desorption under vacuum and the pulp is discharged from the agitated vessels at 140°F with a residual sulfur dioxide content of 0.5 gm/liter. Air is injected into the pulp to remove residual sulfur dioxide and to increase the rate at which molybdenum is absorbed on the charcoal.

Charcoal adsorption is carried out in gently agitated tanks for 12 hr by contact of −8 + 20 mesh activated charcoal with the leach pulp containing 47% solids at 140°F. The charcoal absorbs 96% of the molybdenum blue and other molybdates; it is recycled through the circuit until the molybdenum loading on the charcoal is 8–10 wt. %. Three pounds charcoal/ft^3 pulp gives satisfactory results. A 35-mesh vibrating screen separates the loaded charcoal from the ore. After washing, the loaded charcoal is treated with an air–ammonia mixture to dissolve the adsorbed molybdenum. A high-grade strip liquor is obtained by passing high-purity water or condensate in a countercurrent flow through 2–3 columns of charcoal. Residual ammonia is removed from the charcoal by an acid wash and a portion of the charcoal from every cycle is regenerated by heating at 1475°F for about $\frac{1}{2}$ hr in the presence of combustion gases and steam. The high-grade stripped liquor containing 70 gm/liter molybdenum as ammonium molybdate, plus some sulfate, phosphate, and free ammonia is reacted with a stoichiometric amount of magnesium sulfate to remove the phosphorus as magnesium ammonium phosphate precipitate. A two-effect evaporator–crystallizer is used to produce ammonium paramolybdate crystals. A bleed of liquor from the evaporator–crystallizer is necessary to reduce the sulfur content in the product. The molybdenum content in the bleed stream is precipitated with acid at pH 5 and by solvent extraction treatment of the resultant filtrate. The strip liquor from the solvent extraction is recycled to the phosphate precipitation step. Molybdic oxide (99.8 MoO_3) is produced by calcination in an indirect-fired kiln at 1075°F of the ammonium paramolybdate crystals and the molybdic acid precipitate. The ammonia released during the crystallization and calcination is reprocessed for reuse.

b. Molybdenum Sulfide. The Union Carbide Corporation's operation (M22) in Bishop, California used a process for separating molybdenum from tungsten the flotation concentrate by chemical means. The minerals in the ore consisted of molybdenite, chalcopyrite, covellite, chalcocite, bornite, scheelite, and powellite. Grinding to 85% -65 mesh was necessary to liberate the minerals. A pressure leaching process was used to convert the tungsten and molybdenum into soluble tungstate and molybdate. Molybdenum sulfide was precipitated, filtered off, and roasted to molybdenum trioxide. The tungsten was recovered as calcium tungstate precipitate and was nodulized. The tungsten recovery was 98% and the molybdenum recovery, 95%.

C. NICKEL

The Sherritt Gordon operation in Fort Saskatchewan, Alberta, Canada produces 30×10^6 lb nickel and 10^6 lb of cobalt annually by using both an ammonia leaching process developed by Professor F. A. Forward of the University of British Columbia under the sponsorship of Sherritt–Gordon (F5) and the metal hydrogen reduction technique developed by the Chemical Construction Company (S2). The copper–nickel ore averages 0.83% Ni, 0.49% Cu, and some cobalt. This ore is beneficiated by flotation into a copper concentrate containing 28–30% Cu with less than 1% Ni and a nickel sulfide concentrate containing 10% Ni. 2% Cu, 0.4% Co, 33% Fe, and 30% S. The nickel is mainly present as pentlandite, the copper as chalcopyrite, the iron as a pyrrhotite and pyrite, and the cobalt in pentlandite and as Co–Ni–pyrite. The simplified process flow sheets for the nickel and cobalt refineries are shown in Figs. 12 and 13.

The processing of nickel oxide ores is illustrated by the sulfuric acid leaching of nickeliferous laterites and the precipitation of high-purity nickel and cobalt powder by reaction with hydrogen gas giving ammonium sulfate as by-product. This process is in use at Moa Bay, Cuba.

1. *Beneficiated Concentrate*

a. Nickel Sulfide. In the Sherritt Gordon process (B25), the nickel sulfide flotation concentrate is leached in the presence of air and ammonia in agitated autoclaves operating under a pressure of 100–110 psig and at a temperature of 170–180°F. The leaching is a continuous, two-step, countercurrent operation. The leach liquor from the second set of autoclaves is used to leach the fresh concentrate in the first set of autoclaves, extracting the readily leachable portions. The partially leached concentrate is contacted with fresh leach liquor high in ammonia in the second set of autoclaves to dissolve the less soluble portions of the metal values.

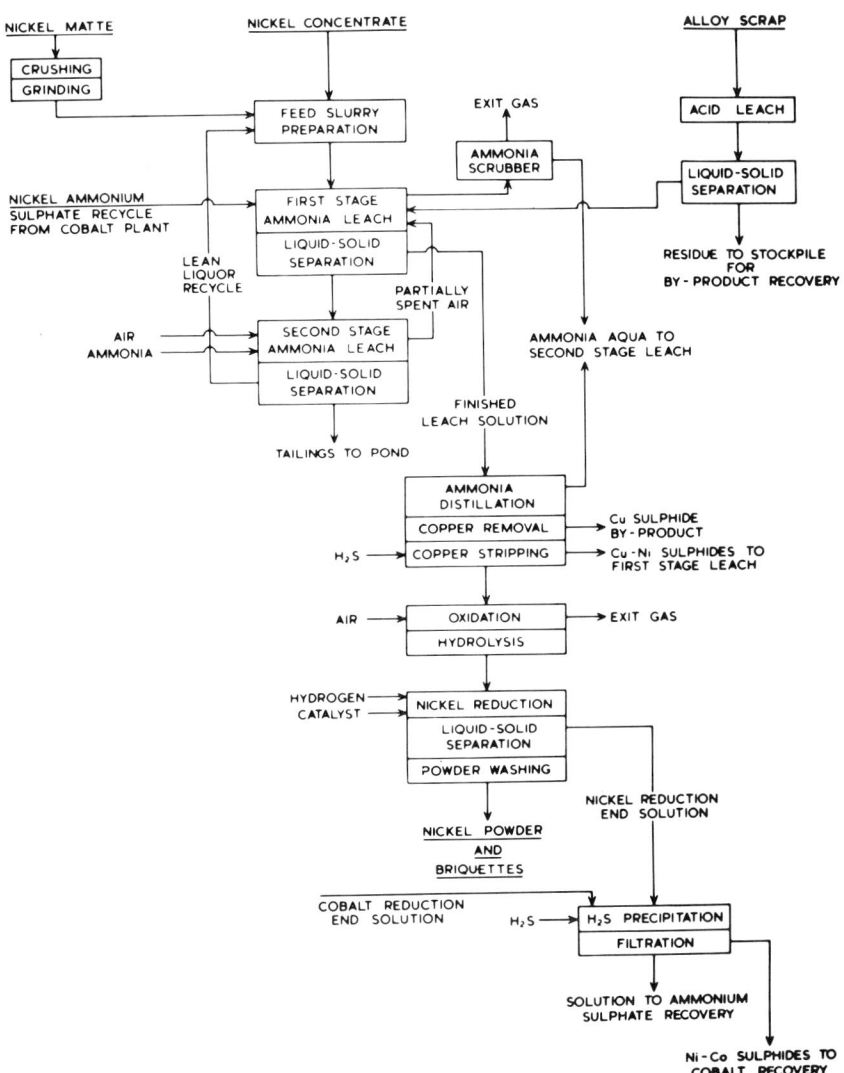

FIG. 12. Sherritt Gordon Mines Limited, simplified nickel refinery flow diagram. Evans (E10).

The leach liquor containing the dissolved metal values is separated from the residue containing iron oxide and other insolubles by thickeners and disc filters. All soluble nickel from the leach residue is recovered by careful washing, repulping, and filtering procedures before being discarded to

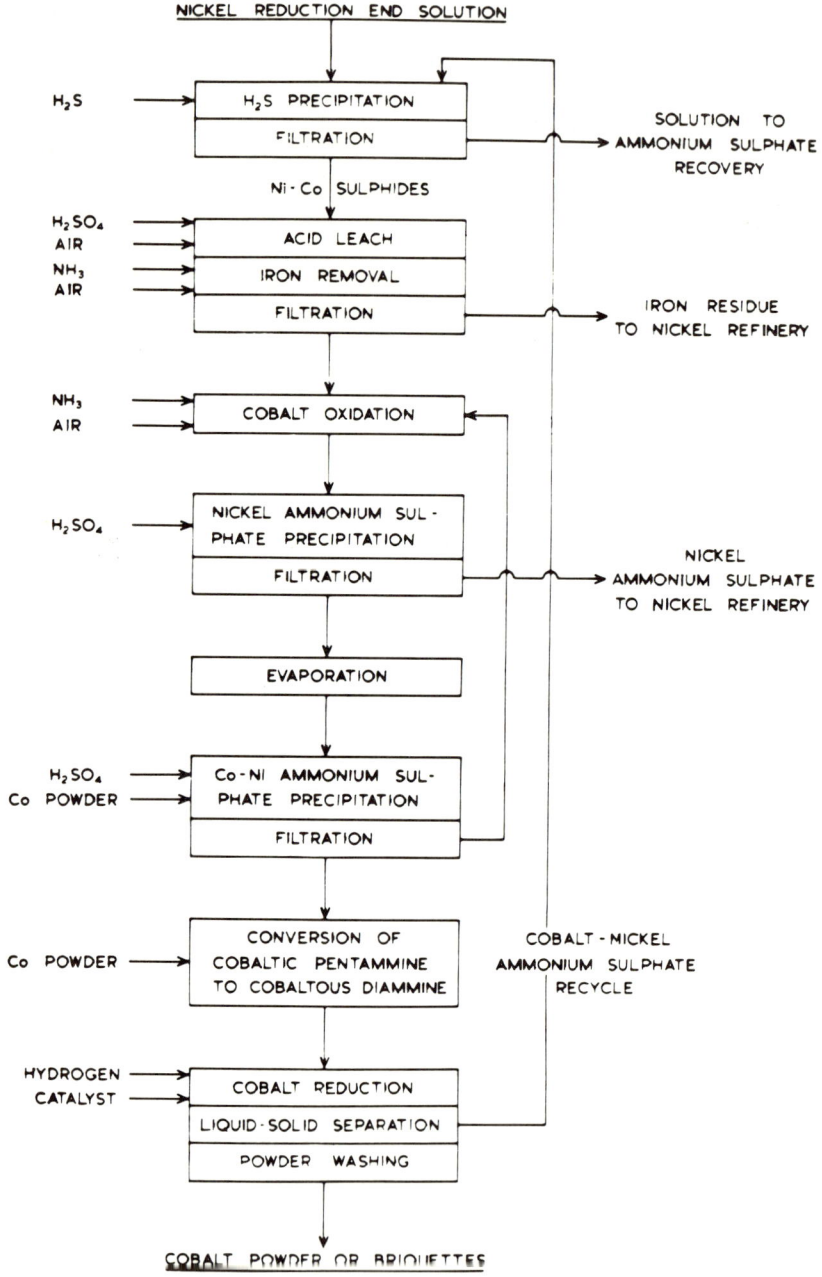

Fig. 13. Sherritt Gordon Mines Limited, simplified cobalt recovery flow diagram. Evans (E10).

the residue ponds. Sulfur is mostly in the form of ammonium sulfate and in a smaller quantity in the form of unsaturated sulfur compounds such as ammonium thiosulfate.

The excess free ammonia in the leach solution is distilled and condensed as 15% aqua which is recycled to the leach circuit. The decrease in ammonia content initiates the precipitation of copper sulfide as a black sludge due to the reaction between copper and the unsaturated sulfur compounds in solution. The precipitation is completed by heating above 250°F. After filtration, copper still remaining in solution is stripped by bubbling with hydrogen sulfide. This copper sulfide precipitate is high in nickel and is recycled to the leaching stage.

The copper-free solution containing nickel, cobalt, ammonium sulfate, ammonium sulfamate, and some unsaturated sulfur compounds undergoes an oxidation–hydrolysis step in an autoclave at 460°F under an air pressure of 600 psig. The unsaturated sulfur compounds are oxidized and sulfamate is hydrolyzed to sulfate. This step prevents the contamination of the fertilizer-grade ammonium sulfate with ammonium sulfamate and of the nickel with sulfur. The oxydrolyzed solution containing about 45 gm/liter Ni, 1 gm/liter Co, and 350 gm/liter $(NH_4)_2SO_4$ is ready for batchwise nickel recovery.

The oxydrolyzed solution is fed into an agitated single compartment autoclave containing a small quantity of fine nickel powder at a temperature of 400°F and under a hydrogen pressure of 500 psig. The nickel in the solution precipitates and no cobalt is precipitated as long as a small amount of nickel is left in solution. At this stage, the agitators are stopped and nickel powder allowed to settle. The depleted solution containing the cobalt is ready for cobalt precipitation. Fresh oxydrolyzed solution is added to the autoclaves and the agitators started to keep in suspension the nickel powder.

After repeating this process about 40–50 times, the solution is drawn off with the agitators running to facilitate the removal of the precipitated nickel. During this step, the molar ratio of ammonia to nickel is maintained at 2:1 to reduce the nickel completely.

The content of the reduction autoclaves is discharged into cone-bottomed flash tanks where separation of the powder from the spent liquor takes place. The nickel metal in slurry form is washed, dried, and packaged as powder or pressed into briquettes, sintered, and packaged ready for shipment. The spent liquor containing about 1 gm/liter Co and 1 gm/liter Ni is precipitated with H_2S, filtered, and sent to the mixed sulfides stage for cobalt recovery.

The mixed sulfides are leached in a six-compartment agitated autoclave at 250°F and 100 psig by the addition of air and sulfuric acid to keep the discharge solution pH at 1.5–2.5. About 97% of the nickel and cobalt are dissolved. The small amount of ferrous and ferric sulfates is removed by agitation at atmospheric pressure with air and enough ammonia to bring the pH to 5.1 in order to oxidize the ferrous salts to the ferric state and precipitate them as hydrated ferric oxide. The clarified solution is mixed with ammonia and oxidized by addition of air in an agitated autoclave at 160°F and under 100 psig to complete the oxidation of cobaltous to cobaltic cobalt to avoid precipitation of cobaltous cobalt with nickel when the solution is acidified.

The oxidized solution pH is adjusted to 2.5 with sulfuric acid. The nickel ammonium sulfate double salt is precipitated, filtered, dissolved in aqueous ammonia solution, and returned to the main leach circuit. The filtrate is evaporated to increase ammonium sulfate and cobalt concentration, and H_2SO_4 is added to remove the residual nickel. The filter residue is returned to the oxidation feed tanks. Cobalt powder is added to the pure cobalt solution to convert the cobaltic to cobaltous, in a process similar to the nickel reduction with hydrogen gas.

The metal-free solution is evaporated to leave behind fertilizer grade ammonium sulfate crystals with a water-soluble nitrogen content of 21% and a sulfur content of 24%.

b. Nickel Oxide. The Freeport Nickel Company (C4, E10, W15) developed a process to treat the nickeliferous laterites at Moa, Cuba in two major parts. The first is the preparation of an upgraded shipping concentrate and the second part is the refining of this concentrate at Port Nickel, Louisiana to high-purity nickel and cobalt powder. The average analysis of the nickeliferous laterite is about 1.35% Ni, 0.13% Co, and 46.5% Fe. The ore is prepared for the leaching plant by slurrying in water and wet screening at 20 mesh and is thickened from 25% solids to 45% solids. The flowsheet for this process is shown in Fig. 14.

The selective leaching of the nickel and cobalt from the gangue material is carried out in Pachuca-type leaching vessels with 98% sulfuric acid at temperatures between 450 and 500°F and with equivalent pressures between 400 and 600 psi. A typical leach liquor contains 6 gm/liter Ni, 0.6 gm/liter Co, 0.6 gm/liter Fe, 2.3 gm/liter Al, and 67 gm/liter SO_4. Each batch is leached for about 90 min to extract 95% of the metal values. Preheating and heating of the ore slurry is by direct steam absorption in the reaction vessel. The barren residue is separated from the leach liquor by countercurrent decantation to a density of 55–60% solids. The solids are washed with very little dilution and about 99.5% of the soluble value are recovered.

The pregnant solution is neutralized to pH 2.5 with specially prepared coral mud containing more than 90% $CaCO_3$. The free acid in the liquor

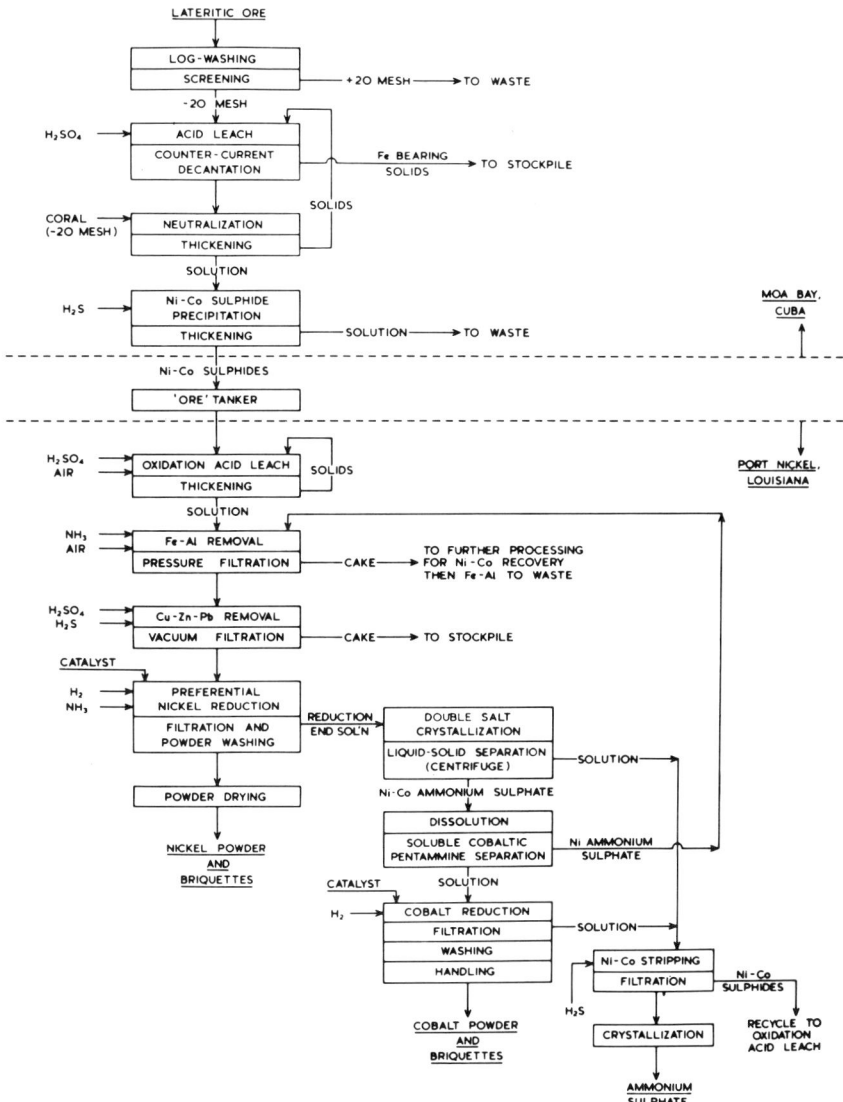

Fig. 14. Freeport Nickel Company, simplified flow diagram for production of nickel and cobalt powders from laterites. Evans (E10).

is readily neutralized due to the high porosity of the coral material. Pretreatment of the pregnant solution with waste hydrogen sulfide eliminates chromate chromium, reduces some ferric iron to ferrous, flocculates any fine ore solids, and reduces the probability of precipitating ferric hydroxide in the circuit. The clarified neutralized liquor is heated to 245°F by direct contact with steam. The precipitation of nickel and cobalt sulfide concentrate is carried out in a bafflled compartmentalized autoclave with high-purity hydrogen sulfide at a total pressure of 150 psia. Over 99% of the Ni and 98% of the cobalt are precipitated as the sulfide. The precipitated sulfides are separated from the excess hydrogen sulfide in flash tanks. The thickened sulfide is washed and prepared for shipment to Port Nickel, Louisiana as a slurry with 65% solids. It contains 54% Ni, 5.3% Co, and 35.2% S.

The nickel–cobalt sulfides as a suspension in dilute sulfuric acid are oxidized in agitated, spherical autoclaves at 350°F in the presence of air at 500 psig and excess sulfides. The unleached solids in the thickener overflow are recycled back to the autoclave and the liquor, containing 50 gm/liter Ni, 5 gm/liter Co as sulfates and small amounts of iron, aluminum, chromium, copper, zinc, and lead, is purified. The iron, aluminum, and chromium are precipitated as the hydroxides by adjusting the pH to 5.5 with ammonia and the copper, lead, and most of the zinc are removed by subsequent addition of sulfuric acid to lower the pH to 1.5 and precipitation of the metals with hydrogen sulfide.

The purified nickel and cobalt sulfates, with low concentrations of ammonium sulfate, are reduced with hydrogen in agitated autoclaves at 375°F and 650 psig. By controlling the pH between 0.9–1.8 by continuous addition of aqueous ammonia, 95% of the Ni is preferentially reduced. The nickel powder is washed, filtered, dried, and packaged or is briquetted and sintered at 1750°F in hydrogen before shipment to alloy manufacturers.

The nickel, cobalt, and zinc in the reduction end solution are precipitated as metal ammonium double salts after solution evaporation to 500 gm/liter ammonium sulfate. The double salts containing the nickel and cobalt centrifuged from the solution are then dissolved in water. Nickel and cobalt are separated by formation of cobaltic pentammine sulfate solution. The cobaltic pentammine solution is reduced at 350°F under hydrogen at 500 psig to produce cobalt powder. The ammonium sulfate by-product is prepared by stripping out the metal values with hydrogen sulfide.

D. Copper–Zinc

The Dowa Mining Company (K9) in Kosaka, Japan has developed a hydrometallurgical process shown in Fig. 15 to treat 2400 metric tons/month of copper–zinc sulfide flotation concentrates. The microscopically fine mixture of copper and zinc sulfides was separated from lead sulfide and barite by flotation. The flotation concentrate analyzed 8.7% Cu,

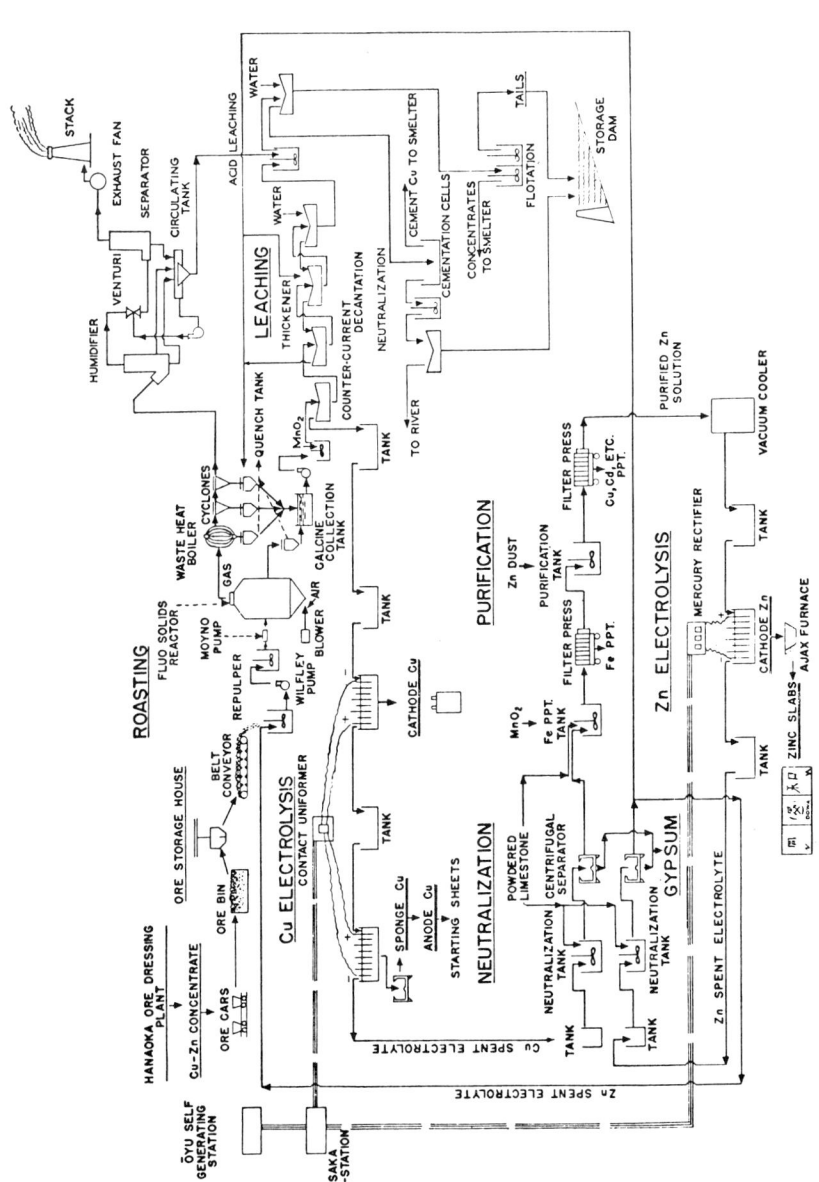

Fig. 15. Flowsheet of the Kosaka hydrometallurgical plant, the Dowa Mining Company. Kurushima and Tsunoda (K9).

15.4% Zn, 21.6% Fe, 2.93% Pb, 32.7% S, and 11.3% insoluble. The concentrates consisting of 80% −200 mesh particles were roasted in a fluidized bed at 700°C with 50% excess air to convert 95% of the copper to copper sulfate, 90% of the zinc to zinc sulfate, and 5% of the iron to ferrous and ferric sulfate. The roaster off gas contains a high SO_2 concentration and is used to produce sulfuric acid. About 95% of the total solids collected from the roaster overflow, waste heat boiler hoppers, and the cyclones were given a 5-hr leach at 70°C with 5–7 gm/liter sulfuric acid solution in agitated wooden vessels. Around 200 lb/hr of pulverized manganese ore containing 60% MnO_2 was added to precipitate the iron. The residue from the primary leach plus the solids collected from the scrubber were leached for 4 hr at 70°C with 50 gm/liter sulfuric acid. The leaching recoveries were 94% for Cu and 88% for Zn. The clarified leach solution from the primary thickener analyzing 54 gm/liter Cu, 100 gm/liter Zn, and 2 gm/liter Fe was sent to the copper tank house. The copper in solution from the single-stage washing thickener was precipitated with iron to produce cement copper which was refined in the smelter. The residue from the single-stage washing thickener was floated to produce a scavenger concentrate containing 20% of the gold, 60% of the silver, 30% of the copper, and 60% of the zinc values from the residue for further recovery of the precious metals and copper in the blast furnace.

The first two stages of copper electrolysis produced cathode copper; and electrolytic stripping of the solution in the shird stage produced sponge copper which was cast into anodes in the pyrometallurgical section. The solution leaving the copper tank house analyzed 1 gm/liter Cu, 100 gm/liter Zn, and 90 gm/liter H_2SO_4. Pulverized limestone was added to neutralize and precipitate iron and the precipitated gypsum was separated by centrifuge. The other metals in solution such as copper, cadmium, nickel, cobalt, etc. were precipitated by zinc powder and β-naphthol. The purified zinc tank house solution was mixed with the cell solution to give 50 gm/liter Zn and 70 gm/liter H_2SO_4 at the start of the electrolysis. This solution was circulated at 3.25 gal/min through the cells until the zinc content decreased to 27 gm/liter and sulfuric acid content increased to 110 gm/liter. The total copper recovery was 93% and zinc recovery was 65%.

VII. Summary

The developments in the field of hydrometallurgy the last thirty years have definitely established its technical and economic potential as a viable alternative to pyrometallurgy in the production of metals. An increasing fraction of the total tonnage of the industrial metals such as copper, nickel, cobalt, chromium, and molybdenum will be produced by this method in the coming years. This will be true especially for copper where research and development work has resulted in a solvent extraction reagent that extracts copper specifically in the presence of other metals in solution. The development of specific reagents for other metals should also be expected during this period. The decreasing reserves of high-grade ores will force the development of the technology needed for large-scale *in situ* leaching of low-grade deposits and closed underground mines where a substantial fraction of the original high-grade ores were left as supports. More of the tailings and waste dumps accumulated over the years will be reworked for their metal contents. Environmental concerns will lead to commercial development of recovering metal values from mine and mill waters as part of their pretreatment prior to being returned to the streams. Altogether, the future of hydrometallurgy as an active field of process engineering is assured.

Acknowledgments

The author would like to acknowledge the secretarial assistance of Mrs. Marcia McCormick and Mrs. Connie McDonnell of the Department of Chemical Engineering, Iowa State University, and the editorial critique given by Mrs. Barbara Warner of the Ames Laboratory, USAEC.

References

A1. "Advances in Extractive Metallurgy," Proc. Symp. Inst. Mining Met., 1967. Inst. Mining Met., London, 1968 (distributed by Elsevier, Amsterdam).
A2. Agers, D. W., and DeMent, E. R., *AIME, SME Prepr.* (1967).
A3. Agers, D. W., Drobnick, J. L., and Lewis, C. J., *AIME, SME Prepr.* (1962).
A4. Agers, D. W., House, J. E., Swanson, R. R., and Drobnick, J. L., *Mining Eng.* (New York) **17**, 12, 76 (1965).
A5. Ambrose, P. M., *U. S., Bur. Mines, Bull.* **630**, 901 (1965).
A6. Anderson, A. E. and Cameron, F. K., *Trans. AIME* **73**, 31 (1926).
A7. Arden, T. V., Davis, J. B., Herwig, G. L., Stewart, R. M., Swinton, E. A., and Weiss, D. E., *Proc. U.N. Int. Conf. Peaceful Uses At. Energy, 2nd, 1958* Vol. 3, p. 396 (1958).
A8. Arehart, T. A., Bresee, J. C., Hancher, C. W., and Jury, S. H., *Chem. Eng. Progr.* **52**, 353 (1956).
B1. Back, A. E., *Trans. AIME* **238**, 12 (1967).
B2. Back, A. E., *J. Metals* **19**, No. 5, 27 (1967).
B3. Back, A. E., Ravitz, S. F., and Tame, K. E., *U.S., Bur. Mines, Rep. Invest.* **4931**, 1–13 (1951).
B4. Baragwanath, J. G., and Chatelain, J. B., *Mining Met.* **26**, 391 (1945).
B5. Baroch, C. J., Smutz, M., and Olson, E. H., *Mining Eng.* (New York) **11**, No. 3, 315 (1959).

B6. Bauer, D. J., Lindstrom, R. E., and Higbie, K. B., *U. S., Bur. Mines, Rep. Invest.* **7100**, 1–12 (1968).
B7. Bautista, R. G., *Philipp. Mining Rec.* **2**, No. 9, 28 (1968).
B8. Bautista, R. G., and Hard, R. A., *Trans. AIME* **227**, 124 (1963).
B9. Bays, C. A., U. S. Patent 2,979,317 (1961).
B10. Beck, J. V., *J. Bacteriol.* **79**, 502 (1960).
B11. Begunova, T. G., *Tsvet. Met.* **30**, No. 7, 14 (1957).
B12. Benedict, C. H., *Mining Sci. Press* **110**, 615 (1915).
B13. Benedict, C. H., U. S. Patent 1,131,986 (1915).
B14. Benedict, C. H., *J. Mining Eng.* **104**, No. 2, 43 (1917).
B15. Benedict, C. H., and Kenny, H. C., *Trans. AIME* **70**, 595 (1924).
B16. Bhappu, R. B., Johnson, P. H., Brierly, J. A., and Reynolds, D. H., *Trans. AIME* **224**, 307 (1969).
B17. Bhappu, R. B., Reynolds, D. H., and Roman, R. J., *J. Metals* **17**, No. 11, 199 (1965).
B18. Bhappu, R. B., Reynolds, D. H., Roman, R. J., and Schwab, D. A., *N. Mex., Bur. Mines Miner. Resour., Circ.* **81**, 1–24 (1965).
B19. Bhappu, R. B., Reynolds, D. H., and Stahmann, W. S., *in* "Unit Processes in Hydrometallurgy" (M. E. Wadsworth and F. T. Davis, eds.), p. 95. Gordon & Breach, New York, 1964.
B20. Bjerrum, J., "Metal Ammine Formation in Aqueous Solution." P. Haase & Son, Copenhagen, Denmark, 1941.
B21. Bjorling, G., *Z. Erzbergbau Metallhuettenw.* **8**, 781 (1954).
B22. Bjorling, G., and Kolta, G. A., "Proceedings of The VIIth International Mineral Processing Congress" (N. Arbiter, ed.), p. 127. Gordon & Breach, New York, 1964.
B23. Black, K. L., Koslov, J., and Moore, J. D., *Proc. U.N. Int. Conf. Peaceful Uses At. Energy, 2nd, 1958* Vol. 3, p. 488 (1958).
B24. Bochinski, J., Smutz, M., and Spedding, F. H., *Ind. Eng. Chem.* **50**, 157 (1958).
B25. Boldt, J. R., Jr., and Queneau, P., "The Winning of Nickel." Longmans Canada Ltd., Toronto, Canada, 1967.
B26. Bonsack, J. P., *Ind. Eng. Chem., Prod. Res. Develop.* **9**, No. 3, 398 (1970).
B27. Bratt, G. C. and Pickering, R. W., *Met. Trans.* **1**, 2141 (1970).
B28. Brenner, A., *in* "Modern Electroplating" (F. A. Lowenheim, ed.), p. 698. Wiley, New York, 1963.
B29. Bridges, D. W., and Rosenbaum, J. B., *U. S., Bur. Mines, Inform. Circ.* **8139**, 1–45 (1962).
B30. Brown, K. B., Coleman, C. F., Crouse, D. J., Blake, C. A., and Ryon, A. D., *Proc. U.N. Int. Conf. Peaceful Uses At. Energy, 2nd, 1958* Vol. 3, p. 472 (1958).
B31. Bryner, L. C., and Anderson, R., *Ind. Eng. Chem.* **49**, 1721 (1957).
B32. Bryner, L. C., Beck, J. V., Davis, D. B., and Wilson, D. O., *Ind. Eng. Chem.* **46**, 2587 (1954).
B33. Bryner, L. C., and Jameson, A. K., *Appl. Microbiol.* **6**, 281 (1958).
B34. Bryner, L. C., and Jones, L. W., *Develop. Ind. Microbiol.* **7**, 287 (1966).
B35. Bryner, L. C., Walker, R. B., and Palmer, R., *Trans. AIME* **238**, 56 (1067).
B36. Burkin, A. R., "The Chemistry of Hydrometallurgical Process." Spon, London, 1966.
B37. Burkin, A. R., *Met. Rev.* **12** (n111), 1 (1967).

B38. Burkin, A. R., *in* "Advances in Extractive Metallurgy," p. 821. Inst. Mining Met., London, 1968.
B39. Butler, J. A., *Eng. Mining J.* **152**, No. 3, 56 (1951).
B40. Butler, J. N., *Nev., Bur. Mines, Rep.* **5**, 1–58 (1963).
C1. Caldwell, N. A., and Frint, W. R., U. S. Patent 3,050,290 (1962).
C2. Callahan, J. R., *Chem. Met. Eng.* **44**, 120 (1937).
C3. Carlson, C. W. and Nielsen, R. H., *J. Metals* **12**, No. 6, 472 (1960).
C4. Carlson, E. T., and Simons, C. S., *J. Metals* **12**, No. 3, 206 (1960).
C5. Carlson, E. T., and Simons, C. S., "Extractive Metallurgy of Copper, Nickel and Cobalt" (P. E. Queneau, ed.). Wiley (Interscience), New York, 1961.
C6. Carman, E. D. H., *J. S. Afr. Inst. Mining Met.* **61**, 647 (1960).
C7. Caron, M. H., U. S. Patent 1,487,145 (1924).
C8. Caron, M. H., *Trans. AIME* **188**, 67 (1950).
C9. Caron, M. H., *Trans. AIME* **188**, 91 (1950).
C10. Casto, M. G., Smutz, M., and Bautista, R. G., *Trans. SME, AIME* **250**, 42 (1971).
C11. Casto, M. G., Hoh, Y. C., Smutz, M., and Bautista, R. G., *Ind. Eng. Chem., Prod. Res. Develop.* **12** (10), (1973).
C12. *Chem. Eng. News* **46**, 44 (1968).
C13. *Chem. Eng. News* **48**, 21 (1970).
C14. *Chem. Eng. News* **48**, 33 (1970).
C15. *Chem. Eng. News* **48**, 35 (1970).
C16. *Chem. Eng. Progr., Symp. Ser.* **50**, No. 14 (1954).
C17. *Chem. Eng. Progr., Symp. Ser.* **55**, No. 24 (1959).
C18. Chilton, C. H., *Chem. Eng. (New York)* **65**, No. 22, 104 (1958).
C19. Clark, J. B., *Trans. AIME* **186**, 1 (1949).
C20. Coffer, L. W., *Chem. Eng. (New York)* **65**, No. 2, 107 (1958).
C21. Cohen, E., and Ng, W. K., *in* "Advances in Extractive Metallurgy," p. 127. Inst. Mining Met., London, 1968.
C22. Colmer, A. R., and Hinkle, M. E., *Science* **106**, 253 (1947).
C23. Colmer, A. R., Temple, K. L., and Hinkle, M. E., *J. Bacteriol.* **59**, 317 (1949).
C24. Colombo, A. F., *U. S., Bur. Mines, Rep. Invest.* **7138**, 1–7 (1968).
C25. Corrick, J. D., and Sutton, J. A., *U. S., Bur. Mines, Rep. Invest.* **5718**, 1–8 (1961).
C26. Corrick, J. D., and Sutton, J. A., *U. S., Bur. Mines, Rep. Invest.* **6714**, 1–21 (1965).
D1. Dasher, J., Gaudin, A. M., and MacDonald, R. D., *Trans. AIME* **209**, 185 (1957).
D2. Davis, J. G., *AIME, SME Prepr.* **70-AS-356** (1970).
D3. Davis, J. G., and Shock, D. A., *Trans. AIME* **247**, 93 (1970).
D4. Davis, M. W., Jr., and Weber, E. J., *Ind. Eng. Chem.* **52**, 929 (1960).
D5. Dean, R. S., U. S. Patent 2,608,463 (1952).
D6. Dean, R. S., *Mining Eng. (New York)* **4**, No. 1, 55 (1952).
D7. Dean, R. S., and Fox, A. L., U. S. Patent 2,621,107 (1952).
D8. Dean, R. S., Leaver, E. S., and Joseph, T. L., *U. S., Bur. Mines, Inform. Circ.* **6770**, 1–92 (1934).
D9. De Cuyper, J. A., *in* "Unit Processes in Hydrometallurgy" (M. E. Wadsworth and F. T. Davis, eds.), p. 126. Gordon & Breach, New York, 1964.
D10. DeMent, E. R., and Merigold, C. R., *AIME, SME Prepr.* **70-B-65** (1970).
D11. Denaro, A. R., "Elementary Electrochemistry." Butterworth, London, 1965.
D12. Dennis, W. H., "Metallurgy of the Non-Ferrous Metals." Pitman, London, 1961.
D13. Donald, M. B., *Trans. Inst. Chem. Eng.* **15**, 77 (1937).

D14. Dorr, J. V. W., "Cyanidation and Concentration of Gold and Silver Ores." McGraw-Hill, New York, 1936.
D15. Downes, K. W., and Bruce, R. W., *Bull. Can. Inst. Mining Met.* **48,** 127 (1955).
D16. Downes, K. W., and Bruce, R. W., *Trans. Can. Inst. Mining Met.* **58,** 7 (1955).
D17. Dresher, W. H., Kaneko, T. M., Fassell, W. M., and Wadsworth, M. E., *Ind. Eng. Chem.* **47,** 1681 (1955).
D18. Dresher, W. H., Wadsworth, M. E., and Fassell, W. M., Jr., *Trans. AIME* **205,** 738 (1956).
D19. Dufour, M. F., and Hills, R. C., *Chem. Ind. (London)* **57,** 621 (1945).
D20. Duggan, E. J., *Eng. Mining J.* **137,** 1008 (1928).
D21. Duncan, D. W., Waldon, C. C., and Trussell, P. C., *Can. Mining Met. Bull.* **59,** 1075 (1966).
D22. Duncan, D. W., Trussell, P. C., and Walden, C. C., *Appl. Microbiol.* **12,** 122 (1964).
D23. Duncan, D. W., Walden, C. C., and Trussell, P. C., *Trans. Can. Inst. Mining Met.* **69,** 329 (1966).
D24. Duncan, D. W., Walden, C. C., Trussell, P. C., and Lowe, E. A., *Trans. AIME* **238,** 1 (1967).
D25. Dunstan, E. T., *J. Chem., Met. Mining Soc. S. Afr.* **31,** No. 7, 190 (1931).
D26. Durie, R. W., and Jessen, F. W., *Trans. AIME* **231,** 275 (1964).
D27. Dutrizac, J. E., MacDonald, R. D., and Ingraham, T. R., *Trans. AIME* **245,** 955 (1969).
D28. Dutrizac, J. E., MacDonald, R. D., and Ingraham, T. R., *Met. Trans.* **1,** 225 (1970).
D29. Dutrizac, J. E., MacDonald, R. D., and Ingraham, T. R., *Met. Trans.* **1,** 3083 (1970).
D30. Dykstra, J., Thompson, B. H., and Clouse, R. J., *Ind. Eng. Chem.* **50,** No. 2, 161 (1958).
E1. Ebner, M. J., *U. S., Geol. Surv., Bull.* **1019**A (1954).
E2. Eddy, L., *Eng. Mining J.* **107,** No. 26, 1162 (1919).
E3. Edwards, J. D., Frary, F. C., and Jeffries, Z., "Aluminum and Its Production." McGraw-Hill, New York, 1930.
E4. Ehrlich, H. L., *Bacteriol. Proc.*, Abstr., 61*st* Annu. Meet., 54 (1961).
E5. Elkin, E. M., and Bennett, P. W., *J. Metals* **17,** No. 3, 252 (1965).
E6. Ellis, D. A., Long, R. S., and Byrne, J. B., *Proc. U.N. Int. Conf. Peaceful Uses At. Energy, 2nd, 1958* Vol. 3, p. 499 (1958).
E7. Elsner, L., *J. Prakt. Chem.* **37,** 441 (1846).
E8. Engel, A. L , and Heinen, H. J., *U. S., Bur. Mines, Rep. Invest.* **5342,** 1–9 (1957).
E9. *Eng. Mining J.* **158,** No. 5, 91 (1957).
E10. Evans, D. J. I., *in* "Advances in Extractive Metallurgy," p. 831. Inst. Mining Met., London, 1968.
E11. Everest, D. A., Napier, E., and Wells, R. A., *Proc. U.N. Int. Conf. Peaceful Uses At. Energy, 2nd, 1958* Vol. 3, p. 396 (1958).
E12. Everest, D. A., and Wells, R. A., *in* "Proceedings of The VIth International Mineral Processing Congress" (A. Roberts, ed.), p. 145. Pergamon, Oxford, 1965.
F1. Falke, W. L., *U. S., Bur. Mines, Rep. Invest.* **6361,** 1–14 (1964).
F2. Fisher, J. R., *Trans. Can. Inst. Mining Met.* **69,** 167 (1966).
F3. Fitzhugh, E. F., Jr., and Seidel, D. C., *Trans. AIME* **238,** 380 (1967).
F4. Fletcher, A. W., and Hester, K. D., *Trans. AIME* **229,** 282 (1964).

F5. Forward, F. A., *Trans. Can. Inst. Mining Met.* **56,** 363 (1953).
F6. Forward, F. A., and Halpern, J., *Trans. Can. Inst. Mining Met.* **56,** 344 (1953).
F7. Forward, F. A., and Halpern, J., *J. Metals* **7,** No. 3, 463 (1955).
F8. Forward, F. A., and Halpern, J., *Inst. Mining Met., Trans.* **66,** 191 (1956).
F9. Forward, F. A., and Mackiw, V. N., *Trans. AIME* **203,** 457 (1955).
F10. Forward, F. A., Samis, C. S., and Kudryk, V., *Trans. Can. Inst. Mining Met.* **51,** 350 (1948).
F11. Forward, F. A., and Veltman, H., *J. Metals* **11,** No. 12, 836 (1959).
F12. Forward, F. A., Veltman, H., and Vizsolyi, H., *in* "Proceedings of The International Mineral Processing Congress," p. 823. Inst. Mining Met., London, 1960.
F13. Forward, F. A., and Vizsolyi, H., *in* "Proceedings of The VIth International Mineral Processing Congress" (A. Roberts, ed.), p. 187. Pergamon, Oxford, 1965.
F14. Forward, F. A., and Warren, I. H., *Met. Rev.* **5,** No. 18, 137 (1960).
F15. Franklin, J. W., *Eng. Mining J.* **159,** No. 4, 85 (1958).
F16. Frisch, N. W., and McGarvey, F. Z., *Chem. Eng. Progr., Symp. Ser.* **55,** 51 (1959).
G1. Gaudin, A. M., *Proc. Int. Conf. Peaceful Uses At. Energy, 1st, 1955* Vol. 8, p. 8 (1956).
G2. Gaudin, A. M., "Flotation." McGraw-Hill, New York, 1958.
G3. Gaudin, A. M., Schuhmann, R., Jr., and Dasher, J., *J. Metals* **8,** No. 8, 1065 (1956).
G4. George, D. R., Crocker, L., and Rosenbaum, J. B., *Mining Eng. (New York)* **22,** No. 1, 75 (1970).
G5. George, D. R., Ross, J. R., and Prater, J. D., *Mining Eng. (New York)* **20,** No. 1, 73 (1968).
G6. George, D. R., Tame, K. E., Crane, S. R., and Higbie, K. B., *J. Metals* **20,** No. 9, 59 (1968).
G7. Gerlach, K. J., and Pawlek, F. E., *in* "Unit Processes in Hydrometallurgy" (M. E. Wadsworth and F. T. Davis, eds.), p. 308. Gordon & Breach, New York, 1964.
G8. Gray, P. M. J., *Rev. Pure Appl. Chem.* **5,** 194 (1955).
G9. Gray, P. M., *Proc. Int. Conf. Peaceful Uses At. Energy, 1st, 1955* Vol. 8, p. 96 (1956).
G10. Grimes, M. E., *AIME, SME Prepr.* **5820A17** (1958).
G11. Grinstead, R. R., Shaw, K. G., and Long, R. S., *Proc. Int. Conf. Peaceful Uses At. Energy 1st, 1955* Vol. 8, p. 71 (1956).
G12. Grunenfelder, J. G., *in* "Copper, The Science and Technology of the Metal, its Alloys and Compounds" (A. Butts, ed.), p. 322. Van Nostrand-Reinhold, Princeton, New Jersey, 1954.
G13. Gruzensky, W. G., and Engel, G. T., *Trans. AIME* **215,** 738 (1959).
H1. Habashi, F., *Mont., Bur. Mines Geol., Bull.* **51** (1966).
H2. Habashi, F., *Trans. AIME* **235,** 236 (1966).
H3. Habashi, F., *Mont., Bur. Mines Geol., Bull.* **59** (1967).
H4. Habashi, F., *Proc. Int. Congr. Ind. Chem., 36th, 1960* Gr.V-S.14–284 (1967).
H5. Habashi, F., *Mining Congr. J.* **55,** No. 6, 38 (1969).
H6. Habashi, F., and Torres-Acuna, N., *Trans. AIME* **242,** 780 (1968).
H7. Halpern, J., *J. Metals* **9,** No. 2, 280 (1957).
H8. Halpern, J., Forward, F. A., and Ross, A. H., *Trans. AIME* **212,** 65 (1958).
H9. Hanson, C., *Brit. Chem. Eng.* **10,** No. 1, 34 (1965).
H10. Hanson, C., *Chem. Eng. (New York)* **75,** No. 18, 76 (1968).

H11. Harada, T., Bautista, R. G., and Smutz, M., *Met. Trans.* **2**, 195 (1971).
H12. Harada, T., Bautista, R. G., and Smutz, M., *in* "Proceedings of The International Solvent Extraction Conference" (J. G. Gregory, B. Evans, and P. C. Weston, eds.), Vol. 2, p. 950. Ind., London, 1971.
H13. Harrison, V. F., Gow, W. A., and Hughson, M. R., *J. Metals* **18**, No. 11, 1189 (1966).
H14. Harrison, V. F., Gow, W. A., and Ivarson, K. C., *Can. Mining J.* **87**, No. 5, 64 (1966).
H15. Haver, F. P., Uchida, K., and Wong, M. M., *U. S., Bur. Mines, Rep. Invest.* **7185**, 1–15 (1968).
H16. Haver, F. P., Uchida, K., and Wong, M. M., *U. S., Bur. Mines, Rep. Invest.* **7360**, 1–13 (1970).
H17. Hayden, W. M., U. S. Patent 2,814,564 (1957).
H18. Hedley, N., and Kentro, D. M., *Trans. Can. Inst. Mining Met.* **48**, 237 (1945).
H19. Hedley, N., and Tabachnick, H., *Amer. Cyanamid Co., Tech. Bull.* **23** (1958).
H20. Hedley, N., and Tabachnick, H., *Trans. AIME* **217**, 19 (1960).
H21. Hedley, N., and Tabachnick, H., *Amer. Cyanamid Co., Mineral Dressing Notes* No. 23 (1968).
H22. Heindl, R. A., Ruppert, J. A., Skow, M. L., and Conley, J. E., *U. S., Bur. Mines, Rep. Invest.* **5124**, 1–98 (1955).
H23. Heindl, R. A., Ruppert, J. A., Skow, M. L., and Conley, J. E., *U. S., Bur. Mines, Rep. Invest.* **5142**, 1–80 (1955).
H24. Heinen, H. J., and Porter, B., *U. S. Bur. Mines, Rep. Invest.* **7250**, 1–5 (1968).
H25. Heister, N. K., and Phillips, R. C., *Chem. Eng. (New York)* **61**, No. 10, 161 (1954).
H26. Helfferich, F., "Ion Exchange." McGraw-Hill, New York, 1962.
H27. Herring, A. P., and Ravitz, S. F., *Trans. AIME* **232**, 191 (1965).
H28. Herzog, E., and Backer, L., *in* "Proceedings of The VIth International Congress of Mineral Processing" (A. Roberts, ed.), p. 171. Pergamon, Oxford, 1965.
H29. Higgins, I. R., and Roberts, J. T., *Chem. Eng. Progr., Symp. Ser.* **50**, 87 (1954).
H30. Hollis, R. F., and McArthur, C. K., *Proc. Int. Con. Peaceful Uses At. Energy, 1st, 1955* Vol. 8, p. 54 (1956).
H31. Howard, E. V., *Mining Eng. (New York)* **20**, No. 4, 70 (1968).
H32. Hudson, A. W., and Van Arsdale, G. D., *Trans. AIME* **69**, 137 (1923).
H33. Hurlbut, C. S., Jr., "Dana's Manual of Mineralogy, 17th ed. Wiley, New York, 1959.
H34. Hurst, F. J., Crouse, D. J., and Brown, K. B., ORNL-TM-2522. Oak Ridge 1st Lab., Oak Ridge, Tennessee, 1969.
H35. Hussey, S. J., *U. S., Bur. Mines, Rep. Invest.* **4374**, 1–34 (1949).
H36. Huttl, J., *Eng. Mining J.* **163**, No. 3, 74 (1962).
I1. Ingraham, T. R., and Kerby, R., *Trans. AIME* **245**, 17 (1969).
I2. Ioannou, T. K., Bautista, R. G., and Smutz, M., *in* "Proceedings of The International Solvent Extraction Conference" (J. G. Gregory, B. Evans, and P. C. Weston, eds.), Vol. 2, p. 957. Soc. Chem. Ind., London, 1971.
I3. Irving, J., *Eng. Mining J.* **113**, No. 17, 714 (1922).
I4. Ivanov, V. I., Nagirnyak, F. I., and Stepanov, B. A., *Mikrobiologiya* **30**, 688 (1961).
I5. Iverson, H. G., and Leitch, H., *J. Metals* **19**, No. 12, 28 (1967).
I6. Iwasaki, I., and Carlson, W. J., Jr., *Trans. AIME* **238**, 439 (1967).
I7. Iwasaki, I., Takahashi, Y., and Kahata, H., *Trans. AIME* **235**, 308 (1966).

J1. Jackson, K. J., and Strickland, J. D. H., *Trans. AIME* **212**, 373 (1958).
J2. Jacobi, J. S., *Mining Eng.* (*New York*) **15**, No. 9, 56 (1963).
J3. Jacobsen, F. M., and Beyer, G. H., *AIChE J* **2**, No. 3, 283 (1956).
J4. Jarman, A., and Brereton, E., LeGay, *Inst. Mining Met., Trans.* **14**, 289 (1904–5).
J5. Joe, E. G., Ritcey, G. M., and Ashbrook, A. W., *J. Metals* **18**, No. 1, 18 (1966).
J6. Johnson, P. H., *Mining Eng.* (*New York*) **17**, No. 8, 64 (1965).
J7. Johnson, P. H., and Bhappu, R. B., *AIME, SME Prepr.* **90-B-70** (1970).
J8. Johnston, W. E., *Can. Inst. Mining Met. Bull.* **26**, No. 254, 224 (1933).
K1. Kajic, J. E., U. S. Patent 3,272,621 (1966).
K2. Kelsall, D. F., *Mining Eng.* (*New York*) **15**, No. 10, 60 (1963).
K3. Kenahan, C. B., and Schlain, D., *U. S., Bur. Mines, Rep. Invest.* **5890**, 1–53 (1961).
K4. Kershner, K. K., and Hoertel, F. W., *U. S., Bur. Mines, Rep. Invest.* **5740**, 1–16 (1961).
K5. Kindig, J. K. and Hazen, W. C., *AIME, SME Prepr.*, (1970).
K6. Kinzel, N. A., *J. Bacteriol.* **80**, 628 (1960).
K7. Kunin, R., Gustafson, E. G., Isacoff, E. G., and Fillus, H., *AIME, SME Prepr.* (1970).
K8. Kunin, R., and Myers, R. Y., "Ion Exchange Resins," p. 135. Wiley, New York, 1950.
K9. Kurushima, H., and Tsunoda, S., *J. Metals* **7**, No. 5, 634 (1955).
K10. Kuznetsov, S. I., Ivanov, M. V., Lyalikova, N. N., Oppenheimer, C. H., and Broneer, P. T., "Introduction to Geological Microbiology." McGraw-Hill, New York, 1963.
L1. Lamborn, R. H., "The Metallurgy of Copper." Lockwood, London, 1875.
L2. Lawrence, H. M., *Eng. Mining J.* **104**, No. 18, 781 (1917).
L3. Leathen, W. W., and Braley, S. A., *Bacteriol. Proc.*, Abstr., 54th Annu. Meet., 44 (1954).
L4. Leathen, W. W., Braley, S. A., and McIntyre, L. D., *Appl. Microbiol.* **1**, 61 (1953).
L5. Leathen, W. W., Kinzel, N. A., and Braley, S. A., *J. Bacteriol.* **72**, 700 (1956).
L6. Leaver, E. S., and Woolf, J. A., *U. S., Bur. Mines, Tech. Pap.* **481**, 1–20 (1930).
L7. Leaver, E. S., and Woolf, J. A., *U. S., Bur. Mines, Tech. Pap.* **494**, 1–63 (1931).
L8. Leaver, E. S., Woolf, J. A., and Karchmer, N. K., *U. S., Bur. Mines, Rep. Invest.* **3436**, 1–25 (1931).
L9. Lemmon, R. J., *Chem. Eng. Mining Rev.* **32**, 103 (1939).
L10. Lemmon, R. J., *Chem. Eng. Mining Rev.* **33**, 227 (1940).
L11. Lewis, C. J., and Crabtree, F. H., *Mining Congr. J.* **44**, No. 1, 65 (1958).
L12. Lewis, C. J., and Drobnick, J. L., *Ind. Eng. Chem.* **50**, No. 12, 534 (1958).
L13. Lewis, C. J., and House, S. E., *Trans. AIME* **220**, 359 (1961).
L14. Lilge, E. O., and Siebert, H., *Trans. AIME* **229**, 325 (1964).
L15. Lodding, W., *Mining Eng.* (*New York*) **19**, No. 3, 60 (1967).
L16. Lower, G. W., and Booth, R. B., *Mining Eng.* (*New York*) **17**, No. 11, 56 (1965).
L17. Lundquist, R. V., *Trans. AIME* **196**, 413 (1953).
L18. Lyons, D. A., and Ralston, O. C., *U. S., Bur. Mines, Bull.* **157**, 1–176 (1918).
M1. MacGregor, R. A., *Trans. Can. Inst. Mining Met.* **69**, 162 (1966).
M2. MacGregor, R. A., *Mining Eng.* (*New York*) **21**, No. 3, 54 (1969).
M3. Mackay, T. L., and Wadsworth, M. E., *Trans. AIME* **212**, 597 (1958).
M4. Mackiw, V. N., *Can. J. Chem. Eng.* **46**, 3 (1968).
M5. Mackiw, V. N., Benz, T. W., and Evans, D. J. I., *Chem.-Ing.-Tech.* **34**, No. 6, 441 (1962).

M6. Mackiw, V. N., Benz, T. W., and Evans, D. J. I., *Met. Rev.* **11** (n109), 143 (1966).
M7. Mackiw, V. N., and Veltman, H., *Can. Inst. Mining Met. Bull.* **60**, No. 657, 80 (1967).
M8. Magner, J. E. and Bailes, R. H., *Proc. U.N. Int. Conf. Peaceful Uses At. Energy, 2nd, 1958* Vol. 3, p. 495 (1958).
M9. Malouf, E. E., Peters, E., and Shoemaker, R. S., "Bio-Extractive Mining." AIME, New York, 1970.
M10. Malouf, E. E., and Prater, J. D., *Mining Congr. J.* **48**, 82 (1962).
M11. Mancantelli, R. W. and Woodward, J. R., *Trans. AIME* **203**, 751 (1955).
M12. Mantell, C. L., "Electrochemical Engineering." McGraw-Hill, New York, 1960.
M13. Marsden, D. D., *J. S. Afr. Inst. Mining Met.* **61**, No. 11, 522 (1961).
M14. Martin, F. S., and Holt, R. J. W., *Quart. Rev., Chem. Soc.* **13**, 327 (1959).
M15. Maschmeyer, S., and Benson, B., *Can. Inst. Mining Met. Bull.* **58**, No. 641, 931 (1965).
M16. Maslenitsky, N. W., and Perlov, P. M., *in* "Proceedings of The International Mineral Processing Congress," p. 839. Inst. Mining Met., London, 1960.
M17. McArthur, J. A., and Leaphart, C., *in* "Extractive Metallurgy of Copper, Nickel and Cobalt" (P. Queneau, ed.), p. 347. Wiley (Interscience), New York, 1961.
M18. McArthur, J. S., Forrest, R. W., and Forrest, W., British Patent 14,174 (1887).
M19. McArthur, J. S., Forrest, R. W., and Forrest, W., U. S. Patent 418,137 (1889).
M20. McCabe, C. L., and Morgan, J. A., *Trans. AIME* **206**, 800A (1956).
M21. McKay, D. R., and Halpern, J., *Trans. AIME* **212**, 301 (1958).
M22. McKinley, H. L., Holmes, T. W., and Sausa, L. E., *Eng. Mining J.* **152**, No. 5, 76 (1951).
M23. McKinney, W. A., and Rampacek, C., *U. S., Bur. Mines, Rep. Invest.* **5629**, 1–16 (1960).
M24. McNeill, R., Swinton, E. A., and Weiss, D. E., *Trans. AIME* **203**, 912 (1955).
M25. Meddings, B. and Mackiw, V. N., *in* "Unit Processes in Hydrometallurgy" (M. E. Wadsworth and F. T. Davis, eds.), p. 345. Gordon & Breach, New York, 1964.
M26. Mellor, J. W., "Comprehensive Treatise on Inorganic and Theoretical Chemistry," Vol. 3, p. 499. Longmans, Green, New York, 1964.
M27. Miller, A., *AIME, SME Prepr.* **67B339** (1967).
M28. Mindler, A. B., and Paulson, C. F., *J. Metals* **5**, No. 8, 980 (1953).
M29. *Mining Eng. (New York)* **16**, No. 6, 72 (1964).
M30. *Mining Eng. (New York)* **18**, No. 12, 88 (1966).
M31. *Mining Eng. (New York)* **19**, No. 11, 66 (1967).
M32. *Mining Eng. (New York)* **20**, No. 10, 92 (1968).
M33. *Mining Eng. (New York)* **20**, No. 10, 100 (1968).
M34. *Mining Eng. (New York)* **20**, No. 10, 104 (1968).
M35. *Mining Eng. (New York)* **20**, No. 10, 114 (1968).
M36. *Mining J. (London)* **258**, 420 (1962).
M37. Mitchell, J. S., *Mining Eng. (New York)* **8**, No. 11, 1093 (1956).
M38. Moison, R. L., and O'Hern, H. A., Jr., *Chem. Eng. Progr., Symp. Ser.* **55**, 71 (1959).
M39. Monninger, F. M., *Mining Congr. J.* **40**, No. 10, 48 (1963).
M40. Morrison, B. H., *in* "Unit Processes in Hydrometallurgy" (M. E. Wadsworth and F. T. Davis, eds.), p. 227. Gordon & Breach, New York, 1964.
M41. Morrow, B. S., and Griswold, G. G., Jr., *Trans. AIME* **27**, 413 (1933).

M42. Musgrove, R. E., Maurer, E. E., and Fischer, R. E., *AIME, SME Prepr.* **5820A16** (1958).
N1. Nadkarni, R. M., Jelden, C. E., Bowles, K. C., Flanders, H. E., and Wadsworth, M. E., *Trans. AIME* **239**, 581 (1967).
N2. Nadkarni, R. M., and Wadsworth, M. E., *Trans. AIME* **239**, 1066 (1967).
N3. Nadkarni, R. M., and Wadsworth, M. E., *in* "Advances in Extractive Metallurgy," p. 918. Inst. Mining Met., London, 1968.
N4. Nash, W. G., *Mining Sci. Press* **104**, 213 (1912).
N5. Nashner, S., *Trans. Can. Inst. Mining Met.* **58**, 212 (1955).
N6. Nolfi, F. V., Shewmon, P. G., and Foster, J. S., *Trans. AIME* **245**, 1427 (1969).
N7. North, A. A., and Wells, R. A., *Inst. Mining Met., Trans.* **74**, 463 (1964–1965).
O1. O'Kane, P. T., *Prepr. Pap., 13th Chem. Eng. Conf., 1963* (1963).
O2. O'Leary, V. D., *Mining Eng. (New York)* **18**, No. 9, 63 (1966).
O3. Ozsahin, S., and Husband, W. H. W., Report No. E-65-1. Eng. Div., Saskatchewan Research Council, 1965.
P1. Painter, L. A., and Izzo, T. F., *Proc. Int. Conf. Peaceful Uses At. Energy, 2nd, 1958* Vol. 3, p. 383 (1958).
P2. Panlasigue, R. A., and Wheelock, T. D., unpublished Ph.D. thesis of R. A. Panlasigue, Iowa State University Library, Ames, 1970.
P3. Pearce, R. F., Warner, J. P., and Mackiw, V. N., *J. Metals* **12**, No. 1, 28 (1960).
P4. Penneman, R. A., and Jones, L. H., *J. Chem. Phys.* **24**, 293 (1956).
P5. Perkins, E. C., *U. S., Bur. Mines, Rep. Invest.* **5341**, 1–16 (1957).
P6. Perry, J. H., "Chemical Engineer's Handbook," 3rd ed., p. 714. McGraw-Hill, New York, 1950.
P7. Peters, E., and Halpern, J., *Trans. Can. Inst. Mining Met.* **56**, 350 (1953).
P8. Peters, F. A., Johnson, P. W., and Kirby, R. G., *U. S., Bur. Mines, Rep. Invest.* **6133**, 1–68 (1962).
P9. Powell, H. E., Smith, L. L., and Cochran, A. A., *U. S., Bur. Mines, Rep. Invest.* **7336**, 1–14 (1970).
P10. Powell, J. E., and Spedding, F. H., *Chem. Eng. Progr., Symp. Ser.* **55**, 101 (1959).
P11. Powell, J. E., and Spedding, F. H., *Trans. AIME* **215**, 457 (1959).
P12. Preuss, A., and Kunin, R., *Proc. Int. Conf. Peaceful Uses At. Energy, 1st, 1955* Vol. 8, p. 45 (1956).
P13. "Proceedings of The International Mineral Processing Congress." Inst. Mining Met., London, 1960.
Q1. Queneau, P., ed., "Extractive Metallurgy of Copper, Nickel, and Cobalt." Wiley (Interscience), New York, 1961.
R1. Rahn, R. W., and Smutz, M., *Ind. Eng. Chem., Process Des. Develop.* **8**, 289 (1969).
R2. Rampacek, C., Fuller, H. C., and Clemmer, J. B., *U. S., Bur. Mines, Rep. Invest.* **5508**, 1–54 (1959).
R3. "Rare Metals Handbook," 2nd ed., p. 582. Van Nostrand-Reinhold, Princeton, New Jersey, 1961.
R4. Ravitz, S. F., Back, A. E., Tame, K. E., Wyman, W. F., and Dewey, J. C., *U. S., Bur. Mines., Tech. Pap.* **723**, 1–45 (1949).
R5. Ravitz, S. F., Wyman, W. F., Back, A. E., and Tame, K. E., *Metals Technol.* **13**, No. 6 (1946).
R6. Razzell, W. E., *Trans. Can. Inst. Mining Met.* **65**, 135 (1962).
R7. Read, F. O., *J. S. Afr. Inst. Mining Met.* **60**, 105 (1959).
R8. Remirez, R., *Chem. Eng. (New York)* **76**, No. 19, 96 (1969).

R9. Ritcey, G. M., *Can. Mining J.* **90**, No. 6, 73 (1969).
R10. Ritcey, G. M., Joe, E. G., and Ashbrook, A. W., *Trans. AIME* **238**, 330 (1967).
R11. Roberts, E. J., *Chem. Met. Eng.* **41**, 242 (1934).
R12. Romankiw, L. T., and de Bruyn, P., *in* "Unit Processes in Hydrometallurgy," (M. E. Wadsworth, and F. T. Davis, eds.), p. 45. Gordon & Breach, New York, 1964.
R13. Rose, D. H., Lessels, V., and Buchwalter, D. J., *Trans. AIME* **238**, 221 (1967).
R14. Rosenbaum, J. B., and McKinney, W. A., *AIME, SME Prepr.* **70-AS-329** (1970).
R15. Ross, J. R., and Schack, C. H., *U. S., Bur. Mines, Rep. Invest.* **6580**, 1–22 (1965).
R16. Rudolfs, W., and Helbronner, A., *Soil Sci.* **14**, 459 (1922).
R17. Ryan, V. H., and Tschirner, H. J., *in* "Proceedings of 13th Annual Meeting, Metal Power Association," p. 25. Metal Powder Ass., Chicago, Illinois, 1957.
R18. Ryon, A. D., Daley, F. E., and Lowrie, R. S., *Chem. Eng. Progr.* **55**, No. 10, 70 (1959).
R19. Ryon, A. D., and Lowrie, R. S., Oak Ridge National Laboratory Report, ORNL-3381. Oak Ridge Nat. Lab., Oak Ridge, Tennessee, 1963.
S1. Saubestre, E. B., *Metal Finish.* **60**, No. 6, 67 (1962); No. 7, 49 (1962); No. 8, 45 (1962); No. 9, 59 (1962).
S2. Schaufelberger, F. A., *Trans. AIME* **205**, 539 (1956).
S3. Schaufelberger, F. A., and Roy, T. K., *Inst. Mining Met., Trans.* **64**, 375 (1955).
S4. Scheiner, B. J., Lindstrom, R. E., and Henrie, T. A., *U. S., Bur. Mines, Tech. Progr. Rep.* **8**, 1–12 (1969).
S5. Sege, G., and Woodfield, F. W., *Chem. Eng. Progr., Symp. Ser.* **13**, Part 3, 179 (1954).
S6. Seidel, D. C., *in* "Unit Processes in Hydrometallurgy" (M. E. Wadsworth and F. T. Davis, eds.), p. 114. Gordon & Breach, New York, 1964.
S7. Seidel, D. C., and Fitzhugh, E. F., Jr., *Trans. AIME* **241**, 261 (1968).
S8. Seraphim, D. P., and Samis, C. S., *Trans. AIME* **206**, 1096 (1956).
S9. Shaw, K. G., and Long, R. S., *Chem. Eng. (New York)* **64**, No. 11, 251 (1957).
S10. Sheffer, H. W., and Evans, L. G., *U. S., Bur. Mines, Inform. Circ.* **8341**, 1–57 (1968).
S11. Sheridan, G. E., and Griswold, G. G., Jr., U. S. Patent 1,427,235 (1922).
S12. Sherman, M. I., and Strickland, J. D. H., *Trans. AIME* **209**, 795 (1957).
S13. Sherman, M. I., and Strickland, J. D. H., *Trans. AIME* **209**, 1386 (1957).
S14. Shoemaker, R. S., and Darrah, R. M., *Mining Eng. (New York)* **20**, No. 12, 68 (1968).
S15. Sievert, J. A., Martin, W. L., and Conley, F. R., *AIME, SME Prepr.* **70-AS-334** (1970).
S16. Silo, R. S., *Trans. AIME* **235**, 225 (1966).
S17. Silverman, M. P., and Lundgren, D. G., *J. Bacteriol.* **78**, 326 (1959).
S18. Sims, C., and Ralston, O. C., *Met. Chem. Eng.* **15**, No. 7, 410 (1916).
S19. Skow, M. L., and Conley, J. E., *U. S., Bur. Mines, Rep. Invest.* **4649**, 1–16 (1950).
S20. Spedden, H. R., Malouf, E. E., and Prater, J. D., *J. Metals* **18**, No. 10, 1137 (1966).
S21. Spence, W. W., and Cook, W. R., *Trans. Can. Inst. Mining Met.* **67**, 257 (1964).
S22. Stanczyk, M. H., and Rampacek, C., *U. S., Bur. Mines, Rep. Invest.* **6038**, 1–12 (1962).

S23. Stanczyk, M. H., and Rampacek, C., *U. S., Bur. Mines, Rep. Invest.* **6808**, 1–13 (1966).
S24. St. Clair, H. W. et al., *U. S., Bur. Mines, Bull.* 577 (1959).
S25. Stephens, F. M., and MacDonald, R. D., *Proc. Int. Conf. Peaceful Uses At. Energy, 1st, 1955* Vol. 8, p. 18 (1956).
S26. Stickney, W. A., *U. S., Bur. Mines, Rep. Invest.* **5499**, 1–22 (1959).
S27. Stickney, W. A., and Town, J. W., *Trans. AIME* **224**, 306 (1962).
S28. Stone, R. L., and Tiemann, T. D., *Trans. AIME* **229**, 217 (1964).
S29. Stone, R. L., Wu, S. M., and Tiemann, T. D., *Trans. AIME* **232**, 115 (1965).
S30. Sullivan, J. D., *U. S., Bur. Mines, Rep. Invest.* **2934**, 1–000 (1929).
S31. Sullivan, J. D., *U. S., Bur. Mines, Inform. Circ.* **6425**, 1–12 (1931).
S32. Sullivan, J. D., *Trans. AIME* **106**, 515 (1933).
S33. Sullivan, J. D., and Bayard, K. O., *U. S., Bur. Mines, Rep. Invest.* **3073**, 1–43 (1931).
S34. Sulman, H. L., *Inst. Mining Met., Trans.* **14**, 363 (1904–5).
S35. Swanson, R. R., and Agers, D. W., *AIME, SME Prepr.* (1964).
S36. Swanson, R. R., Dunning, H. N., and House, J. E., *Eng. Mining J.* **162**, No. 10, 110 (1961).
T1. Taylor, J. H., and Whelan, P. F., *Inst. Mining Met., Trans.* **52**, 35 (1942).
T2. Temple, K. L., and Colmer, A. R., *J. Bacteriol.* **62**, 605 (1951).
T3. Temple, K. L., and Delchamps, E. W., *Appl. Microbiol.* **1**, 255 (1953).
T4. Thomas, R. W., *Mining Met.* **19**, 481 (1938).
T5. Thornhill, E. B., *Trans. AIME* **52**, 165 (1915).
T6. Thornhill, E. B., *Mining Sci. Press* **110**, 873 (1915).
T7. Tiemann, T. D., *Trans. AIME* **223**, 173 (1962).
T8. Tiemann, T. D., *Trans. AIME* **229**, 258 (1964).
T9. Tiemann, T. D., *Trans. AIME* **232**, 307 (1965).
T10. Tilley, G. S., Miller, R. W., and Ralston, O. C., *U. S., Bur. Mines, Bull.* **267**, 1–85 (1927).
T11. Todd, D. B., *Chem. Eng. Progr.* **62**, No. 8, 119 (1966).
T12. Tolun, R., *Proc. Int. Conf. Peaceful Uses At. Energy, 2nd, 1958* Vol. 3, p. 502 (1958).
T13. Tougarinoff, B., Van Goetsenhoven, F., and Dewulf, A., *in* "Advances in Extractive Metallurgy," p. 741. Inst. Mining Met., London, 1968.
T14. Town, J. W., Link, R. F., and Stickney, W. A., *U. S., Bur. Mines, Rep. Invest.* **5748**, 1–39 (1961).
T15. Tremblay, R., and Bramwell, P., *Trans. Can. Inst. Mining Met.* **62**, 44 (1959).
T16. Treybal, R. E., *Ind. Eng. Chem.* **54**, No. 5, 55 (1962).
T17. Trussell, P. C., Duncan, D. W., and Walden, C. C., *Can. Mining J.* **85**, No. 3, 46 (1964).
T18. Tschirner, H. J., and Williams, L. A., U. S. Patent 3,127,264 (1964).
T19. Turner, T. L., and Swift, J. H., *U. S., Bur. Mines, Rep. Invest.* **6368**, 1–33 (1964).
U1. United Nations, "Proceedings of International Conferences on the Peaceful Uses of Atomic Energy," 1st, 2nd, and 3rd. United Nations, New York, 1956, 1958, and 1965.
V1. Van Arsdale, G. D., "Hydrometallurgy of Base Metals." McGraw-Hill, New York, 1953.

V2. van Zyl, J. J. E., Marsden, D. D., Finkelstein, N. P., and Douglas, W. D. *J. S. Afr. Inst. Mining Met.* **67**, No. 5, 241 (1966).
V3. Vedensky, D. N., *Eng. Mining J.* **147**, No. 7, 58 (1946).
V4. Veltman, V., and O'Kane, P. T., *AIME, Prepr.* (1968).
V5. Veltman, H., Pellegrini, S., and Mackiw, V. N., *J. Metals* **19**, 21 (1967).
V6. Vermeulen, T., and Hiester, N. K., *Chem. Eng. Progr., Symp. Ser.* **55**, 61 (1959).
V7. Vizsolyi, A., and Forward, F. A., in "Proceedings of The VIIth International Mineral Processing Congress" (N. Arbiter, ed.), p. 147. Gordon & Breach, New York, 1964.
V8. Vizsolyi, A., Veltman, H., and Forward, F. A., *Trans. AIME* **227**, 215 (1963).
V9. Vizsolyi, A., Veltman, H., and Forward, F. A., in "Unit Processes in Hydrometallurgy" (M. E. Wadsworth and F. T. Davis, eds.), p. 326. Gordon & Breach, New York, 1964.
V10. Vizsolyi, A., Veltman, H., Warren, I. H., and Mackiw, V. N., *J. Metals* **19**, No. 11, 52 (1967).
V11. von Hahn, E. A., and Ingraham, T. R., *Trans. AIME* **236**, 1098 (1966).
V12. von Hahn, E. A., and Ingraham, T. R., *Trans. AIME* **237**, 1895 (1967).
W1. Wadsworth, M. E., *Trans. AIME* **245**, 1381 (1969).
W2. Wadsworth, M. E., and Davis, F. T., eds., "Unit Processes in Hydrometallurgy." Gordon & Breach, New York, 1964.
W3. Wadsworth, M. E., and Wadia, D. R., *J. Metals* **7**, No. 6, 755 (1955).
W4. Waksman, S. A., and Joffe, J. S., *J. Bacteriol.* **7**, 239 (1922).
W5. Walden, S. J., Hendriksson, S. T., Arbstedt, P. G., and Mioen, T., *J. Metals* **11**, No. 8, 528 (1959).
W6. Warren, I. H., *Aust. J. Appl. Sci.* **9**, 36 (1958).
W7. Wartman, F. S., and Roberson, A. H., *U. S., Bur. Mines, Rep. Invest.* **3746**, 1–16 (1944).
W8. Weed, R. C., *Mining Eng. (New York)* **8**, No. 7, 721 (1956).
W9. Welsh, J. Y., and Peterson, D. W., *J. Metals* **9**, 762 (1957).
W10. Whatley, M. E., U. S. Patent 2,754,179 (1956).
W11. Whitaker, J. F., *Wire Wire Prod.* **36**, No. 10, 1346 (1961).
W12. White, P. A. F., and Smith, S. E., *J. Brit. Nucl. Energy Soc.* **8**, No. 2, 93 (1969).
W13. Williams, L. M., *Eng. Mining J.* **157**, No. 10, 75 (1956).
W14. Wilson, D. A., and Sullivan, P. M., *U. S., Bur. Mines, Rep. Invest.* **7042**, 1–21 (1967).
W15. Wilson, F., *Mining Eng. (New York)* **10**, No. 5, 563 (1958).
W16. Winchell, H. V., *Mining Sci. Press* **104**, 314 (1912).
W17. Windolph, F. J., unpublished report, Climax Molybdenum Division, AMAX, Climax, Colorado, 1970.
W18. Woodcock, J. T., *Australas. Inst. Mining Met., Proc.* **224**, 47 (1967).
Y1. Yurko, W. J., *Met. Soc. Conf. [Proc.]* **49**, 55–91 (1966).
Y2. Yurko, W. J., *Chem. Eng. (New York)* **73**, No. 18, 64 (1966).
Z1. Zakarias, M. J., and Cahalan, M. J., *Inst. Mining Met., Trans., Sect. C* **75**, C245 (1966).
Z2. Zimmerley, S. R., Wilson, D. G., and Prater, J. D., U. S. Patent 2,829,964 (1958).
Z3. Zubryckyj, N., Evans, D. J. I., and Mackiw, V. N., *J. Metals* **17**, No. 5, 478 (1965).

DYNAMICS OF SPOUTED BEDS

Kishan B. Mathur and Norman Epstein

Department of Chemical Engineering
University of British Columbia
Vancouver, Canada

I. The Phenomenon of Spouting 111
II. Location of Spouting in the Gas–Solids Contacting Spectrum 115
III. The Onset of Spouting 117
    A. Mechanism 117
    B. Minimum Spouting Velocity 123
    C. Pressure Drop. 131
IV. Flow Patterns. 140
    A. Flow Pattern of Gas. 140
    B. Flow Pattern of Solids 144
    C. Mixing Characteristics 158
V. Bed Structure. 163
    A. Spout Shape 163
    B. Voidage Distribution. 169
VI. Spouting Stability 173
    A. Effect of Various Parameters 174
    B. Maximum Spoutable Bed Depth 180
    Nomenclature. 187
    References. 188

I. The Phenomenon of Spouting

Fluidization is a well-established method for contacting and agitating particulate solids with a fluid. With coarse and uniform-sized particles, however, the full benefits of fluidization are not always realized because of the growth of large bubbles in the bed; these cause a tendency toward slugging. This limitation of fluidized beds can be overcome if the gas enters the bed through a small opening at the center of a conical base, instead of through a uniform distributor. The high-velocity gas jet causes a stream of solids to rise rapidly in a hollowed central core or spout within the bed. The particles, having reached somewhat above the bed level, fall back onto the annular space between the spout and the container wall and travel downward as a packed bed. Thus, the bed becomes a composite of a central spout, in which the particles move upward in a dilute phase, and a dense-phase downward-moving annulus with countercurrent percolation of gas. A systematic cycle pattern of solids movement is established with effective contact between the gas and the solids, giving rise to a unique hydrodynamic system, which is more suitable for certain applications than more conventional fluid–solids configurations.

The name "spouted bed" or "spouting" was given to this technique by Gishler and Mathur (G5) who developed it in 1954 at the National Research Council of Canada. In addition to drying of granular materials to which spouting was initially applied (M11, P2), it has since attracted wide attention for a variety of apparently unrelated processes involving coarse solids, such as cooling (F1), blending (B8), and coating (H2, S1) of various materials, drying of pastes (R2), granulation of fertilizers and other products (B5, B6, V4), pyrolysis of shale (B7), and carbonization of coal (B2, B9). Some of these applications have reached the commercial stage, and industrial spouted bed units are in operation in Canada, the United States, Britain, France, the Soviet Union, and India (M9).[1]

The growing acceptance of spouted beds lies primarily in the specific fluid and particle dynamics associated with them. This is the aspect with which the present review is concerned. Some of the more important features of spouted beds, which distinguish it from other similar phenomena, are the following:

(1) Spouting action can be achieved with either a liquid or a gas as the jet fluid. Liquid spouting, however, has not attracted much interest, possibly because it does not offer any obvious advantage over particulate fluidization, which, unlike aggregrative fluidization, is as effective for coarse and uniform-sized solids as for fine materials.

(2) The minimum particle diameter for which spouting is practical appears to be 1–2 mm. Finer materials tend to fluidize rather than spout. The upper limit of particle size for good fluidization is generally regarded as 10–20 mesh (D2). Thus with respect to particle size, spouting takes over where fluidization leaves off.

(3) For a given solid material, column diameter, and fluid inlet diameter, a "maximum spoutable bed depth" exists beyond which the spouting action degenerates into poor-quality fluidization.

(4) At depths below the maximum for spouting, there exists an upper limit of gas-flow rate for stable spouting, above which the systematic move-

---

[1] A detailed account of these developments has been given recently by the present authors (Mathur, K. B. and Epstein, N., Developments in spouted bed technology, paper presented at 4th Internat. Cong. Chem. Eng., Prague, Czechoslovakia, September 1972).

ment of solids tends to become disorganized and eventually gives way to slugging (Fig. 1).

(5) The spouting vessel is commonly either cylindrical or conical in shape (Fig. 2). With the former it is preferable to have a short conical base tapering down to the inlet orifice, so that the solids in the annulus can easily slide into the gas jet region without forming any dead zone at the base. Most of the work outside the Soviet Union has been carried out in this type of vessel, while spouting in conical vessels has received particular attention in the Soviet Union, and has been extensively discussed in a book by Romankov and Rashkovskaya (R4). The general discussion of spouted bed behavior in this review is in the context of a cylindrical column.

(6) A typical spouted bed has a substantial depth, which in the case of a cylindrical vessel is usually at least of the order of two column diameters, measured from the inlet orifice to the surface of the annulus. If the bed is much shallower, the system becomes hydrodynamically different from true spouting. The situation in this respect is similar to that for gas-fluidized beds, where the generally formulated principles of fluidization are not applicable to very shallow beds.

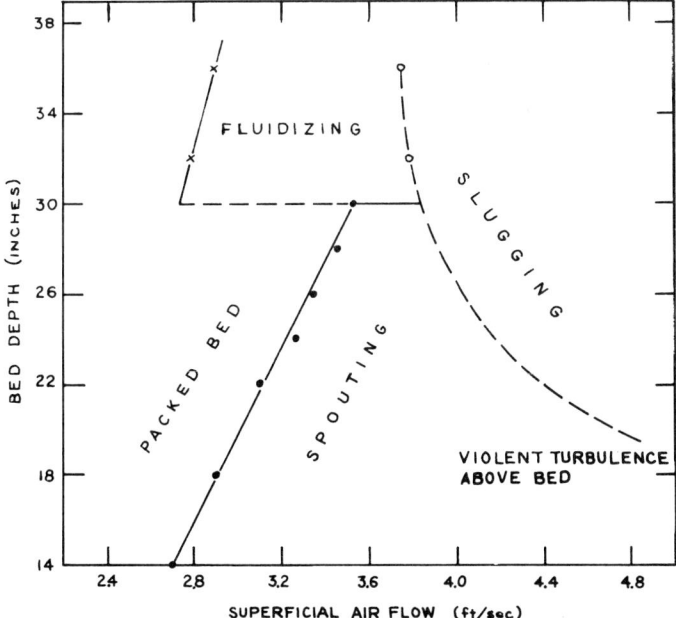

Fig. 1. A typical phase diagram (M10). Material; wheat; D_c = 6 in.; D_i = 0.5 in.; d_p = 3.2–6.4 mm.

(7) The pressure gradient (dP/dH) along the height of a spouted bed is small near the base and increases to a maximum value at the bed surface. In contrast to this, the pressure gradient in fluidization is constant, even in a conical–cylindrical bed.

(8) The total pressure drop across a spouted bed is always lower than that required to support the weight of the bed (or the fluidization pressure drop). While this is also the case in a channeling-fluidized bed, the alleged similarity between spouting and channeling (L3, Z1) is somewhat misleading. Channeling is an undesirable feature of a fluidized bed associated with either uneven distribution of the gas or with the use of very fine particles (B1, M8), and involves the passage of gas through part of the bed without inducing much movement in the surrounding particles. In spouting, on the other hand, agitation of the entire bed is achieved by means of the gas jet and, in addition, intimate contact between the particles and the gas occurs both in the dense-phase region and in the jet.

(9) Solids can be added into and withdrawn from a spouted bed so that,

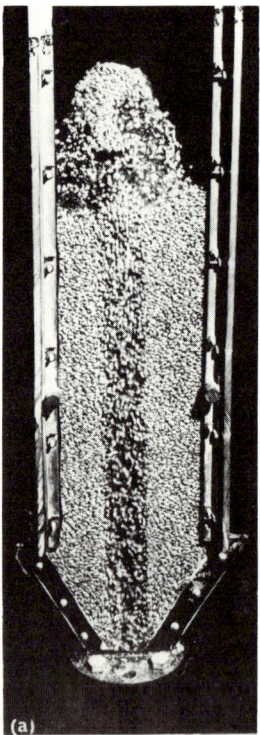
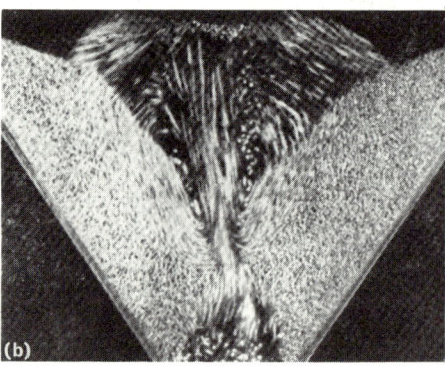

FIG. 2. (a) Spouting in a cylindrical vessel (M10); (b) spouting in a conical vessel (G7).

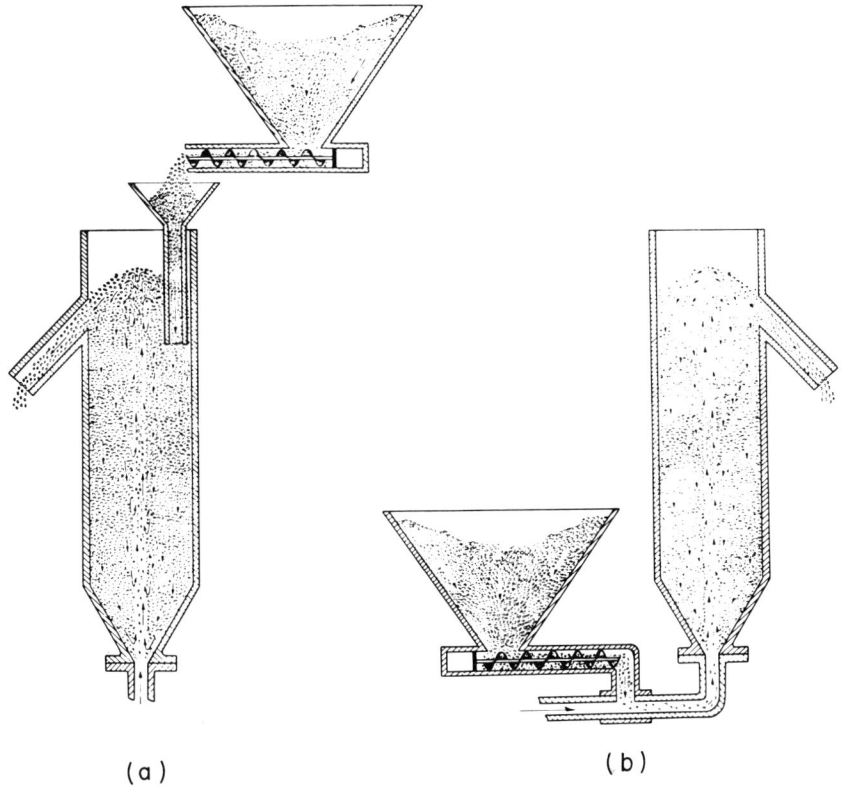

FIG. 3. Continuous spouting operation. *Note:* Solids can also be discharged from the base of the column. (a) Solids fed into annulus (after Mathur and Gishler, M11); (b) solids fed with incoming air (after Manurung, M7).

like fluidization, spouting lends itself to continuous operation. The solids may be fed into the bed either at the top near the wall so that they join the downward-moving mass of particles in the annulus, or with the incoming gas (Fig. 3). Since the annular solids are in an aerated state, material discharges readily through an overflow pipe located in the column wall, or through an outlet near the base of the bed.

II. Location of Spouting in the Gas–Solids Contacting Spectrum

The more common gas–solids contacting systems may be broadly classified as (a) nonagitated, (b) mechanically agitated, and (c) gas-agitated.

Fixed and moving packed beds, which fall in the first category, are applicable to processes that do not call for high rates of heat and mass transfer between the gas and the coarse solids, and in which uniformity of conditions in different parts of the bed is either not critical or not desirable. In a fixed bed, solids cannot be continuously added or withdrawn, and treatment of the gas is usually the main objective. A moving bed does provide continuous flow of solids through the reaction zone. Its application therefore extends to solids treatment, in such uses as roasting of ores, calcining of limestone, and drying and cooling of pellets and briquettes. In both fixed and moving beds, gas movement is close to "plug flow;" this feature is advantageous for certain exothermic chemical reactions where it is desired to have the temperature increase as the reaction proceeds, i.e., as the driving force diminishes.

Limited agitation can be imparted to the solids by mechanical means, either by movement of the vessel itself, as in rotary dryers and kilns, or by the use of internal agitators. In either case, most of the material at any instant is still maintained in a packed-bed condition; but the relative movement of particles improves contacting effectiveness since fresh surface is continually exposed for action by the gas. Also, the blending of solids by agitation levels out interparticle gradients in composition and temperature. Mechanical systems are mainly used for processes involving solids treatment such as drying, calcining, cooling, but are obviously unsuitable for processes that require the gas to be uniformly treated.

In gas-agitated systems such as fluidized and suspended beds, a more intense form of agitation is imparted to each solid particle by the action of the gas stream. In dense-phase fluidization, the bed of fine solids often behaves as a smoothly circulating body of liquid, and the high solids surface involved gives rise to high particle-to-gas heat- and mass-transfer rates. A stream of solids can be easily added to and withdrawn from the bed. Because of these basic features, the fluidized bed has emerged as the preferred contacting method for a number of processes which include gas-phase chemical reactions (both catalytic and noncatalytic) as well as treatment of solids. In fluidization, the presence of a coherent bed allows the solids flowing through the system to spend a certain time in the reaction zone, which can be controlled by adjusting the ratio between the feed rate and the weight of the bed. The bed also acts as a buffer to dampen out any instabilities that arise during continuous operation.

Dilute-phase gas–solids contacting (a suspended bed or transport system) may be more suitable than fluidization, in other process situations. In such systems the contact time between a given particle and the gas is very short—no more than a few seconds—because of very high gas velocities. Intense turbulence makes for high coefficients of heat and mass transfer, but the extents of heat and mass transfer for the particles are not high because of their small residence time in the reaction zone. Dilute-phase systems are suitable for processes in which the gas–solid interaction is surface-rate controlled rather than diffusion controlled. Examples of established applications are combustion of pulverized coal, flash roasting of metallic sulfides, and drying of sensitive materials which can tolerate exposure to heat for only a few seconds.

In Perry's "Handbook" (P1) spouted beds are placed in the category of moving beds, presumably because the annular region that contains most of the solids does constitute a moving bed with countercurrent plug flow of gas. However, inasmuch as the bed solids are well agitated (i.e., recirculated many times), the operation comes closer to fluidization in its characteristics. While it was originally suggested that "spouting appears to achieve the same purpose for coarse particles as fluidization does for fine materials" (M10), it now seems also that the systematic cyclic movement of particles in a spouted bed, as against the more random motion in fluidization, is a feature of critical value for certain applications (for instance, in granulation or coating processes). In the spectrum of contacting systems, then, spouted beds occupy a rather complex position, overlapping fluidized and moving beds to some extent, but at the same time having a place of their own by virtue of certain unique characteristics.

Apart from achieving gas–solids contact, spouting is also useful as a means of agitating coarse particles with a gas, the contact between the gas and solid particles being incidental. When a spouted bed is compared to conventional mechanical agitators, the solids turnover rates in a spouted bed are seen to be high, yet the equipment is much simpler and the energy requirement lower. Spouting has thus proved successful for blending of plastic granules on an industrial scale (I1).

III. The Onset of Spouting

A. MECHANISM

The mechanism by which a bed of particulate solids transforms to the spouted state is best explained by following the changes that occur in the pressure drop across the bed, with varying rate of gas flow (Fig. 4).

At low flow rates, the gas simply passes up without disturbing the particles, the pressure drop rising with flow rate as in a packed bed (along AB).

At a certain flow rate, the jet velocity becomes sufficiently high to push back the particles from the immediate vicinity of the orifice, forming a relatively empty cavity. The particles surrounding the cavity are compressed against the material above, forming a compacted arch which offers a greater resistance to flow (see Fig. 5a). Therefore, despite the existence of a hollow cavity, the total pressure drop across the bed continues to rise.

With further increases in gas flow, the cavity elongates to an internal spout (Fig. 5b). The arch of compacted material still exists above the internal spout, so that the pressure drop across the bed rises further until it reaches a maximum value (Fig. 4, point B).

As the flow rate is increased beyond point B, the height of the relatively hollow internal spout becomes large in comparison with that of the packed material above the spout. The pressure drop, therefore, begins to decrease along BCD. By point C, enough solids have been displaced from the central core to cause a noticeable expansion of the bed, so that the decrease in ΔP is arrested over a small region of gas-flow represented by CD (Fig. 5c).

With only a slight increase in flow rate beyond point D, which is called the point of incipient spouting, the internal spout breaks through the bed

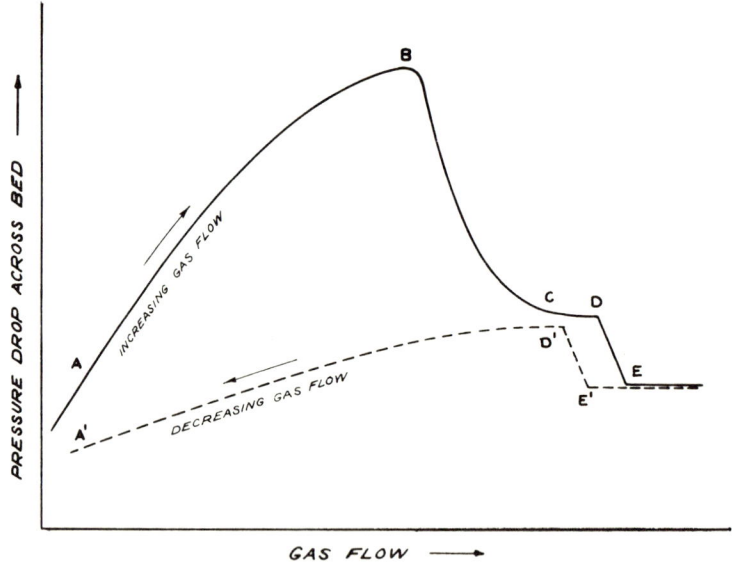

FIG. 4. Typical pressure drop–flow rate curve [D, incipient spouting; E, onset of spouting; E', minimum spouting] (after Thorley et al., T2).

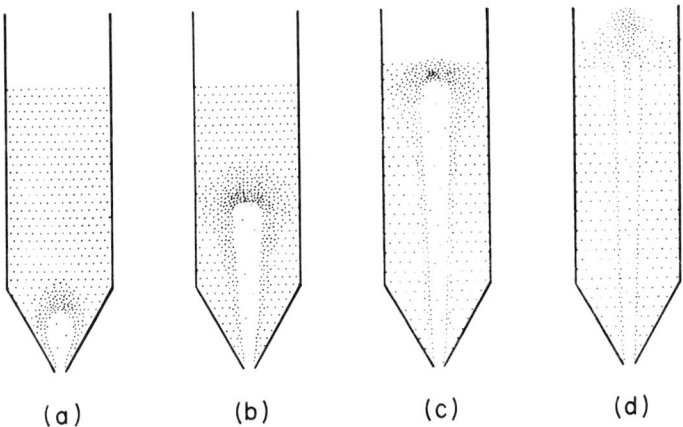

FIG. 5. Development of the spout: (a) cavity forms; (b) cavity elongates (internal spouting); (c) bed expands (incipient spouting); (d) steady spouting sets in.

surface (Fig. 5d). When this happens, the solids concentration in the region directly above the internal spout decreases abruptly, causing a sharp reduction in pressure drop to point E (Fig. 4), when the entire bed becomes mobile and steady spouting sets in. The region between C and D is an unstable one during which the internal spout can be observed to expand and contract alternately as it attempts to pierce through the bed. The incipient spouting velocity (D) and the onset of spouting (E), being bed-history dependent, are not exactly reproducible.

With still further increase in gas flow, the additional gas simply passes through the spout region, which is now established as the path of least resistance, causing the spout to shoot up higher without any significant effect on the total pressure drop. The pressure drop beyond point E, therefore, remains substantially constant.

If the gas flow rate is now slowly decreased, the bed remains in the spouted state until E'; a slight reduction in flow rate at this stage causes the spout to collapse and the pressure drop to rise suddenly to D'. Point E' represents the minimum spouting condition. Once the spout has collapsed, the pressure drop decreases steadily with decreasing flow rate; however, the curve now falls lower than for increasing flow since the energy required by the gas jet to penetrate the solids is no longer expended during the collapse of the spout. Even at low flow rates, when the bed has reverted back to a packed condition, the pressure drop remains lower than previously, the packing of the particles now being looser.

The mechanism just described is based on the work reported by Mathur and co-workers (M10, T1, T2), and Madonna and Lama (M2), who all used cylindrical vessels with either a flat or a short conical base. For observing the physical behavior, they employed half-round sectional columns in which the axial zone of the bed could be seen against the flat transparent face.

Goltsiker (G7) has reported the opposite direction of disturbance spreading for a conical bed, using a two-dimensional column, even though the pressure-drop behavior with flow was of the same type as shown in Fig. 4. In his case the first disturbance of particles occurred not near the gas inlet, but at the surface of the bed, causing an upward deformation of the top layer of particles in the axial region. With increasing gas flow, the deformation spread downward, and the cavity at the gas inlet did not appear until after all the layers of particles constituting the bed had been deformed. On the pressure-drop diagram (Fig. 4) the first appearance of the cavity corresponded to point C rather than to some point preceding B, though the onset of spouting still occurred at point E. The spreading of the deformation of layers of particles from top to bottom was confirmed by Goltsiker in vessels of different sizes and with different solid materials, and has been put forward by him as a general feature of the mechanism of onset of spouting in conical vessels.

It is, however, difficult to see how this difference from cylindrical-bed behavior can be attributed to the vessel shape. A more plausible explanation possibly lies in the small bed depths used by Goltsiker (about 8 in.) since in the absence of a substantial resistance due to the weight of the bed above the inlet region, any disturbance caused by the incoming gas would be easily transmitted to the bed surface.

A more detailed study of the sequence of transition to the onset of spouting has been reported by Volpicelli and Raso (V1), who carried out instantaneous pressure measurements with an electric transducer near the base of the bed at varying gas flow rates in a two-dimensional column. They also recorded the development of the internal spout and the particle trajectories with a motion-picture camera. The average pressure, together with records of instantaneous pressures published by these workers, are reproduced in Fig. 6. Their results broadly confirm the mechanism already described (Figures 4 and 5), and provide the following additional information:

(1) The internal spout consists of two cavities which are symmetrically located about the gas orifice axis, and are delimited by two rings of moving particles, the particle flow lines being similar to those for a pair of vortices in a continuous medium.

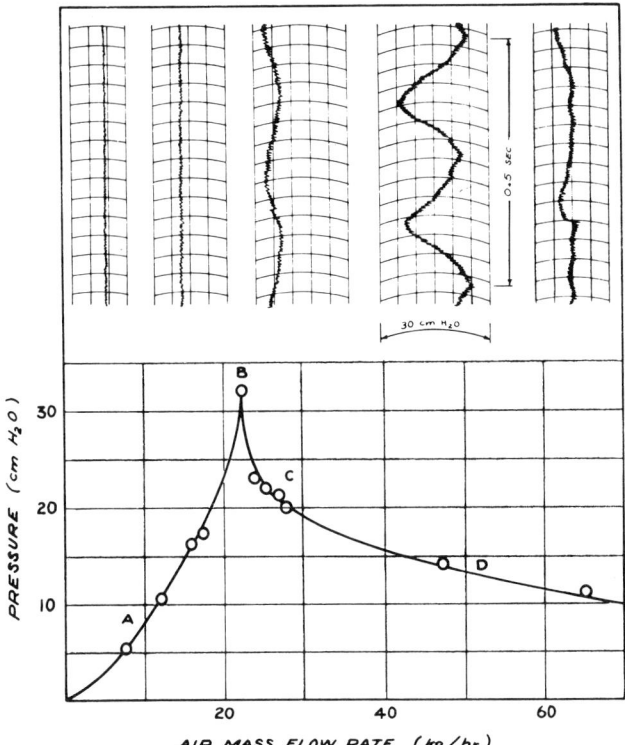

FIG. 6. Instantaneous and average pressures measured at $z = 4$ cm on the nozzle axis. Glass spheres, 3.1 mm diameter; bed depth, 22 cm (Volpicelli and Raso, V1).

(2) The mean pressure near the base of the bed continues to rise with increasing gas flow after the internal spout comes into being and until it reaches a certain critical size. Until this point is reached (point B in Figs. 4 and 6), the instantaneous pressure does not fluctuate. Beyond point B the gas jet penetrates through a zone of low resistance in the compacted arch, forming a somewhat winding channel through the bed and causing the pressure to drop sharply. At this stage, the path forced open in the arch can easily get blocked again by the solids, causing the through-spout to retract back to the original internal spout. Hence, this state is marked by a periodic succession of penetrations and retractions of the spout, with corresponding fluctuations in the pressure (region BD, Fig. 6).

(3) Beyond point D the gas–solids jet becomes fully developed, blockage of the channel no longer occurs, and therefore pressure fluctuations even out.

Volpicelli and co-workers (V1, V3) have attempted to provide support for the physical mechanism described above, by calculating the distribution of local pressures in the bed during the transition stage in which the internal spout has not yet developed to its critical size, using certain assumptions which follow from the observed physical behavior, and matching the calculated values against measurements carried out in a two-dimensional bed (20 cm × 1.55 cm, rectangular cross section) of 3.1-mm-diameter glass beads. Their theoretical analysis is valid only for region AB in Fig. 6, so that the method provides support for only this part of the transition mechanism.

It was assumed that the permeation of gas across the wall of the internal spout or cavity into the surrounding packed solids occurs in accordance with

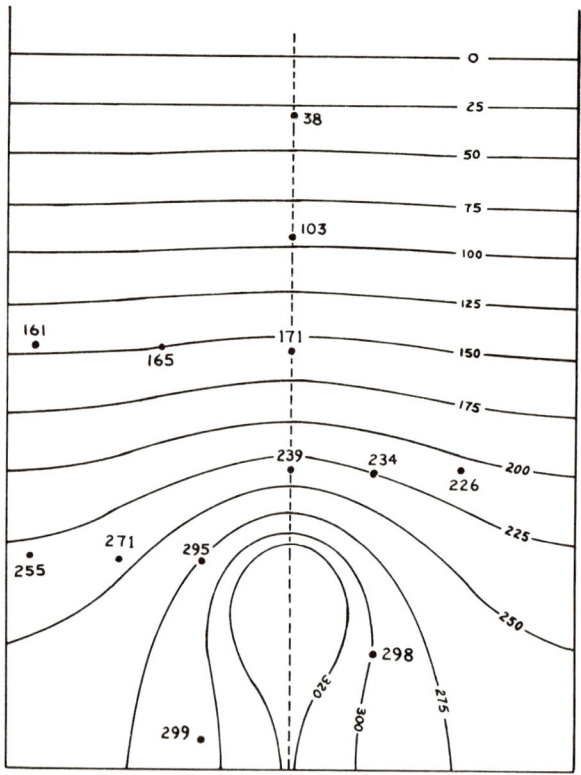

FIG. 7. Pressure distribution during transition. Observed (●) and calculated values, millimeters of water (Volpicelli *et al.*, V3). Glass spheres, 3.1 mm diameter; bed depth, 22 cm.

Darcy's law,
$$\nabla P = KU \quad (1)$$
Neglecting any density variation of the gas and assuming isotropy of the packed solids led to the Laplace equation
$$\nabla^2 P = 0 \quad (2)$$
This equation was integrated by finite differences using relaxation methods, the field of integration being delimited by (a) the measured value of pressure along the contour of the cavity or internal spout, which is regarded as an isobar, as the lower boundary and (b) the surface of the bed, where the pressure is atmospheric, as the upper boundary. The observed and calculated pressure distributions, shown in Fig. 7, are in fairly good agreement, and tend to confirm that over the stable part of the transition, corresponding to the growth of the internal spout and prior to penetration of the gas jet through the arch of solids surrounding the spout (region AB in Fig. 6), the bed solids outside the spout behave substantially as a packed bed.

B. Minimum Spouting Velocity

The minimum fluid velocity at which the bed would remain in the spouted state has been the subject of over a dozen investigations (B3, C1, G3, G7, G8, K2, M1, M5, M7, M10, N1, S2, T2, T4). No satisfactory general correlation for predicting this important parameter has emerged since the minimum spouting velocity (U_{ms}), in addition to being dependent on the properties of the solids and the fluid, is also a function of the bed geometry. Although the problem is inherently more complex than in either fluidization or pneumatic transport, some of the correlations that have been proposed are reasonably general, and it is useful to examine their validity against the wide range of experimental data now available.

1. *Cylindrical Vessels*

a. *Mathur and Gishler* (M10). From results obtained with a number of solid materials in cylindrical columns ranging in size from 3 to 12 in. in diameter, using air as well as water as the spouting fluid, the above workers correlated the minimum superficial fluid velocity for spouting with the variables of the system by the following dimensionally consistent equation:

$$U_{ms} = (d_p/D_c)(D_i/D_c)^{1/3}[2gH(\rho_s - \rho_f)/\rho_f]^{1/2} \quad (3)$$

As shown in Table I, this correlation has proved valid over a wide range of variables.

TABLE I

Range of Conditions Over Which Eq. (3) Has Proved To Be Valid Within ±10% (Spouting Fluid–Air)

Source	D_c (in.)	θ (°)	Solids	$d_p{}^a$ (mm)	ρ_s (lb/ft³)	D_i (in.)	H (in.)	U_{ms} (expt.) (ft/sec)
			A. Closely sized solids					
Mathur and Gishler (M10)	6	85	Wheat	3.2	85.9	0.49–2.0	12–30	2.44–3.52
			Mustard seed	2.2	75.2	0.49	12–30	1.42–2.54
			Rapeseed	1.8	68.9	0.49	12–30	1.15–2.02
			Peas	6.4	86.6	0.49	8–12	4.15–5.31
			Ottawa sand	0.6	145.0	0.49	12–27	0.55–0.75
			Gravel	1.0–3.5	164.0	0.49	24–48	1.72–4.67
	9	85	Wheat	3.2	85.9	0.5–2.0	25–60	2.30–3.80
	12	85	Wheat	3.2	85.9	0.5–3.0	25–75	2.00–3.20
Thorley et al. (T2)	24	45, 60, 85	Wheat	3.6	85.9	1.87–4.0	36–73	1.67–2.48
Manurung (M7)	6	60	Millet	1.3	80.8	0.36–0.60	8–36	0.93–1.59
			Nylon flakes	2.8	68.6	0.5	10–36	1.95–3.62
			Polystyrene	1.3–2.6	65.5	0.36–0.60	8–30	0.89–2.66

Reference		Material					
Smith and Reddy (S2)	6	Coal, rounded	0.9–2.5	89.0	0.5	9–40	0.76–3.22
	4	Coal, irregular	1.0–2.6	89.0	0.36–0.60	8–40	0.56–2.38
Kugo et al. (K2)	60	Alundum	0.3	246.3	0.37	13–23	0.47–0.59
	60	Coke	1.2–4.0	88.7	0.41	4–10	1.01–5.03
		Wheat	3.1	84.7	0.3–0.6	4–9	1.88–3.84

B. Mixed-size solids

Reference		Material					
Manurung (M7)	60	Coal, rounded (13 mixtures)	1.0–2.2	89.0	0.5	≃11	0.87–1.87
		Coal, irregular (4 mixtures)	1.2–1.5	89.0	0.5	9–30	0.89–1.93
		Polystyrene (3 mixtures)	1.5–2.2	65.5	0.5	9–30	0.91–2.12
Smith and Reddy (S2)	6	Alundum (11 mixtures)	0.7–1.7	246.3	0.375–0.75	8–39	0.77–4.21
		Silica sand	1.0	165.0	0.75	8–37	1.13–1.83
		Crystolon	0.7	199.6	0.75	9–35	0.91–1.55
		Polystyrene (5 mixtures)	1.5–2.4	65.8	0.375–0.75	10–38	1.18–2.85

[a] Smaller dimension in case of wheat. Arithmetic-mean diameter is used in Part A; reciprocal-mean diameter in Part B.

Although no theoretical basis for it was claimed by the authors, Ghosh (G3) subsequently arrived at an analogous relation by equating the momentum gained by the particles to that lost by the entering fluid. He assumed that (a) a particle at the bottom must attain a velocity equal to $(2gH)^{1/2}$ in order to reach the top of the spout, and (b) that the number of particles following each other in succession per unit time is proportional to v/d_p, since each particle requires a time interval of d_p/v to vacate its position. Considering that n particles can enter the spout through its periphery at any instant, the total number of particles accelerated per unit time becomes $n(v/d_p)$. Hence the momentum gained by the particles (assuming spherical shape) per unit time is

$$M = (nv/d_p)(\pi d_p^3/6)(\rho_s - \rho_f)(2gH)^{1/2} \tag{4}$$

where $v = (2gH)^{1/2}$. The momentum lost by the fluid jet is proportional to its initial momentum and is given by

$$M = \tfrac{1}{4}K\pi D_i^2 u_i^2 \rho_f \tag{5}$$

where u_i is the jet velocity through the orifice at minimum spouting. Solving Eqs. (4) and (5) for u_i, and converting to U_{ms} by multiplying both sides by $(D_i/D_c)^2$,

$$U_{ms} = (2n/3K)^{1/2}(d_p/D_c)(D_i/D_c)[2gH(\rho_s - \rho_f)/\rho_f]^{1/2} \tag{6}$$

Aside from the numerical constant, the only point of difference between Eqs. (6) and (3) is the exponent to the group D_i/D_c, its value being $\tfrac{1}{3}$ in the empirical equation as against unity in the theoretical. The difference can be explained by the fact that a substantial portion of the fluid entering the orifice flares out into the annulus immediately above the orifice (see Section IV,A), and therefore the momentum lost by the fluid jet is more weakly related to its initial momentum than assumed by Ghosh in writing Eq. 5.

The choice of a suitable particle diameter is important in testing the validity of the equation against experimental results. Mathur and Gishler, who worked mostly with closely sized materials, used the arithmetic mean of screen apertures (except in the case of wheat, for which they took the smaller dimension on the grounds that wheat particles were observed to align themselves vertically in the spout). Subsequent data for closely sized solids, including materials other than those used by Mathur and Gishler, showed good agreement with their equation (M5, M7, T2).

Manurung (M7) used the reciprocal mean diameter

$$d_p = 1/\sum (r_i/(d_p)_i)$$

as d_p in Eq. (3), and reported good agreement of his data for mixed sizes of coal and polystyrene, for a 2.5-fold size-spread. The reciprocal mean diameter is the same as the volume–surface mean diameter if one assumes that particle shape is invariant with particle size.

Serious deviations from Eq. (3) were reported by Smith and Reddy (S2), who worked with solid materials having a wide spread of particle size (up to sixfold) and used the weight-mean diameter $[d_p = \sum x_i(d_p)_i]$. When recalculated in terms of the reciprocal mean diameter (S3), these data show much better agreement with the equation (see Table IB). Hence, Eq. (3) appears to predict the spouting velocity for materials having a wide range of particle size if the reciprocal mean diameter is used.

Another question which has been raised concerning Eq. (3) is the dependence that it shows of U_{ms} on fluid density. Charlton et al. (C1), who used air, carbon dioxide, and helium as spouting fluids ($\rho_f = 0.000165 - 0.00182$ gm/cm³), found the effect of fluid density to be much smaller for their data than given by Eq. (3) and suggested that the effect attributed to density could be due to the difference in viscosity of air and water. Charlton et al. in their experiments used conical vessels with shallow beds of very-high-density solids, conditions which are completely outside the range of parameters in which Eq. (3) is applicable. The effect of fluid properties given in Eq. (3)—if the fluid is a gas other than air—is supported by Ghosh's theory, but may still require experimental verification.

The included angle of the conical base, which varied between 30° and 85° for the data in Table I, did not significantly affect the spouting velocity for columns up to 12 in. in diameter. In a 24-in. column, however, Thorley et al. found the spouting velocity for wheat to be about 10% higher with an 85° cone than with a 45° cone. They initially proposed (T1) the use of a variable exponent to the ratio D_i/D_c in Eq. (3) for a 24-in. column (0.23 for 45° and 60° cone angle, 0.13 for 85°), but later felt that the effect of cone angle was not large enough to justify this added complication (T2).

b. *Becker* (*B3*). Based on extensive data for spouting of various uniform-sized materials with air in 6–24-in. diameter columns, Becker found that

$$U_{ms}/U_m = 1 + s \ln(H/H_m) \tag{7}$$

where $U_m = U_{ms}$ at H_m. Experimental results for U_m, which is independent of column geometry, were separately correlated in terms of drag coefficient and particle Reynolds number:

$$C_D(\psi) = 22 + 2600/\text{Re}_m \tag{8}$$

where $C_D = 4d_v g(\rho_s - \rho_f)/3\rho_f U_m^2$ and d_v is particle size as diameter of an equivolume sphere. Curves of C_D versus Re_m for the different materials ran parallel to each other. A shape factor ψ was therefore introduced, the values assigned to it (1.0 for spheres, 0.62 and 0.76 for wheat, 0.35 for flax seed, etc.) being such that the data for particles of different shapes were brought together. Since at a depth just above the maximum spoutable, a spouted bed changes into a fluidized bed (Fig. 1), U_m is similar to the minimum fluidization velocity, and hence Eq. (8) is, in effect, an alternative equation for U_{mf}. The column geometry variables are incorporated in coefficient s [Eq. (7)], for which the following empirical relationship was determined:

$$s = 0.0071(D_c/D_i) \operatorname{Re}_m^{0.295} \psi^{2/3} \qquad (9)$$

For calculating H_m, which is required in Eq. (7), Becker proposed a separate empirical equation:

$$\left(\frac{H_m}{d_v}\right)\left(\frac{d_v}{D_c}\right)^{1.76}\left(\frac{12.2 D_i}{D_c}\right)^{1.6 \exp(-0.0072 \operatorname{Re}_m)} \left(22 + \frac{2600}{\operatorname{Re}_m}\right) \psi^{2/3} \operatorname{Re}_m^{1/3} = 42 \qquad (10)$$

Thus, the minimum spouting velocity for a given material, column size, inlet size, and bed depth can be obtained by combining Eqs. (7)–(10). Calculation by this method is valid for H/D_c greater than 1, Re_m of 10–100, and D_i/D_c less than 0.1.

The above equations were developed by Becker on the basis of not only his own data but also the previous data of Mathur and Gishler. It should, however, be noted that the equations are entirely empirical, and that since Becker's data covered more or less the same range of variables as those of the previous workers, the generality of his calculation method is no better than that of Eq. (3). As for accuracy, a detailed comparison of Becker's equations with Eq. (3) by Manurung (M7), using his own data obtained for 6 in. diameter beds as well as previous data for the same bed diameter (M10), did not clearly establish which of the two equations is the more reliable. The considerable complexity of Becker's correlation, therefore, does not seem to have yielded any compensating benefits, except for the prediction of H_m (see Section VI,B).

c. *Manurung* (M7). Manurung confined his experimental work to a 6-in. diameter column with a 60° cone, but studied a wide variety of materials, consisting of both close fractions and mixed sizes (coal of six different sizes, polystyrene, rapeseed, millet, $d_p = 1$–4 mm, $\rho_b = 0.48 - 0.75$ gm/cm³). Supplementing his own data with the larger-column results of Mathur and Gishler and of Thorley *et al.*, he developed the following equation:

$$U_{ms} = 7.73(\tan \alpha)^{0.79}(d_p/D_c)^{0.69}(D_i/D_c)^{0.155 \tan \alpha}(HU_m g)^{1/0} \qquad (11)$$

The coefficient of internal friction, $\tan \alpha$, was introduced to allow for the effect of widely different surface characteristics of his materials; the value of tan α, measured according to the method of Zenz and Othmer (Z2, p. 75), varied between 1.25 for rapeseed and 3.2 for coal. For estimating U_m in Eq. (11), Manurung compared Becker's equation [Eq. (8)] with Ergun's equation as applied to incipient fluidization [Eq. (81)] against experimental data, and selected the latter, but found it necessary to introduce a shape factor and a surface-roughness factor in the Ergun equation to deal with the rough, irregularly shaped coal particles. Despite these refinements which gave better predictions for coal mixtures, Manurung found that the general accuracy of his equation over the entire range of conditions was no better than that of the previous two equations.

d. Smith and Reddy (S2). These workers determined the minimum air velocity required to spout materials containing a distribution of particle sizes in a 6-in. column, using several inlet-orifice sizes. The angle of the conical base of the column was kept constant at 60°. On testing their data against the Mathur–Gishler equation, they found that, while the effect of particle size and solids density was correctly given by this equation, the dependence of U_{ms} on bed depth was more complex; the value of n in the relationship $U_{ms} \propto H^n$ varied depending on the orifice size. Representing the particle size of the material in terms of the surface-length mean diameter $(d_p = \sum [x_i/(d_p)_i]/\sum [x_i/(d_p)_i^2])$ as calculated from sieve analysis data, they correlated their data by the following equation, after dimensional analysis:

$$U_{ms} = d_p \left[\frac{g(\rho_s - \rho_f)}{\rho_f D_c}\right]^{1/2} \left[0.64 + 26.8 \left(\frac{D_i}{D_c}\right)^2\right] \left[\frac{H}{D_c}\right]^{0.50 - 1.76 D_i/D_c} \quad (12)$$

Since column diameter was not varied in their experiments, they wrote Eq. (12) for a 6-in. column in the following form, which is dimensional,

$$U_{ms} = [d_p(0.905 + 152 D_i^2)/(2H)^{3.52 D_i}][2gH(\rho_s - \rho_f)/\rho_f]^{1/2} \quad (13)$$

the linear unit being feet. While the inadequacy of a simple relationship between U_{ms} and H, brought out by Smith and Reddy, is also supported by Manurung's data for mixed-size materials, the exponent on H (in $U_{ms} \propto H^n$) depends not only on D_i but also on the size-spread and to some extent on the nature of the solid material. Hence Eq. (13), though satisfactory for most mixtures (including the coal mixtures of Manurung), gives generally poor results for closely sized materials, even for a 6-in. diameter column. When tested against the 24-in. column data of Thorley *et al.* for wheat, the performance of Eq. (12) was found to be even worse, which is perhaps not surprising considering that D_c was included in Eq. (12) without any experimental support, to make it dimensionless.

TABLE II

MINIMUM SPOUTING VELOCITY CORRELATIONS FOR CONICAL VESSELS (SPOUTING FLUID–AIR)

Investigators	Correlation	Bed geometry	Solids used
Nikolaev and Golubev (N1)	$(Re)_{ms} = 0.051(Ar)^{0.59} (D_i/D_c)^{0.10} (H/D_c)^{0.25}$	Cone plus short cylinder of diameter $D_c = 4.7$ in. $D_i = 0.8$–2.0 in. $H = 3.5$–6.0 in. θ not given	Spherical particles of five different sizes $d_p = 1.75$–5.6 mm
Gorshtein and Mukhlenov (G8)[a]	$(Re_i)_{ms} = 0.174 \dfrac{(Ar)^{0.50}}{(\tan \theta/2)^{1.25}} \left(\dfrac{D_b}{D_i}\right)^{0.85}$	Conical vessel with a short cylindrical upper part $D_i = 0.4$–0.5 in. $H = 1.2$–6.0 in. $\theta = 12°$–$60°$	Quartz, sand, millet, aluminum silicate, $d_p = 0.5$–2.5 mm $\rho_s = 0.98$–2.36 g/cm³ $\rho_b = 0.70$–1.63 g/cm³
Tsvik et al. (T4)	$(Re_i)_{ms} = 0.4(Ar)^{0.52} (H/D_i)^{1.24} (\tan \theta/2)^{0.42}$	$D_i = 0.8$–1.6 in. $H = 4$–20 in. $\theta = 20°$–$50°$	Fertilizer fractions, $d_p = 1.5$–4.0 mm $\rho_s = 1.65$–1.70 g/cm³ $\rho_b = 0.78$–0.84 g/cm³
Goltsiker (G7)	$(Re_i)_{ms} = 73(Ar)^{0.14} (\rho_s/\rho_s)^{0.47} (H/D_i)^{0.9}$	$D_i = 1.6$–4.8 in. $H = 2.0$–12.2 in. $\theta = 26°$–$60°$	Fertilizer and silica gel closely sized $d_p = 1.0$–3.0 mm Equation proved valid for antimony sand with a size-spread between 0.63 and 5.0 mm, using the reciprocal mean diameter as d_p

[a] Two modified versions of this correlation have been subsequently published by the same workers (M16, M17).

In summary, then, the simplest and most reliable method for estimating the minimum spouting velocity for common materials over a wide range of practical conditions is by use of Eq. (3), provided that in the case of mixed-size particles the reciprocal mean diameter is employed. The equation has been validated for column diameters up to 2 ft, not only for wheat but also for a coarse grade of ammonium nitrate (I2), but industrial data for larger units have unfortunately not been released.

2. Conical Vessels

If either the spouting vessel is conical in shape, or in a conical–cylindrical vessel the bed is so shallow that it remains mostly in the conical part, the gas flow for spouting can no longer be conveniently expressed in terms of a fixed superficial velocity. Several Soviet investigators, who worked with such beds, therefore used the gas velocity through the inlet orifice for correlating the minimum flow required for spouting. The starting point for this group of equations is the method commonly used in the Soviet Union for correlating minimum fluidization velocity data (for cylindrical columns with a uniform gas distributor) in the form

$$\mathrm{Re} = f(\mathrm{Ar}) \tag{14}$$

where $\mathrm{Re} = d_p U_{mf} \rho_f / \mu$ and Ar (Archimedes number) $= g d_p^3 \rho_f (\rho_s - \rho_f) / \mu^2$. For spouted beds, the above relationship has been modified by introducing bed-geometry parameters, using dimensional analysis. Table II shows different correlations which have been developed in this manner from experimental data, covering the range of variables indicated. The Reynolds number is based on the gas velocity through the inlet orifice and the diameter of the particle (except in the equation of Nikolaev and Golubev, who have not defined their Reynolds number clearly). The equations appear to be of limited applicability, and a comparative evaluation is unlikely to be useful. It should moreover be noted that the equations show widely different effects of variables such as cone angle and particle diameter on minimum spouting velocity.

C. Pressure Drop

Referring to Fig. 4, the pressure-drop values of practical interest are those corresponding to B and E, namely the peak pressure drop attained prior to the onset of spouting (ΔP_m) and the pressure drop at steady spouting (ΔP_s). The former would be encountered when starting up a spouted unit and must be allowed for in designing the gas-delivery system, while the latter would determine the operating power requirement.

1. *Peak Pressure Drop*

The high peak in ΔP that occurs just before spouting sets in is not specifically a feature of a spouted bed, but is associated generally with the entry of a high-velocity gas jet into a bed of solids. The occurrence of a similar peak has been reported by Gelperin et al. (G1) for the case of fluidization in a conical vessel, as well as in a conical–cylindrical vessel (Fig. 8). In both these situations the gas jet must penetrate the solids in the lower region of the bed, as it does in spouting, before it can cause movement of the solids in the upper part. Even in a cylindrical fluidized bed which the gas enters through a uniform distributor, the same phenomenon occurs; but the excess pressure drop attained prior to fluidization is only slight since each small gas jet that enters the solids through numerous orifices in the distributor can penetrate only a few layers of the particles before losing its identity by breaking into bubbles (Z2, p. 282). Therefore, the occurrence of a peak in the curve of pressure drop versus flow rate, prior to the onset of both spouting and fluidization, can be attributed to the energy required by the gas jet to rupture the packed-bed structure and to

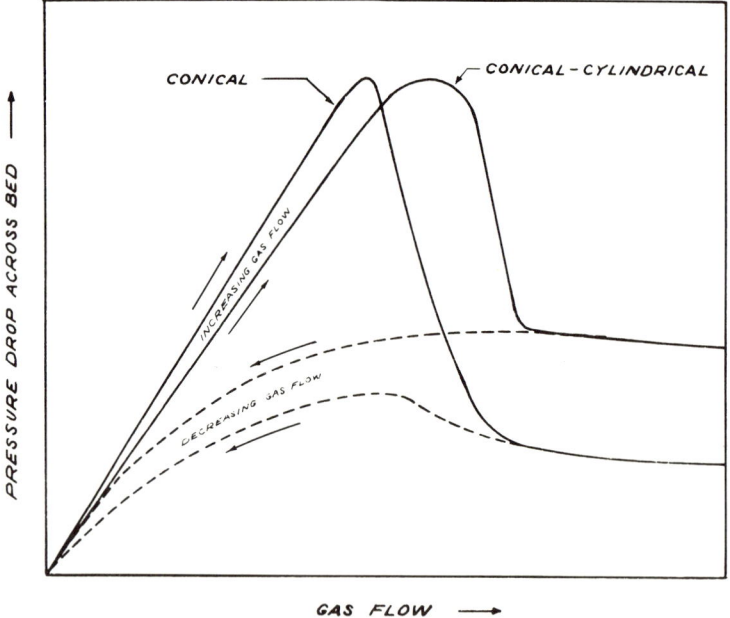

FIG. 8. Pressure drop curves for fluidization in conical and conical–cylindrical vessels (Gelperin et al., G1).

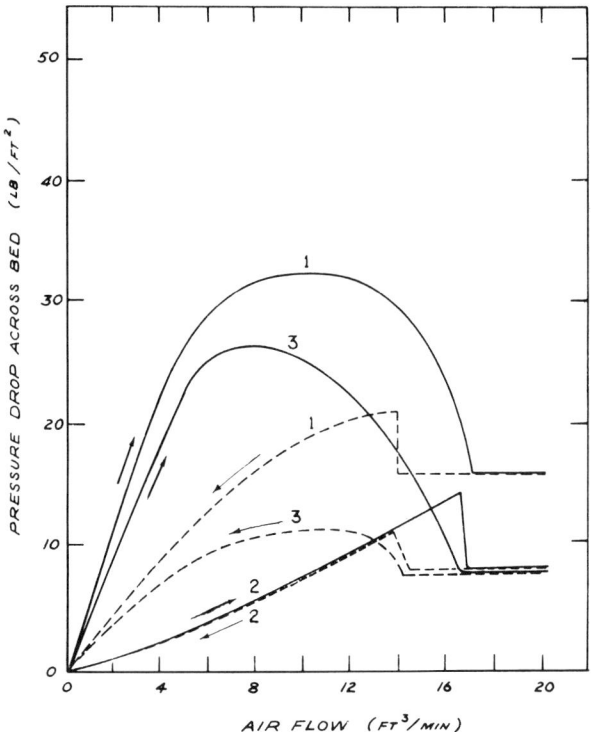

Fig. 9. Pressure drop–flow rate curves of Manurung (M7). 1, ΔP total; 2, ΔP upper; 3, ΔP lower; solid lines, increasing flow; dashed lines, decreasing flow.

form an internal spout in the lower part of the bed. Whether this internal spout subsequently develops into a through-spout or gives rise to fluidization will depend on whether the critical conditions such as particle size, orifice diameter, bed depths, etc., required for the spouting action are satisfied or not.

This explanation for the existence of a peak pressure drop is supported by experimental results obtained by Manurung (M7), who measured pressure drops separately across the upper cylindrical part and the lower conical part of the bed contained in a conical–cylindrical column as a function of both increasing and decreasing air flow. It is seen in Fig. 9 that the pressure drop across the upper part of the bed, up to the point at which the spout breaks through, corresponds to that in a packed bed and remains the same irrespective of whether the flow is increasing or decreasing. A peak well before the onset of spouting occurs only in the curve for the lower part of the bed, which falls much lower for decreasing than for increasing flow since the rupture energy is no longer required.

Manurung considered ΔP_m as being composed of a rupture pressure drop and a frictional pressure drop. The latter is the total pressure drop for decreasing gas flow, while the former is given by the difference between the total pressure drops for increasing and decreasing gas flow. From experimental results for a variety of materials in a 6-in. column with a 60° conical base, he derived empirical relationships for the two ΔP components separately, which on combining gave the following correlation:

$$\Delta P_m = [(6.8/\tan \alpha)(D_i/D_c) + 0.8]H\rho_b - 34.4 d_p \rho_b \qquad (15)$$

The data supporting the above equation were obtained for the same range of solid materials as those for Eq. (11).

The cone angle does not appear in Eq. (15), nor was it varied by Manurung in his experiments. The cone angle, unless it is very small, should not have any pronounced effect on jet penetration immediately above the inlet orifice. This view is supported by the close agreement obtained between ΔP_m predicted by Eq. (15) and the data reported by Lefroy and Davidson (L2) for kale-seed beds spouted with air in a flatbase column (see Table III). However, when checked against a wide range of data reported by different workers, the equation does not prove to be consistently reliable (Table III).

Madonna and Lama (M2) attempted to correlate ΔP_m values by a packed-bed equation, allowing for the "rupture" pressure drop in the numerical value of the constant. The constant, however, was found to vary widely with column size as well as with particle properties; moreover, their equation did not allow for the effect of orifice diameter. Malek and Lu (M3) obtained fresh data for several solid materials in columns of 4, 6, and 9 in. diameters, and arrived at the simple relationship that the maximum pressure drop approximately equals the weight of the bed, regardless of the size of the inlet orifice used, provided that the ratio of bed height to column diameter is greater than unity. A check against some of the data reported by other workers (Table III) shows that the simple relationship suggested by Malek and Lu is not generally valid, the ratio between ΔP_m and bed weight being often much higher than unity. Comparison of the results of different workers is, however, complicated by the fact that the exact location of the pressure tap, which could have a considerable effect on the observed value of ΔP_m due to the temporary pressure drop immediately downstream of an orifice plate, was not the same in every case. Thus, the generally lower results of Malek and Lu could be accounted for by the fact that these workers, unlike the others, located their lowest pressure tap at a point 1 in. above the inlet orifice.

TABLE III
Peak Pressure Drop Data

Source	D_c (in.)	D_i (in.)	θ (°)	Solids	d_p mean (mm)	α (°)	ρ_b (lb/ft³)	H (in.)	ΔP_m observed (lb/ft²)	ΔP_m by Eq. (15) (lb/ft²)	Bed weight $= \rho_b H$ (lb/ft²)	$\dfrac{(\Delta P_m)_{obs}}{\rho_b H}$
Kugo et al. (K2)	4	0.48	60	Wheat	4.1	55[a]	55	7.1	57.8	20.2	32.7	1.77
Manurung (M7)	6	0.50	60	Millet	1.3	54.5	48	13.4	56.8	57.6	53.6	1.06
								30.7	134	138	120	1.11
				Polystyrene	1.8	59.0	41	15.4	54.6	52.5	52.5	1.04
								25.3	91.2	89.7	85.7	1.06
				Coal	1.8	67.5	46	16.9	61.0	58.0	64.7	0.94
								23.4	93.0	86.9	92.6	1.00
								36.6	146	133	137	1.07
Lefroy and Davidson (L2)	12	1.0	180	Kale seed	1.7	52[b]	36	5.9	19.5	15.2	17.8	1.10
								11.8	29.7	37.5	35.6	0.83[d]
								23.6	77.8	82.1	71.2	1.09
								47.2	157.7	171.3	142.4	1.11
								70.9	245.8	261.0	213.9	1.15
Thorley et al. (T2)	24	4.0	45	Wheat	4.8	55[a]	56	24.0	203	148	112	1.81
								70.5	604	494	329	1.84
Malek and Lu (M4)[c]	6	0.375	60	Wheat	3.7	55[a]	52	10.9	44.8	30.0	46.9	0.96
								28.0	126.5	102.0	120.4	1.05
		1.0						13.9	56.6	38.3	59.8	0.95
								20.5	88.4	119.2	88.2	1.00
		1.5						11.0	46.0	72.6	47.3	0.97
								16.5	70.4	119.7	71.0	0.99

[a] From Lama (L1).
[b] Assumed same as for Manurung's rapeseed.
[c] Pressure tap 1 in. above orifice.
[d] 1.05 if ΔP_m is based on Eq. (15).

In experiments with conical vessels, Gelperin et al. (G2) obtained values of ΔP_m which in some cases were two to three times the bed weight. Their study was primarily concerned with fluidization of relatively fine materials, but their findings are considered to be relevant to spouting since the magnitude of the pressure peak should not depend on whether the bed would subsequently fluidize or spout. Expressing their ΔP_m results in terms of an excess pressure drop, over and above the weight of the bed, they proposed the following empirical correlation:

$$\Delta P_e/\rho_b H = 0.062(D_b/D_i)^{2.54}(\tan \theta/2)^{-0.18}[(D_b/D_i) - 1] \quad (16)$$

where θ is the included cone angle and D_b is the diameter of the upper surface of the bed. The range of variables covered was $\theta = 10°–60°$, $d_p = 0.16–0.28$ mm, $D_b/D_i = 1.3–6.8$, and $H = 4–10$ in. The inlet diameter remained constant at 2 in. and quartz was the only solid material used. With larger particles (3.2-mm diameter), for which distinct spouting action was obtained beyond the point of peak pressure drop, Goltsiker et al. (G6) obtained values of ΔP_m which were lower than those predicted by Eq. (16). From a theoretical derivation for pressure drop across a packed bed of conical shape, these workers showed that omission of particle diameter in Eq. (16), though justified for laminar flow, is no longer valid with larger particles of the order of a few millimeters.

Mukhlenov and Gorshtein (M16), who also worked with conical vessels, argued that the ratio between the peak pressure drop and the pressure drop at steady spouting should bear a relationship to the geometry of the system, and to the properties of the gas and the solids. Starting with dimensional analysis, they proposed the following empirical correlation:

$$\frac{\Delta P_m}{\Delta P_s} = \frac{6.65(H/D_i)^{1.2}(\tan \theta/2)^{0.5}}{(Ar)^{0.2}} + 1 \quad (17)$$

where ΔP_s, the spouting pressure drop, is given by Eq. (29) and Ar is the Archimedes number. The data that support the above equation (within 10%) were obtained in vessels having angles of 12°, 30°, 45°, and 60°, with several materials: $d_p = 0.5–2.5$ mm and $\rho_s = 0.978–2.36$ gm/cm³. Four orifice sizes in the range 0.4–0.5 in. were used, with bed depths varying between 1.2 and 6.0 in.

2. Spouting Pressure Drop

In the spouted state, the pressure drop across the bed arises out of two parallel resistances, namely that of the spout, in which dilute phase transport of particles is occurring, and that of the annulus, which is a downward-moving packed bed with countercurrent flow of gas. Since the gas entering the base flares out radially from the axial zone into the annulus as it travels upward, the vertical pressure gradient increases from zero at the base to a maximum at the bed top. The total pressure drop across the bed can therefore be obtained by integrating the longitudinal pressure-gradient profile over the height of the bed. Since the fluidization pressure gradient is approached only in the upper part of a deep bed, the total pressure drop across a spouted bed is always less than the pressure drop which would arise if the same material were fluidized. It has been shown from theoretical considerations, by two different methods (L2, M6), that for a deep bed, the spouting pressure drop bears a fixed ratio to the fluidization pressure drop.

An expression for the cumulative pressure drop profile in a spouted bed was derived by Mamuro and Hattori (M6) from a consideration of the balance of forces acting on a differential height dz of the annular solids (Fig. 10):

$$dP_b = -(\rho_s - \rho_f)(1 - \epsilon_a)(g/g_c)\, dz + (-dP_f) \tag{18}$$

where P_b is the downward force per unit cross-sectional area of the annulus.

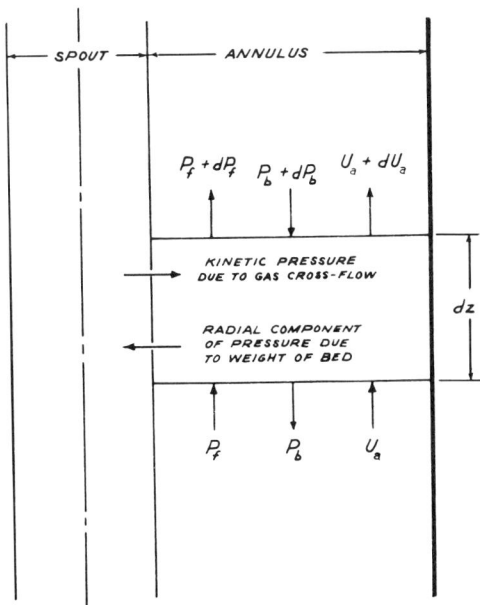

FIG. 10. Model for forces acting in the annulus (Mamuro and Hattori, M6).

Since the interface between the spout and the annulus is maintained steady, the kinetic pressure due to cross-flow of gas from the spout to the annulus must be in balance with the radial component of the annular bed pressure. Assuming, by analogy with Janssen's assumption for powders in hoppers (B10), that the vertical component of this pressure is proportional to the radial component, it is contended by these authors that:

$$P_b = k\rho_f U_r^2 / 2g_c \tag{19}$$

The change in static pressure of the gas passing through the annulus at low particle Reynolds number should be in accordance with Darcy's law since the annular solids are essentially in the packed-bed condition. Therefore

$$-dP_f/dz = KU_a \tag{20}$$

and from a fluid mass balance over the height dz,

$$dU_a/dz = \pi D_s U_r / A_a \tag{21}$$

Equations (18)–(21), together with the boundary conditions,

$$U_a = 0 \quad \text{at} \quad z = 0$$

$$U_a = U_{mf} \quad \text{at} \quad z = H = H_m$$

give rise to the following expression for the cumulative pressure drop profile:

$$(\Delta P_s)_{0 \to z} = \int -dP_f = \frac{KU_{mf}H_m}{4} \left(\frac{z}{H_m}\right)^2 \left[\left(\frac{z}{H_m}\right)^2 - 4\left(\frac{z}{H_m}\right) + 6\right] \tag{22}$$

Therefore, for a bed of height H_m,

$$\Delta P_s = \tfrac{3}{4} K U_{mf} H_m \tag{23}$$

while for the same bed, using Eq. (20),

$$\Delta P_{mf} = K U_{mf} H_m \tag{24}$$

Hence the ratio between the spouting and fluidization pressure drop equals 0.75.

The starting point for the second method mentioned is the generalization suggested by Lefroy and Davidson (L2), that the pressure in a spouted bed (measured near the spout) is distributed vertically according to a quarter cosine curve i.e.,

$$P_a = \Delta P_s \cos(\pi z / 2H) \tag{25}$$

At the top of height H_m, the pressure gradient is sufficient to support the annulus solids; hence, at the top,

$$dP_a/dz = -(\rho_s - \rho_f)(g/g_c)(1 - \epsilon_a) \quad (26)$$

Differentiating Eq. (25) with respect to z and substituting the result in Eq. (26), after putting $z = H = H_m$, gives

$$\Delta P_s = 2H_m(\rho_s - \rho_f)(g/g_c)(1 - \epsilon_a)/\pi \quad (27)$$

which shows the ratio between spouting and fluidization pressure drop to be $2/\pi = 0.64$.

The two derivations quoted above rely on the boundary condition that the gas flow through the annulus near the top of the bed is sufficient to fluidize the solids, and they therefore are valid only for beds of maximum spoutable depth. The derived values of $\Delta P_s/\Delta P_{mf}$ thus represent the upper limit which would be approached with increasing bed depth for a given system. This is borne out by the experimental results plotted in Fig. 11, which cover different materials as well as column geometries. The maximum $\Delta P_s/\Delta P_{mf}$ ratios attained are seen to be in remarkably good agreement with the predicted values of 0.64–0.75.

In the case of more shallow beds, as also for conical vessels, theoretical analysis becomes more complex because of the difficulty of selecting a suitable boundary condition for integrating the differential pressure gradient. The following empirical correlations have been proposed:

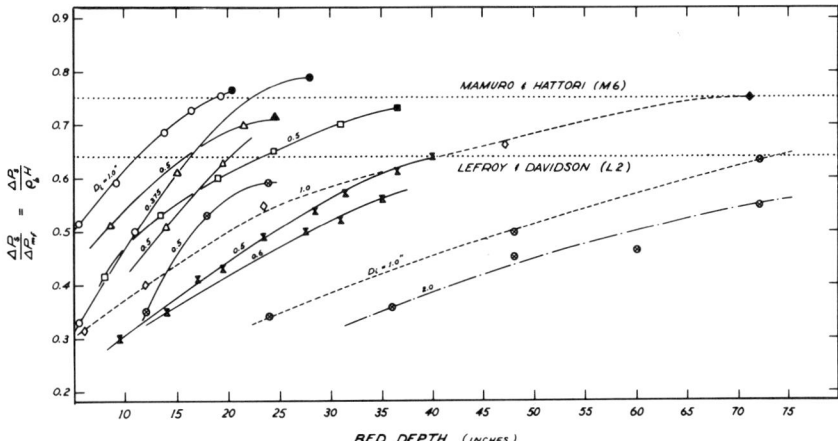

FIG. 11. Spouting pressure drop data. ⊗ Wheat (T2); ○ wheat (M3); △ rapeseed (M7); □ millet (M7); ◇ kale seed (L2); ⊠ coal (M7); —— 6 in.; ---- 12 in.; —·— 24 in. Solid points are data for maximum spoutable bed depths.

For cylindrical vessels (Manurung, M7):

$$\Delta P_s/H\rho_b = 1/[1 + K(D_c/H)] \quad (28)$$

where

$$K = [0.81(\tan \alpha)^{1.5}/\bar{\psi}^2](D_c d_p/D_i^2)^{0.78}$$

The range of data from which the above equation was derived is the same as for Eq. (11). The shape factor $\bar{\psi}$ was taken as the ratio between the reciprocal mean diameter d_p by screen analysis and the diameter of an equivolume sphere, its value being 0.81–1.00 for the materials tested.

For conical vessels (Mukhlenov and Gorshtein, M15, M17):

$$\frac{\Delta P_s}{H\rho_b} = \frac{7.68(\tan \theta/2)^{0.2}}{\mathrm{Re}_i^{0.2}(H/D_i)^{0.33}} \quad (29)$$

Cone angle θ varied from 12° to 60°, H from 1.2 to 6 in., D_i from 0.4 to 0.5 in., and ρ_b from 0.7 to 1.6 gm/cm³. Re_i is based on the diameter of the particle and the air velocity through the orifice.

IV. Flow Patterns

A. Flow Pattern of Gas

The distribution of gas between spout and annulus is important in assessing the effectiveness of gas–solids contact. Qualitatively the flow pattern in a spouted bed is obvious: the gas jet flares out as it travels upward, causing the gas flow rate in the spout to decrease, and the flow rate in the annulus to increase, with increasing distance from the inlet orifice. Attempts to quantify this pattern and to relate it to the variables of the system have been made, both theoretically and experimentally.

Mamuro and Hattori (M6) extended their analysis of the balance of forces acting on the spout–annulus interface (Fig. 10) to derive an expression which enables estimation of gas distribution between the annulus and the spout at various levels in the bed. Substitution of Eqs. (19) and (20) into Eq. (18) gives

$$-(\rho_s - \rho_f)(1 - \epsilon_a)(g/g_c) + KU_a = (k\rho_b U_r/g_c)(dU_r/dz) \quad (30)$$

Differentiating Eq. (30) with respect to z and combining the result with Eq. (21) gives the differential equation

$$(\pi D_s K g_c/k\rho_f A_a)U_r = d(U_r\, dU_r/dz)/dz \quad (31)$$

The solution of the above equation is

$$U_r = C_1(z + C_2)^2 \tag{32}$$

which on substitution into Eq. (21) leads to

$$dU_a/dz = (\pi D_s/A_a)C_1(z + C_2)^2 \tag{33}$$

Since $U_a = 0$ at $z = 0$, integration of Eq. (33) gives

$$U_a = (\pi D_s C_1/3A_a)[(z + C_2)^3 - C_2^3] \tag{34}$$

Assuming that the gas velocity at the top of the annulus is sufficient to fluidize the solids (i.e., that the bed is at the maximum spoutable depth H_m), Eq. (32) together with Eqs. (30) and (20) gives rise to

$$(U_r \, dU_r/dz)_{z=H_m} = 2C_1^2(H_m + C_2)^3 = 0 \tag{35}$$

Therefore $C_2 = -H_m$. On substituting this result in Eq. (34), with the boundary condition $U_a = U_{mf}$ at $z = H_m$,

$$U_a/U_{mf} = 1 - (1 - z/H_m)^3 \tag{36}$$

Mamuro and Hattori further suggest that for depths below the maximum spoutable, where the value of U_a at $z = H$ (i.e., U_{aH}) does not reach U_{mf}, the above equation should be modified to the form

$$U_a/U_{aH} = 1 - (1 - z/H)^3 \tag{37}$$

The experimental results which can be used to verify the preceding theory have been obtained by two different techniques. Mathur and co-workers (M10, T1, T2), treating the annulus as similar in voidage to a loosely packed stationary bed, determined the superficial gas velocity in the annulus from measurements of vertical static-pressure-drop profiles. The experimental investigation was carried out in columns ranging in diameter from 6 to 24 in., with different cone angles, bed depths, and orifice sizes. The system studied was air–wheat. All observations were made in the region above the cone, where the pressure across a horizontal section of the bed was found to be uniform. The volumetric flow in the annulus at different levels was calculated from the known cross-sectional area of the annulus and the corresponding superficial velocities, and the flow through the spout was obtained by difference. As pointed out by these workers, the assumption that the annular solids behave hydrodynamically as a loosely packed stationary bed is only an approximation. While the assumption was supported by pressure drops measured for the same solids in a moving bed, which is known to display the same voidage as a loosely packed stationary bed (H1) and which visually resembles the spouted-bed annulus, the particles near the bottom were in fact observed to be somewhat more tightly packed than those at the top of the annulus. This inhomogeneity would introduce some error in this method of estimating the gas distribution, especially at the extremities of the cylindrical portion of the bed.

Results of direct pitot-tube measurements of the local upward gas velocity, both in the spout and in the annulus, have been reported by Becker (B3) and by Mamuro and Hattori (M6). The accuracy of such measurements in the annulus is questionable because of the low velocities involved and the possible disturbance of local porosity by the introduction of the pitot tube. Becker reports the measured value of the local air velocity near the top of a wheat bed in a 6-in. diameter column to be about 24 ft/sec, in the annulus as well as in the spout. Since the superficial velocity through the column was only about 3 ft/sec, the annulus result calls for either a randomly packed bed voidage of 12–13% or a very high degree of particle orientation. Both situations are well nigh inconceivable in a well-aerated moving mass of particles. On the other hand, a local velocity of 24 ft/sec, in the spout seems reasonable, being consistent with the superficial velocity of 12 ft/sec obtained from pressure-drop data in a similar bed by Mathur and Gishler. The range of variables covered by Mathur and co-workers (M10, T2), in general, encompasses that of Becker.

Mamuro and Hattori (M6) dismissed their annulus data and used only the local air-velocity results obtained in the spout to calculate the air distribution between the spout and the annulus. The calculation, however, requires separate knowledge of spout diameter and of the vertical voidage profile in the spout. While the former could be directly measured, for the latter they had to rely on the values estimated by Mathur and Gishler for a wheat bed 25 in. deep and 6 in. in diameter with the further assumption that the same profile (voidage versus reduced bed level) is valid, not only for wheat beds of different depths, but also for soma sand. This assumption is speculative and weakens the reliability of their calculated air-distribution results as a means for verifying Eq. (37). The earlier experimental results based on pressure-drop measurements (M10, T1), on the other hand, are free from the above objection, and these provide some support for the theoretical derivation of Mamuro and Hattori (Fig. 12), especially for Eq. (36), although less so for its arbitrary extension to Eq. (37).

It should be noted further that Eq. (37) does not enable one to predict theoretically the gas-distribution profile for a given bed since the value of U_{aH} must be independently known. The figures in Table IV show that U_{aH} and the ratio it bears to U_s (superficial spouting velocity, corresponding to total flow through the column) are dependent on bed geometry. Hence

TABLE IV

EXPERIMENTALLY DETERMINED VALUES OF U_{aH}/U_s FOR AIR–WHEAT

Source	D_c (in.)	D_i (in.)	θ (°)	H (in.)	U_{aH} (ft/sec)	U_s (ft/sec)	U_{aH}/U_s
Mathur and Gishler (M10)	6	0.5	85	25	2.66	3.72	0.72
Mathur and Gishler (M11)	12	2	85	45	2.69	2.99	0.90
Thorley et al. (T2)	24	2	45	72	1.88	2.02	0.93
		2	60	36	1.26	1.24	1.0
		2	60	48	1.52	1.52	1.0
		2	60	60	1.67	1.79	0.93
		2	60	72	1.83	1.96	0.93
		3	60	72	2.05	2.27	0.91
		4	60	72	2.12	2.38	0.89
		2	85	72	1.67	1.81	0.92
		2	85	72	1.74	1.93	0.90
		2	85	72	1.70	2.14	0.80

Note: The spouting velocities used in the above experiments are roughly 10% above U_{ms} except for the last two results of Thorley et al. (20 and 30% above).

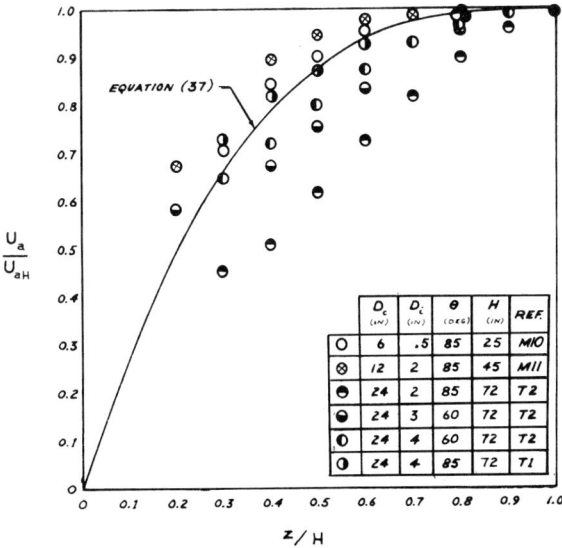

FIG. 12. Vertical profile of air distribution. Comparison of experimental results for air–wheat with Eq. (37). $H \approx H_m$ only in the case of the 6-in. column data.

the vertical profile of the percentage of total gas flowing through the annulus also depends on bed geometry, the percentage at corresponding fractional bed levels being generally higher for high values of U_{aH}/U_s. From the values of this ratio shown in Table IV, it is seen that the maximum proportion of gas passing through the annulus is obtained with large column diameter, large orifice size, small cone angle, bed operation near the minimum spouting velocity, and, surprisingly, with shallow bed depth.

The effect of solids properties on the air-flow pattern has not been studied in detail, but Mamuro and Hattori, on the basis of limited results for soma sand (d_p = 1.18 mm, ρ_s = 2.61 gm/cm³) in a 6-in. column, suggest that the pattern described by Eq. (37) applies irrespective of the solids properties, as is implicit in the derivation of the equation.

B. Flow Pattern of Solids

The solids movement in a spouted bed is initiated by the interaction between the particles and the high-velocity gas jet, so that particle flow in the spout region shapes the entire solids-flow pattern. While a mutual dependence between the solids flow in the spout and in the annulus is inherent to a spouted bed, it is nevertheless convenient to discuss the flow in the spout and in the annulus separately.

1. *In the Spout*

A particle starting from the base of the bed first accelerates from rest to a peak velocity, and then decelerates until it again reaches zero velocity at the top of the spout, where it reverses its direction of movement. In addition, a radial profile of longitudinal velocity exists at a particular level, the velocity at the axis of the spout being higher than at other radial positions. Velocity profiles, both in the vertical and the radial directions, have been measured by several investigators, with various solid materials and bed geometries. Two different experimental techniques have been used: visual observation and particle-impact measurement.

a. Direct Observation. It is not possible to observe the particle motion in the spout in an ordinary spouted column since the spout, being surrounded by the annular solids, is not open to view. However, if the bed is contained in a column of semicircular cross section, with matching conical base and inlet orifice, the axial region of the bed becomes visible against the flat transparent face of the column. The question as to whether or not a half-column truly represents the behavior of a spouted bed in a whole column has been experimentally investigated by obtaining comparative data on various aspects of the behavior in the two types of columns. While Mathur and Gishler (M10), who first used a half-column, had expressed some reservation on this point, subsequent evidence (L2, M13) has shown that any difference which exist are not significant for most purposes.

 b. *Piezoelectric Technique.* This technique enables the spout particle velocities to be measured in a whole column. The method, originally developed by Gorshtein and Soroko (G10), and subsequently used for

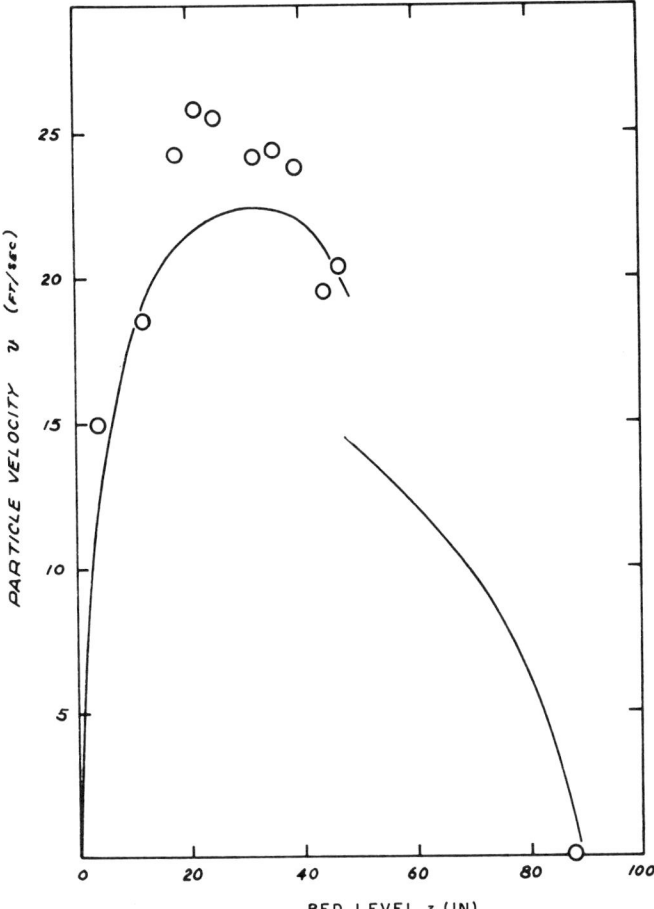

Fig. 13. Vertical profile of upward particle velocity in the spout (after Thorley *et al.*, T2). Air–wheat; $D_c = 24$ in.; $D_i = 4$ in.; $H = 48$ in.; $u_i = 80.6$ ft/sec; — theoretical curves; ○ experimental data.

spout-velocity measurements by other Soviet workers, is based on the piezoelectric effect, namely, that crystals of certain materials such as piezoquartz, barium titanate, Rochelle salt, etc., when subjected to a short-term mechanical shock, generate an electric charge on the surface of the crystal, the magnitude of the charge being a function of the force with which the crystal is hit. A piezo-crystal, of a size similar to that of the bed particles, is used as the sensing element, and is mounted to serve as a probe for measuring velocity profiles in the spout. This is connected to an amplifying device and then to an oscilloscope which records the charge generated. Prior calibration of the instrument, with the same solid particles, is necessary to convert the electric charge data into particle velocities. The lower limit of sensitivity of the technique has been quoted as 10–20 cm/sec for particles of 2-mm diameter.

c. Vertical Profile. A typical profile of the particle velocity along the spout height is shown in Fig. 13. The particles, having attained a peak value about half way up the bed, decelerate rapidly until they reach the level of the bed surface and then more slowly in the fountain above the bed proper. In an attempt to explain the observed velocity profile from theoretical considerations, Thorley *et al.* (T2) divided the vertical height between the inlet orifice and the spout top into three regions (Fig. 14) and considered the force balance on a particle in each region separately. In the lower part of the spout (region 1), the accelerating force due to frictional drag on the particles by the high-velocity gas jet dominates. In region 2, where the gas velocity is much lower, the drag force becomes unimportant in comparison with the decelerating effects due to gravity and due to the particles entering from the annulus through the side of the spout. In region 3, the latter effect disappears, the main force now acting being that of gravity. Based on this simplified analysis, Thorley *et al.* applied the following force balance equation to all three regions:

$$\frac{dv}{dt} = \frac{3C_D \rho_f (u_s - v)^2}{4 d_p \rho_s} - \left(\frac{\rho_s - \rho_f}{\rho_s}\right) g \qquad (38)$$

The above equation together with the relationship

$$dz/dt = v \qquad (39)$$

can be used to calculate the radial average vertical particle velocity at various heights, provided that the corresponding values of u_s are known. These workers carried out the calculation for region 1 with the assumption that u_s changes linearly with bed level and extended the calculation to region 2 by resorting to the further assumption that the decelerating force

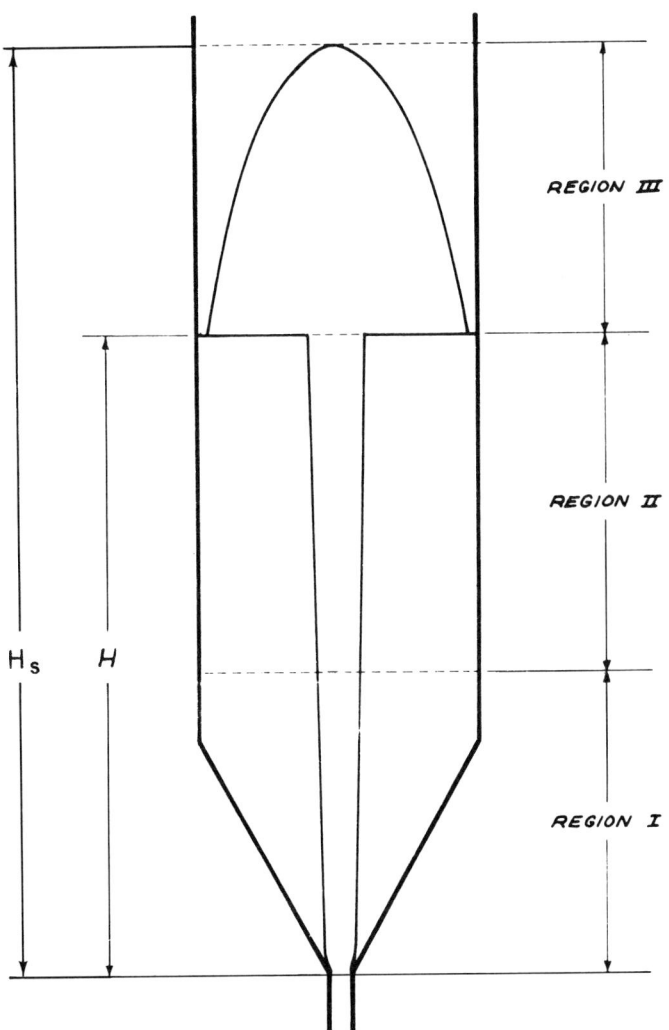

FIG. 14. Model of Thorley et al. (T2) for calculating the vertical particle velocity profile in the spout.

due to cross-flow of particles from the annulus into the spout is negligible. Though both these assumptions (particularly the latter) grossly oversimplify the true behavior, they nevertheless produced some measure of agreement with the observed results (Fig. 13). For estimating the velocity in region 3, Thorley et al. neglected the first term on the right-hand side of

Eq. (38) and integrated Eqs. (38) and (39) together using the boundary condition $v = 0$ at $z = H_s$, with the result

$$v = \{2[(\rho_s - \rho_f)/\rho_s]g(H_s - z)\}^{1/2} \tag{40}$$

The curve based on Eq. (40) is shown in Fig. 13, but no measurements were made in this region.

A more rigorous analysis of the problem has recently been made by Lefroy and Davidson (L2). This analysis takes into account the vertical gas velocity and voidage profiles in the spout, as well as the effect of particle cross-flow from the annulus. Disregarding the radial variations across any horizontal section of the spout, the following equation can be written for continuity of gas flow and solids flow, respectively:

$$d(\epsilon_s u_s)/dz + 4U_r/D_s = 0 \tag{41}$$

$$d[(1 - \epsilon_s)v_m]/dz - 4v_c/D_s = 0 \tag{42}$$

From a momentum balance on the gas,

$$\rho_f \, d(\epsilon_s u_s^2)/dz = -\epsilon_a \, dP_a/dz - \beta(u_s - v_m)^2 \tag{43}$$

and from a momentum balance on the solids,

$$\rho_s \, d[(1 - \epsilon_s)v_m^2]/dz = -(1 - \epsilon_a) \, dP_a/dz$$
$$- (\rho_s - \rho_f)g(1 - \epsilon_s) + \beta(u_s - v_m)^2 \tag{44}$$

β in Eqs. (43) and (44) represents forces of interaction between the gas and the particles. Lefroy and Davidson derived an expression for β, using the Richardson–Zaki (R3) equation for fluidized beds in conjunction with Eq. (43). The latter equation, when applied to a fluidized bed, becomes

$$-dP_a/dz = \beta u^2/\epsilon \tag{45}$$

Combining Eq. (45) with the condition that pressure drop in a fluidized bed equals bed weight per unit area, we get

$$\beta u^2/\epsilon = (\rho_s - \rho_f)g(1 - \epsilon) \tag{46}$$

According to Richardson and Zaki

$$U = u\epsilon = v_t \epsilon^{2.39} \tag{47}$$

at high Reynolds numbers. Since at high Reynolds numbers, the single-particle drag coefficient is practically constant at 0.44, the following expression may be written for the particle terminal velocity:

$$v_t^2 = (\rho_s - \rho_f)g \, d_p/0.33\rho_f \tag{48}$$

Substituting Eqs. (47) and (48) into Eq. (46), we get

$$\beta = 0.33\rho_f(1 - \epsilon)/d_p\epsilon^{1.78} \qquad (49)$$

A somewhat different expression for β can be derived by solving Eq. (45) with the Ergun (E3) packed-bed equation, but since the spout is a high-voidage system, the use of Eq. (49) is more appropriate.

For describing the gas-velocity profile in the spout, Lefroy and Davidson started with the assumption, already discussed in Section IV,C, that the pressure variation with height in a spouted bed follows a quarter-cosine curve:

$$P_a = (2/\pi)B\rho_s(g/g_c)H(1 - \epsilon_0)\cos(\pi z/2H) \qquad (50)$$

Equation (50) combines Eqs. (25) and (27) with an empirical constant B, the value of which is 1.0 for $H = H_m$ and less than 1.0 for $H < H_m$. For the latter case, B must be evaluated from the measured vertical-pressure profile. An expression for the vertical-velocity profile in the annulus was then derived from Eq. (50), assuming that the flow of gas from the spout to the annulus occurs in accordance with Darcy's law, and that for practical purposes, the pressure across the horizontal section of the bed is uniform:

$$U_a = BU_{mf}\sin(\pi z/2H) \qquad (51)$$

The above expression, combined with Eq. (41) and with a gas balance,

$$U_r = [(D_c^2 - D_s^2)/4D_s](dU_a/dz) \qquad (52)$$

gives

$$u_i - \epsilon_s u_s = BU_{mf}[(D_c^2/D_s^2) - 1]\sin(\pi z/2H) \qquad (53)$$

After substituting β from Eq. (49), Eqs. (43) and (44) can be numerically integrated with the aid of Eq. (50) to give $\epsilon_s u_s^2$ and $(1 - \epsilon_s)v_m^2$ respectively, at various values of z, while Eq. (53) can be solved for corresponding values of $\epsilon_s u_s$. The lower boundary conditions for the integration are at $z = 0$,

$$\epsilon_s = 1.0, \qquad v_m = 0, \qquad u_s = u_i$$

Thus vertical profiles of particle velocity, as well as of voidage and gas velocity, can be calculated for a given system, using experimentally determined values of D_s, B, and u_i $(= U_s D_c^2/D_i^2)$.

Lefroy and Davidson did not make a direct check of their theory with measured particle velocity profiles. Instead they used the calculated values of ϵ_s and v_m, in conjunction with Eq. (42), to compute the rate of cross-flow of solids into the spout as a function of bed level. The results thus obtained agreed approximately with predictions made from an independent particle-collision model for solids cross-flow (discussed in the following section).

However, particle-velocity profiles for the system of Thorley *et al.*, calculated by the present authors from the equations of Lefroy and Davidson, fail to give quantitative agreement with the experimental results in Fig. 13, irrespective of the value of B assumed.

Particle-velocity profiles for several solid materials spouted in a 3.7-in. conical–cylindrical column, measured by the piezoelectric technique, have been reported by Mikhailik and Antanishin (M14, millet, silica gel, poly-

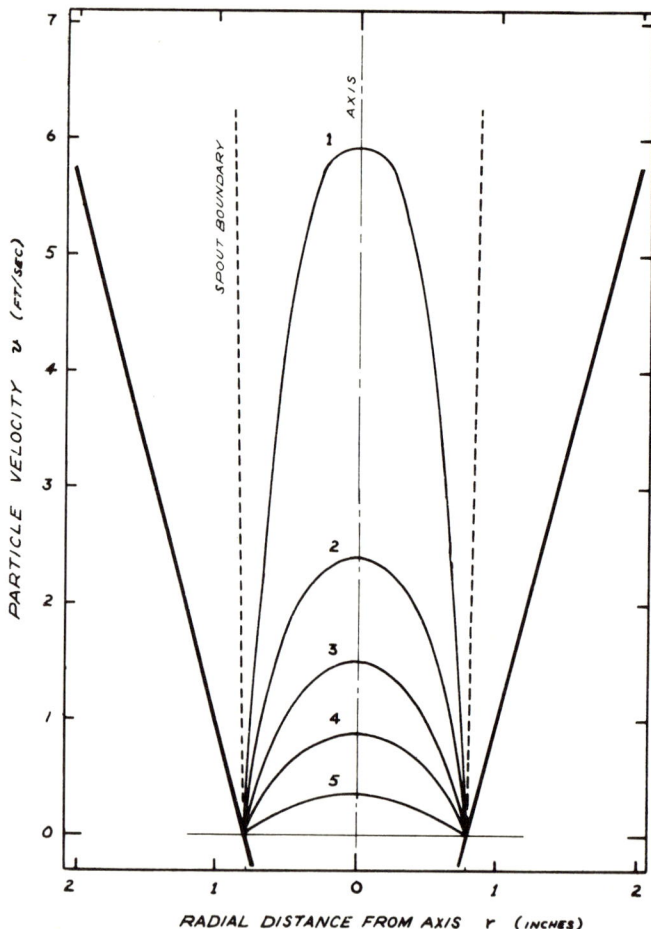

FIG. 15. Radial profile of particle velocity in the spout (Gorshtein and Mukhlenov, G9). Conical vessel $\theta = 30°$, $D_i = 1.6$ in., $H = 4.7$ in.; catalyst marbles, $d_p = 1.5$ mm, $u_i = 11.4$ ft/sec; z (inches) = (1) 1.7; (2) 2.1; (3) 2.8; (4) 4.0; (5) 5.2.

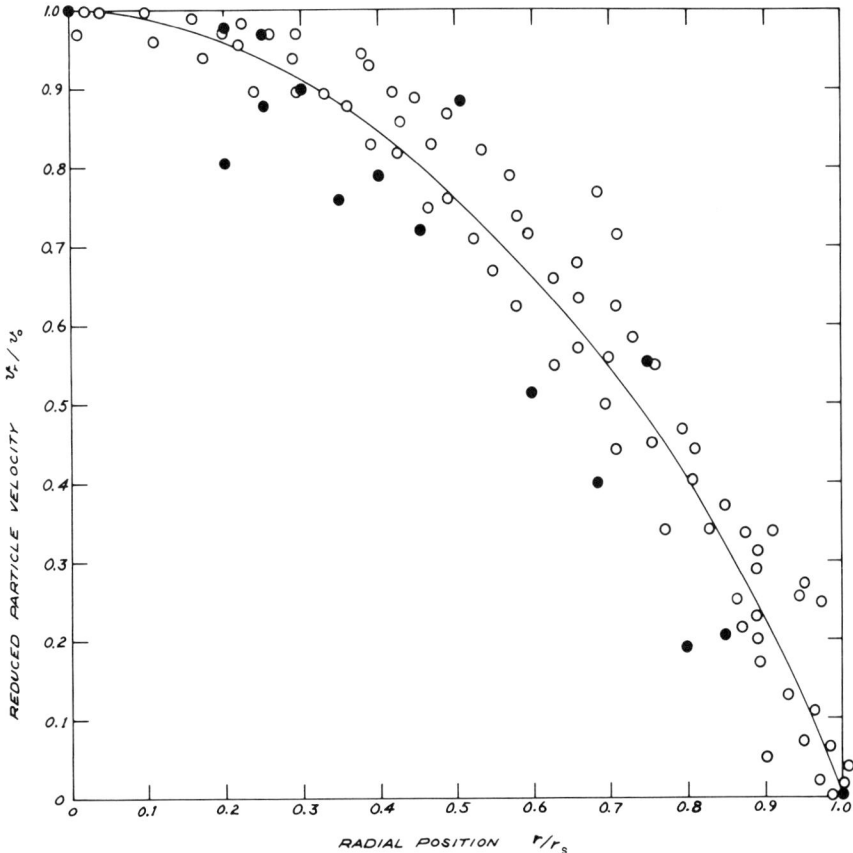

Fig. 16. Normalized plot of spout particle velocities. ○ Gorshtein and Mukhlenov (G9): conical vessel; catalyst marbles, $d_p = 1.5$ mm; $\theta = 20°$–$60°$; $H/D_i = 1.5$–6.0; $z = 2$ and 4 in. ● Lefroy and Davidson (L2): cylindrical vessel, 12-in. diameter; polyethylene chips, $d_p = 3.5$ mm; $\theta = 180°$; $H/D_i = 39$; z not given.

styrene, $d_p = 1.5$–7.0 mm) and for a conical vessel by Gorshtein and Mukhlenov (G9, catalyst beads, $d_p = 1.5$ mm, Fig. 15). The former workers obtained velocity profiles similar in shape to those of Fig. 13, and found that for a given air velocity, smaller particles reached a higher peak value and took a longer distance to do so than larger particles. Empirical correlations of their respective data have been presented by both groups of workers, with separate equations for the acceleration and deceleration zones.

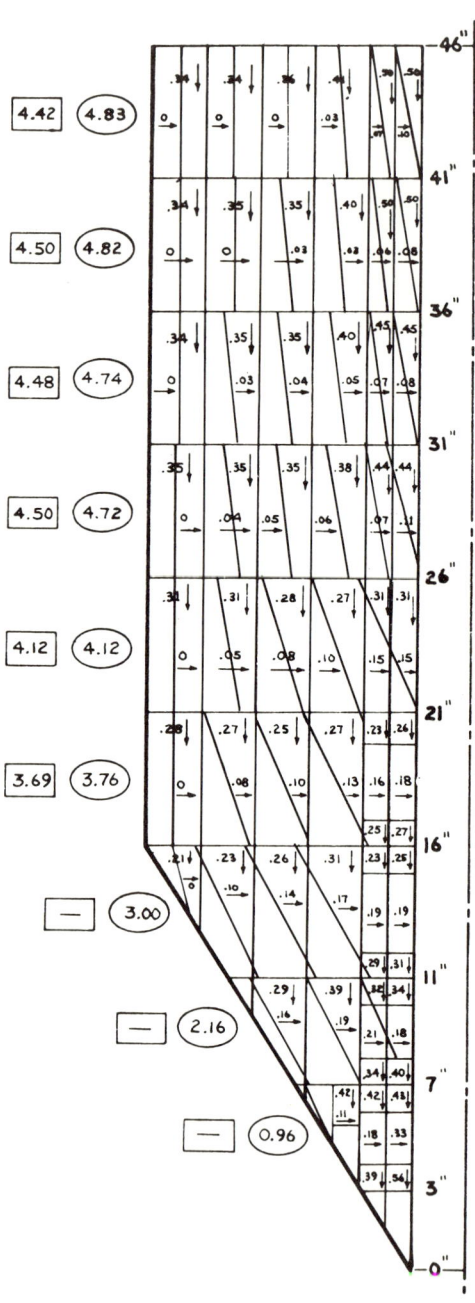

d. Radial Profile. Experimental data on radial particle-velocity profiles at different levels in the spout have been reported for various materials and column geometries. There is general agreement that the radial profile at any level is parabolic, as shown in Fig. 15 by the results of Gorshtein and Mukhlenov (G9), which are typical. The normalized plot of these workers (Fig. 16), which brings together their results obtained with different cone angles, bed depths, and orifice diameters, shows that at any level in the spout of radius r_s, the relationship between the particle velocity at the radial position r, and the maximum velocity at the axis v_0 is described by the equation for a parabola,

$$v_r = v_0 [1 - (r^2/r_s^2)] = 2v_m [1 - (r^2/r_s^2)] \tag{54}$$

The results of Lefroy and Davidson obtained in a 12-in. flat-based column for polyethylene chips are seen in Fig. 16 to obey the same relationship, which appears to be a universal one.

2. *In the Annulus*

The individual particles in the annular part of the bed move vertically downward and radially inward, describing an approximately parabolic path. Detailed measurements of the vertical and horizontal velocity components in different parts of the annulus have been reported by Thorley *et al.* (T1), who traced particle-flow lines by observation against the flat wall of a 24-in. diameter half-column (see Fig. 17). The main point which emerges from the results in Fig. 17 is that, although the vertical velocities near the spout are somewhat higher than those near the column wall, the gradient across any given horizontal section is not large. This finding simplifies the study of annular solids flow since it means that the particle velocity at the column wall, which is convenient to measure, gives an approximately correct indication of the flow rate across the entire cross section of the annulus, at any given level.

This contention is supported by the values of solids flow rate shown in Fig. 17, obtained by (a) integrating the flow-rate profile across the annulus

FIG. 17. Solids flow pattern in the annulus (Thorley *et al.*, T1). Column diameter, 24 in.; air-inlet diameter, 4 in.; cone angle, 60°; bed depth expanded, 48 in.; air flow, 210 cfm; minimum spouting condition; ↓ vertical component of particle velocity (in./sec); → horizontal component of particle velocity (in./sec); data in ovals give solids flow (lb/sec) based on resultant particle velocity; data in rectangles give solids flow (lb/sec) based on particle velocity at column wall.

The figures for solids flow in pounds per second are calculated for full and not half-column. The air-flow rate and particle velocity values refer to the half-column actually used.

cross section at each level, and (b) calculating the total flow rate directly from the wall velocities. In the lower conical region, however, the flow lines are deflected by the sloping wall of the cone, and the velocity at the wall cannot be relied upon to indicate the total solids flow in this region. For the region above the cone, the particle velocity at the wall should decrease with decreasing bed level as a reflection of the cross-flow of solids from the annulus into the spout. The data of Thorley *et al.* for wheat beds in 6 and 24-in. columns show that while this is in fact true for the small column, the vertical velocity gradient in the large column is not a sensitive indication of cross-flow since the ratio between the rate of cross-flow per unit height to the total downward flow rate in the annulus decreases rapidly with increasing column diameter (E2).

Thorley *et al.* (T2) studied the effect of column diameter, cone angle, orifice size, bed depth, and gas flow on the particle velocity at the wall for wheat in air. Their data consist of observed particle velocities at the top of the annulus, and the vertical velocity gradients, from which the total solids flow rates at the top and the cross-flow rates can be calculated. A selection from the extensive data reported by these workers is shown in

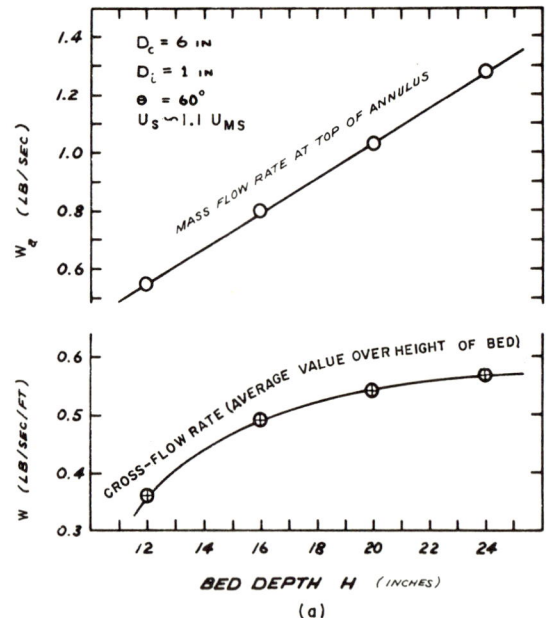

FIG. 18 (pp. 154–156). Effect of design variables on annular solids flow calculated from the data of Thorley *et al.* for wheat (T1, T2). (a) Effect of bed depth; (b) effect of air flow rate; (c) effect of column diameter; (d) effect of orifice diameter; (e) effect of cone angle.

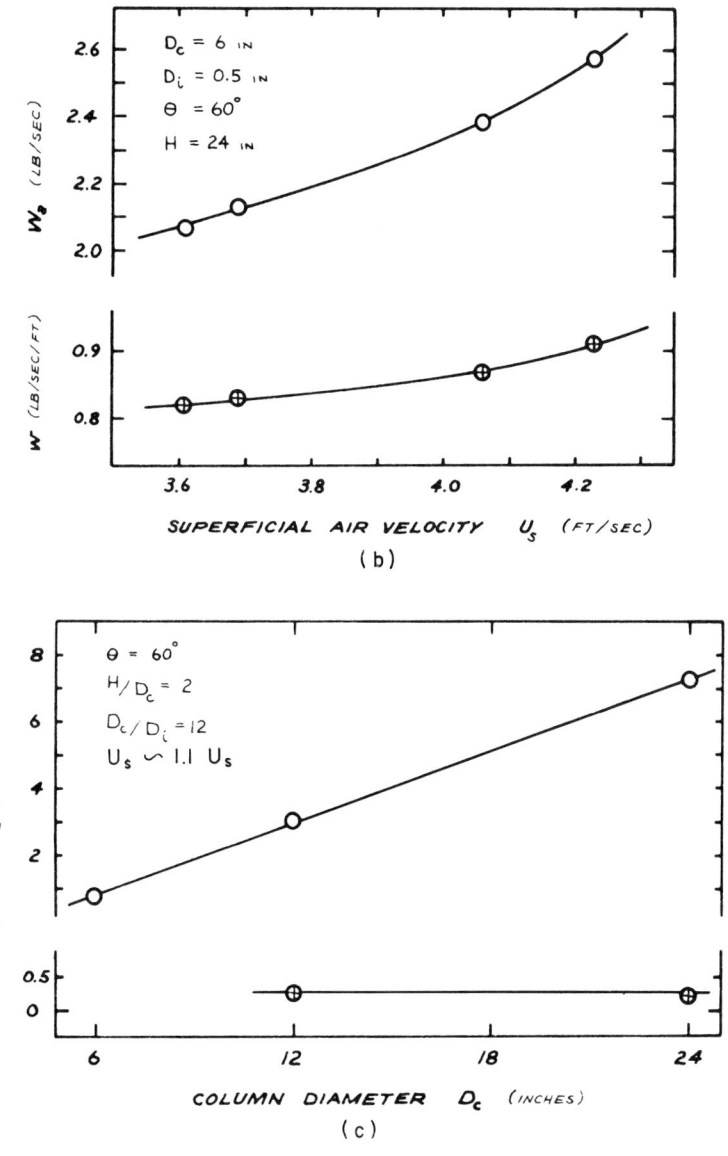

Fig. 18b and c.

Fig. 18 to demonstrate the following trends:

(a) Higher total solids flow rates, as well as higher cross-flow rates, were obtained with deeper beds (Fig. 18a), and for a given bed, with higher air flows (Fig. 18b).

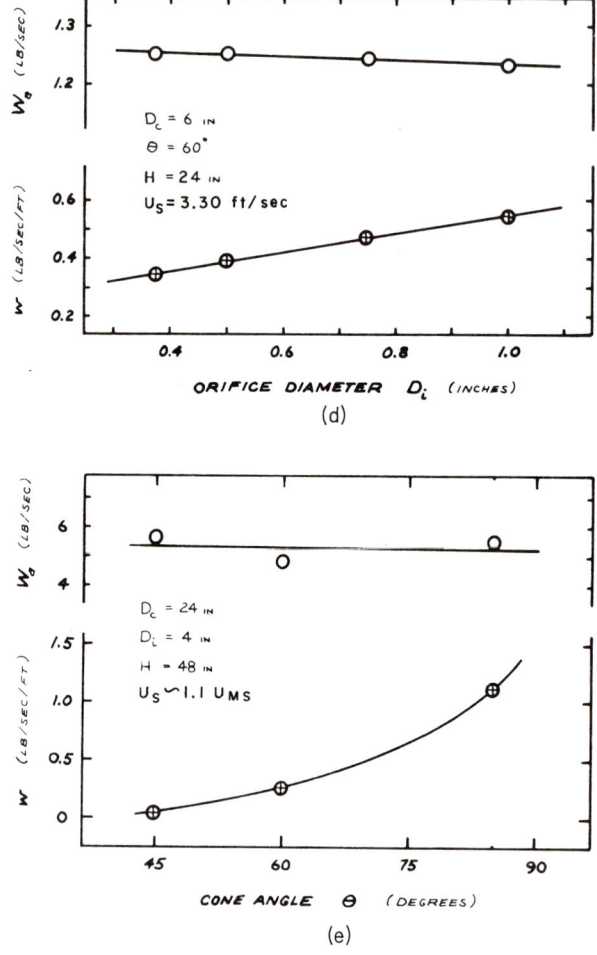

FIG. 18d and e. For caption see p. 154.

(b) For similar H/D_c ratios, the total solids flow rate increased with column diameter, while the cross-flow rate remained unchanged (Fig. 18c).

(c) At a given air flow, higher cross-flow rates were obtained with increasing orifice size, but the total solids flow was not significantly affected (Fig. 18d).

(d) Increasing the cone angle from 45° to 85° increased the cross-flow rate, but again with little influence on solids flow rate (Fig. 18e). The latter result is somewhat at odds with the results of Elperin et al. (E1), obtained with beds which remained wholly or mainly in the conical section of the spouting vessel. For such beds, a cone angle of 40°–45° has been established by these workers as the optimum for maximizing solids circulation rate.

A theoretical model that relates the rate of cross-flow along the spout height with the upward particle velocity in the spout has been proposed by Lefroy and Davidson (L2). The starting point of their analysis is the observation originally made by Thorley *et al.* (T2), from high-speed motion pictures of the spout, that near the base of the bed, the spout wall continually collapses and particles are being swept into the high-velocity jet; while further up the bed, lateral transfer of particles across the spout boundary appears to occur as a result of collisions between the upward-moving particles in the spout against the particles forming the spout wall. The model is developed from an analysis of the mechanism of particle-to-particle collisions, depicted in Fig. 19, and therefore concerns only the upper part of the bed. A fast-moving particle (1) is assumed to collide with a stationary particle (2), which is thereby thrown against the wall of particles formed by the annulus. Particle 2 then rebounds toward the spout, into which it may be entrained if space is available. Thus, a relatively smaller fraction of the collisions are successful in causing entrainment in the upper part of the spout, where the voidage is smaller than in the lower part. From an analysis of these impacts, the authors derived the following equation, relating the entrainment velocity v_{3r} (Fig. 19) to the original velocity of the spout particle, v_a:

$$v_{3r} = \pi^2 v_a e (1 + e)/64 \tag{55}$$

where e is a coefficient of restitution (the normal relative velocity after

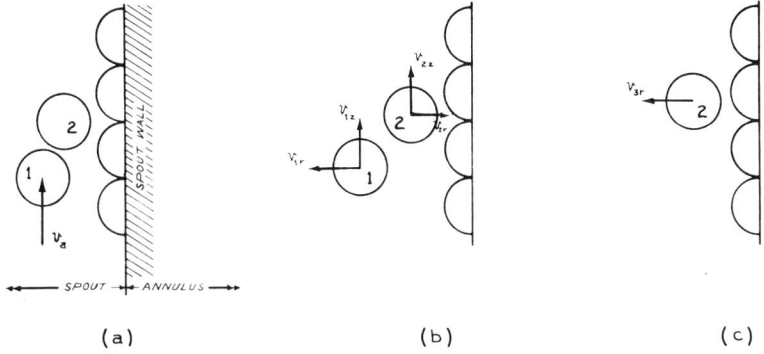

Fig. 19. Collision model of Lefroy and Davidson (L2). v_a velocity of particle No. 1 prior to collision; v_{1r}, v_{2r}, v_{3r} radial velocities due to collisions; v_{1z}, v_{2z} vertical velocities due to collisions.

impact will be e times the relative velocity before impact). If it is assumed that the center of particle 1 in Fig. 19 is one particle diameter away from the spout wall before collision, and that $d_p \ll D_s$, v_a in Eq. (55) can be expressed in terms of radial mean particle velocity in the spout by use of Eq. (54), to give

$$v_a/v_m = 8d_p/D_s \qquad (56)$$

Combining Eqs. (55) and (56), we get

$$v_{3r}/v_m = \pi^2 e (1 + e) d_p/8D_s \qquad (57)$$

In order to obtain the rate of cross-flow per unit height from Eq. (57), allowance must be made for the variable voidage along the spout height. Taking spout voidage ϵ_s as being uniform across the spout, the rate of lateral movement of type-2 particles per unit of spout height can be expressed as

$$w_{\text{type 2}} = \rho_s(1 - \epsilon_s)v_{3r}\pi D_s \qquad (58)$$

But only a fraction of these particles will be entrained, while the remainder will rebound due to impact with particles already in the spout. If it is further assumed that the number of type-2 particles per unit spout height is the same as the number of type 1, the fraction entrained will equal $(\epsilon_s - \epsilon_a)/(1 - \epsilon_a)$. Combining this condition with Eq. (58) gives the rate of actual cross-flow from the annulus into the spout:

$$w = \rho_s[(1 - \epsilon_s)(\epsilon_s - \epsilon_a)/(1 - \epsilon_a)]v_m(v_{3r}/v_m)\pi D_s \qquad (59)$$

The term v_{3r} can be eliminated by combining Eq. (59) with Eq. (57), but the resulting equation still does not lend itself to direct verification against experiment in the absence of specific data on ϵ_s and v_m.

Lefroy and Davidson compared v_{3r}/v_m obtained by Eq. (59) using values of dw/dz and ϵ_s computed independently from their momentum balances [Eqs. (43) and (44)] against results from Eq. (57), and obtained good agreement between the two methods in the upper 80% of the bed. However, the validity of their mathematical models remains largely unproven since the procedure of matching the results obtained from one theoretical derivation against those from another, with each one involving simplifying assumptions, can at best provide only weak support for the underlying models.

C. Mixing Characteristics

The speed with which intermixing of solids occurs in a spouted bed is obviously dependent on the solids flow pattern, but since intermixing is a bulk property of the bed, it is difficult to relate it quantitatively to the

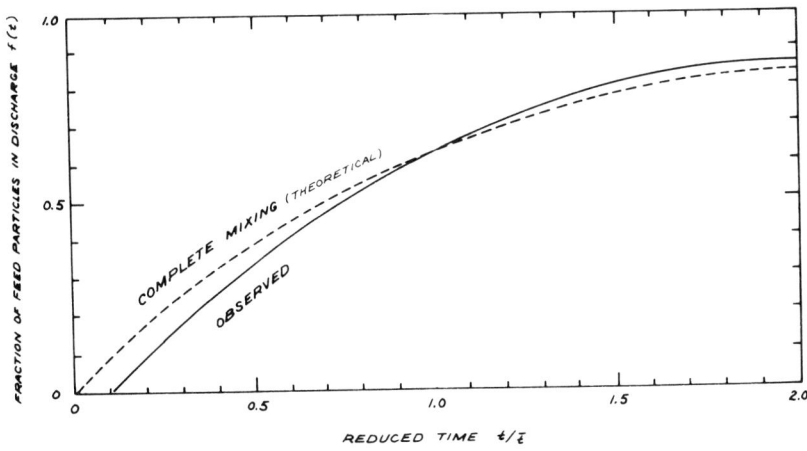

Fig. 20. Mixing curve of Kugo et al. (K2).

movement of individual particles in the spout or the annulus. The mixing characteristics have, therefore, been directly investigated.

From a practical point of view, interest primarily lies in knowing how rapidly a continuous stream of solids fed into a spouted bed will mix with the bed solids. This aspect has been experimentally studied by Kugo et al. (K2), using a system in which fresh feed was continuously introduced into the annular region at the top of a spouted wheat bed, and an equivalent quantity of solids was discharged through an opening in the conical base so as to maintain a constant bed weight. The normal feed was suddenly replaced by colored material, and the response of this stepwise change was determined in terms of the concentration of colored particles in the discharge at 1-min intervals. A typical set of results obtained by these workers in a 6-in. column is shown in Fig. 20 as fraction of feed particles in the discharge, $f(t)$, against reduced time $t/\bar{t}$, $\bar{t}$ being the mean residence time (i.e., feed rate divided by weight of bed). From mixing curves, such as Fig. 20 ("F-Curves" of Danckwerts, D1), they calculated a holdback value, which expresses the approach to perfect mixing:

$$\text{holdback} = \bar{t}^{-1} \int_{t=0}^{t=\bar{t}} f(t)\, dt \qquad (60)$$

where for plug flow, holdback = 0, and for perfect mixing, holdback = $1.0/e$.[2] Over the range of variables studied, which included mean residence

---

[2] Kugo et al. (K2) omitted the e in their paper, but it is apparent from their F-curves and holdback values that e was incorporated in their calculation of holdback. Their statement that holdback = 1.0 for perfect mixing is incorrect; rather holdback = 1.0 for total deadwater (D1).

times of 25–40 min, bed depths of 3–5 in., orifice diameters of 0.24–0.66 in., and air velocities up to 20% above U_{ms}, the holdback values obtained were all within the range $(0.84-0.93)/e$. Kugo et al. therefore concluded that mixing in a spouted bed is nearly perfect, regardless of operating conditions, thus confirming a previous finding of Becker and Sallans (B4) which was based on limited tracer experiments in a 9-in. diameter × 36-in. deep wheat bed.

A similar investigation with coal beds ($-6 + 10$ mesh) has been reported by Barton et al. (B2), with the difference that in their case the product was discharged through an overflow pipe located diametrically opposite the feed inlet. As the tracer, they used a high-volatiles coal, the bed material being low-volatiles coke. Thus by monitoring the volatile matter in the product stream, they obtained F-diagram data similar to those of Kugo et al. For interpreting their experimental results, Barton et al. used the concept of a mixed model (C3, L4), regarding the spouted bed as a backmix régime composed of a perfectly mixed volume V_b and a deadwater volume, V_d. This overall backmix régime, which has a relatively long recirculation time, is considered to include a plug-flow volume V_p as a lag element. The response to a step input is related to the effective volumes of these three regions in the following manner:

$$c/c_0 = \exp[-(V/V_b)(T - V_p/V)] \qquad (61)$$

where $V = V_b + V_d$, $T = Ft/V$ = dimensionless time, F is the volumetric feed rate of solids, and t is time. The experimental data, when plotted as c/c_0 versus T, give a straight line with a mean deviation of $\pm 10\%$ (Fig. 21), which is described by the following equation:

$$c/c_0 = \exp[-(1/0.92)(T - 0.10)] \qquad (62)$$

Barton et al. originally interpreted Eq. (62) as signifying that the total spouted bed volume is 92% perfectly mixed and 8% deadwater with no by-passing, and that "included in this volume is a plug flow region with a lag corresponding to an effective volume of 10% of the total," this plug flow region representing the upper part of the annulus. Subsequently, however, they revised the above interpretation (Q1) to one which simply states that 8–10% of the total volume is in plug flow while the remainder is perfectly mixed, deadwater and by-passing being negligible, which is more correct when viewed in the light of the originally formulated mixed models (C3, L4).

Barton et al. have pointed out that the above picture of a spouted bed is valid for the particular apparatus used in their experiments since any change in bed design could lead to changes in residence time distribution.

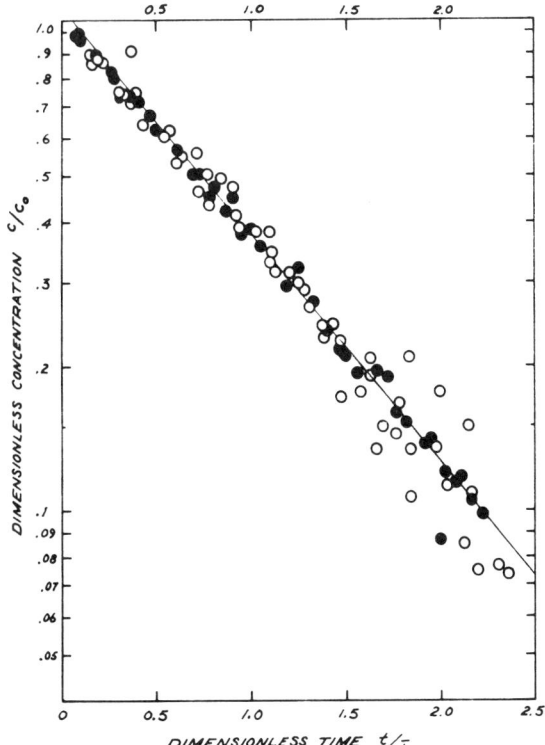

FIG. 21. Correlation of mixing data; ○ Barton et al. (B2); ● Kugo et al. (K2).

On the other hand, the data of Kugo et al., which cover a range of experimental conditions including orifice size and bed depth, show remarkably good agreement with the correlating line in Fig. 21. Therefore Eq. (62) would seem to represent a fairly general description of the mixing characteristics of a continuously operating spouted bed, as long as the feed and discharge points are so located that any obvious short-circuiting of the solids is avoided.

A somewhat different, though related, aspect of solids mixing has been investigated by Chatterjee (C2), who studied the rate of intermixing of solids in the bed itself, rather than mixing between bed solids and fresh feed in a continuous system. He applied a technique, previously used for similar studies in fluidized beds (Z2, p. 290), in which a bed initially composed of two regions of identifiably different particles was maintained in the spouted state, and the rate at which blending of the solids in the two regions proceeded was determined from measurements of concentration x of tracer in the initially traced bed region versus time t. From material-balance equations for the movement of solids from one region to the other, it can be shown that

$$\ln \frac{x_n}{x - x_t} = W \left[\frac{1}{w_t} + \frac{1}{w_n} \right] t \qquad (63)$$

where

$$x_n = w_n/(w_t + w_n), \qquad x_t = w_t/(w_t + w_n)$$

w_t is the weight of initially traced bed region and w_n is the weight of initially untraced bed region. Thus, the overall circulation rate W, which is really the flow of solids across the boundary between the two regions, can be evaluated by plotting the experimental results on semilog coordinates.

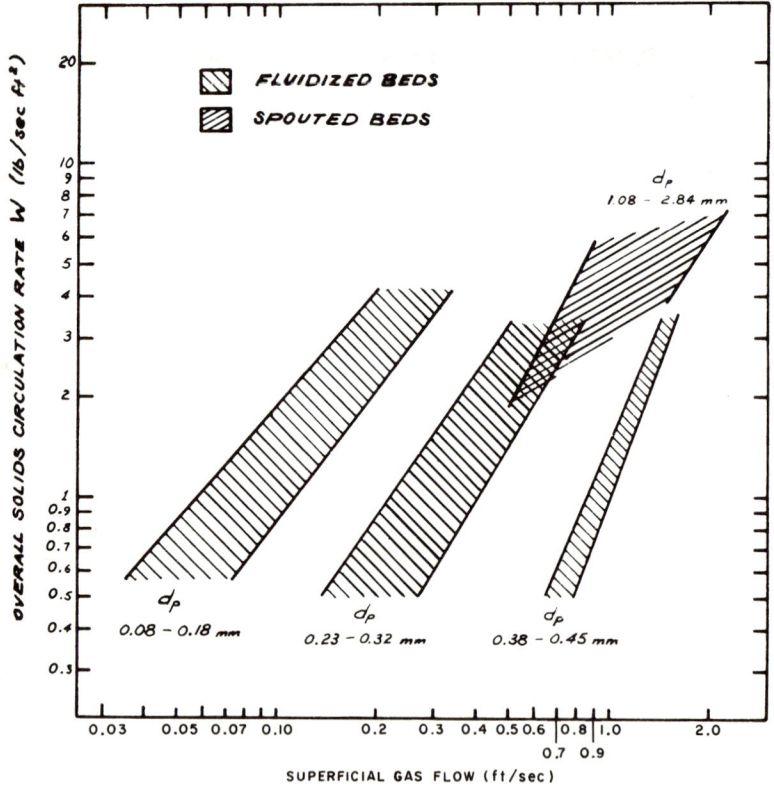

FIG. 22. Solids circulation in fluidized (Z2, p. 296) and spouted beds (C2).

The circulation rates thus obtained in a 6-in. column for beds of sand, mustard seed, sago, and coal ($d_p = 1\text{–}3$ mm, $\rho_s = 1\text{–}3$ gm/cm³) ranged between 0.6 and 1.2 lb/sec. The data were correlated by the following empirical equation:

$$W = 870(G/G_{ms})(d_p^{0.27}/\rho_s^{0.17}) \tag{64}$$

where W is in kilograms per hour, d_p in millimeters, and ρ_s in grams per cubic centimeter. Equation (64) shows that, while particle size and density have some effect on circulation rate, the dominating variable is gas flow rate.

Spouted-bed circulation rates obtained by Chatterjee are considerably higher than those reported for fluidized beds using a similar technique (Fig. 22). This comparison supports Chatterjee's view that spouting is a suitable operation for solids-blending processes.

The mixing characteristics of the fluid in a spouted bed of solids have not been studied. Since the particle Reynolds numbers involved in both the annulus and the spout are relatively high, the extent of any axial mixing experienced by the fluid in either channel is likely to be small, especially for the more common case of gas spouting. For the bed as a whole, however, it could be significant due to the very uneven velocity distribution as between spout and annulus, which would cause an effect akin to Taylor diffusion.

V. Bed Structure

A. Spout Shape

The fact that in a steadily spouting bed the spout assumes a stable shape implies the existence of a state of dynamic equilibrium among the various forces acting on the spout–annulus interface. Since these forces arise from the movement of both solids and gas, the shape of the spout provides a valuable clue to an understanding of the entire dynamics of spouted beds. The spout shape has, therefore, received considerable attention.

Spout-shape observations have usually been made against the flat transparent face of sectional columns, either semicircular or "two dimensional." Doubt concerning distortion of the spout in a half-sectional cylindrical column by the flat face has been dispelled by Mikhailik (M13), who made parallel measurements in half and full columns, using the piezoelectric technique mentioned in Section IV,B for the latter. The absolute significance of measurements made in two-dimensional columns remains open to question, but qualitative trends of variation of spout diameter with bed level observed in such columns are probably acceptable for generalization. A variety of spout shapes have been observed under different experimental conditions and these are illustrated in Fig. 23, which shows notional sketches of the different types of spout shapes reported in the literature. The experimental conditions associated with each type are listed in Table V. All one can infer from these results is that the pattern of change of diameter with bed level is not related to the inlet orifice diameter, or indeed to any other variable of the system, in any simple manner.

Lefroy and Davidson (L2), in an extension of the theoretical analysis described in Section IV,B, have attempted to explain the spout shape from a consideration of the balance of forces acting on an element of particles in the spout wall (see Fig. 24). F_z and F_r are due to percolation of air through the particles in the annulus, and are calculable from the air-velocity distribution given by Eqs. (51) and (52), plus the conditions that the drag force on an element at minimum fluidization will equal its weight, and that the shear stress τ depends on the angle of repose ϕ of the solid material. The force balance, $F_z + \tau = mg$, can then be expressed as

$$\sin(\pi z/2H) + [\pi(D_c^2 - D_s^2)/8D_sH]\tan\phi\cos(\pi z/2H) \geq 1/B \quad (65)$$

At $H = H_m$, $B = 1$; hence for a bed of maximum spoutable depth, Eq. (65) becomes

$$\frac{\pi(D_c^2 - D_s^2)}{8D_sH_m}\tan\phi \geq \frac{1 - \sin(\pi z/2H_m)}{\cos(\pi z/2H_m)} \quad (66)$$

From a comparison of the theoretical spout diameters required to satisfy the above equation against experimentally measured values, Lefroy and Davidson conclude that a constant spout diameter is possible for the upper half of the bed, but that in the lower half, the spout must taper downward to less than half its upper diameter, as for example in Fig. 23a and b.

By combining the force balance Eq. (66) with the spout momentum balance Eqs. (43) and (44) and the particle entrainment Eq. (59), Lefroy and Davidson were able to derive the following expression for spout diameter:

$$D_s = 1.07 D_c^{2/3} d_p^{1/3} \quad (67)$$

(The coefficient they gave is 1.06, but the arithmetic they used corresponds more closely to the shown value of 1.07.) The above equation, though based in part on first principles, is really semiempirical, since certain approximations based on experimental observations had to be made in its derivation, the range of variables covered being $d_p = 1.7$–9.0 mm and $D_c = 7.6$ and 30.5 cm. Because of oversimplification, spout diameters

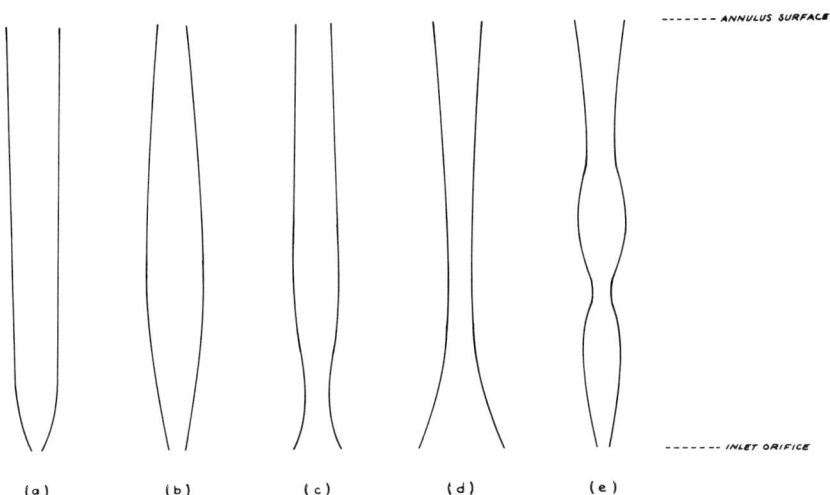

FIG. 23. Observed spout shapes: (a) diverges continually; (b) expands then tapers; (c) necks, expands, then tapers slightly; (d) necks, then expands; (e) necks twice.

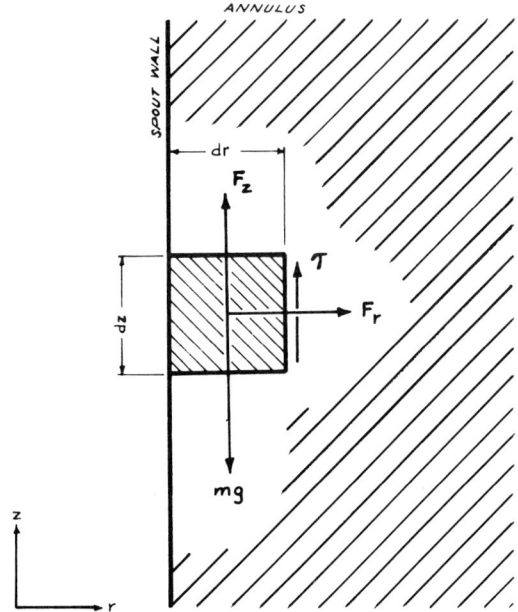

FIG. 24. Force balance model of Lefroy and Davidson (L2).

TABLE V
EXPERIMENTAL CONDITIONS FOR FIG. 23

Spout type	Column type	D_c (in.)	D_i (in.)	θ (°)	H (in.)	Solids	Comments	Reference
a	Semicircular	6	0.49	85	25	Wheat	Spout diverges at angle of $\simeq 18°$ (see Fig. 2a)	M9
	Conical	—	0.8–1.6	20–50	4–20	Fertilizer fractions	Spout diverges at angles between 10° and 19°	T3
	Conical	—	—	20–70	—	—	Angle of spout divergence remains at 5–7° over a wide range of conditions	M16
	Semicircular	6	0.37–0.75	85	12–24	Wheat		T2
	Semicircular	6	0.5	60	7–10	Wheat	Spout diverges quickly to nearly constant diameter	M4
	Circular	3.7	0.59	45	<7.9	Wheat, millet, polystyrene, silica gel		M13

	Shape					Material	Notes	Ref
b	Semicircular	6	0.5	60, 90, 180	7–10	Polystyrene	—	H3
	Semicircular	24	2–3	60	36–72	Wheat	Note that with a larger orifice, spout shape changed to type c	T2
c	Circular	3.7	0.59	45	>7.9	Wheat, millet, polystyrene, silica gel	Tapering of spout dia. near top occurred only for $H > 7.9$ in.	M13
	Semicircular	12	0.63–3.74	180	23–71	Kale seed, polyethylene, peas	Spout walls near base were continually collapsing	L2[a]
d	Semicircular	24	4	60	72	Wheat	With wheat, maximum spout diameter attained at $z \simeq 0.4H$	T2
	Two-dimensional with conical face	—	≃2.4	70	≃8	"Beads" of 1.0–1.5 mm diameter	See Fig. 2b. Angle of the upper diverging part of the spout varied between 8.5° and 18°, depending upon air flow rate	G7
e	Two-dimensional with rectangular face	0.63 × 7.9	0.16 × 0.63	180	8–12	Ceramic chips, glass beads	Disturbances developed at the bottom of the spout and moved upward like ripples	V2

[a] However, in these experiments upper spout diameter exceeded D_i, as in spout type a.

calculated from Eq. (67) show only a rough agreement with the experimental results on which it is based, and hardly any agreement with the data of other investigators (viz. M3, T2); nevertheless, the dependence of spout diameter on column diameter and particle size and its independence of orifice size, as indicated by the equation, are at least qualitatively correct.

Spout pinching at a short distance above the gas inlet rather than flaring out is predicted by Volpicelli *et al.* (V2), who applied Helmholtz instability analysis for the growth of a disturbance at the interface between two flowing fluids to determine the stable spout shape. This approach to the problem arose out of their observation that disturbances developed at the bottom of the spout and moved upward like ripples. The analysis led Volpicelli *et al.* to conclude that self-adjustment of spout diameter in the region above the orifice occurs so as to keep the lateral velocity of the downward moving annular solids below a certain maximum value, analogous to the velocity in gravity flow of solids through the bottom of a converging bin. It should be pointed out that this analysis is valid for the lower region of the bed only, and therefore neither supports nor contradicts the force-balance analysis of Lefroy and Davidson applicable to the upper part of the bed.

Under most three-dimensional conditions, the major adjustment in the spout diameter occurs in the region immediately above the inlet orifice, variation in diameter further up the bed being relatively small. Malek *et al.* (M4) found that in 4- and 6-in. semicircular columns, spout diameters measured at various levels starting from 1 in. above the orifice were generally within 10% of the mean value. Mean spout diameters observed by these workers for eight different solid materials, using different orifice sizes and air-flow rates, as well as the results reported by Thorley *et al.* (T2) for 24-in. diameter wheat beds, were correlated by the following empirical equation:

$$D_s = (0.115 \log D_c - 0.031) G^{1/2} \qquad (68)$$

where D_s and D_c are in inches and G in pounds per hour per square foot. Thus the gas flow rate reflects the effects of particle properties, bed depth, and orifice size; but column diameter, which is likewise related to gas-flow rate, also appears in the equation. The data used by Malek *et al.* in arriving at Eq. (68) were for columns with a 60° cone as the base, but the cone angle was found to be unimportant by Hunt and Brennan (H3), whose results for a variety of materials spouted in a 6-in. column, with cone angles of 30°, 60°, 90°, and 180° (flatbase), agreed with Eq. (68). Mikhailik (M13), who worked with a 3.7-in. column, also reported agreement with Eq. (68), but only for solids of densities up to 1.5 gm/cm³. For heavy materials such as pig-iron pellets and slag beads ($\rho_s = 7.8$ gm/cm^3), the mean spout diameter was overestimated by Eq. (68), the data being correlated by the following modified form of this equation:

$$D_s = 10(0.115 \log D_c - 0.031)(G/\rho_s)^{1/2} \qquad (69)$$

in which conversion to units of inches, pounds per hour per square foot, and pounds per cubic foot has been made by the present authors.

Mukhlenov and Gorshtein (M16, M17) maintain that the spout shape is determined in accordance with the principle of least resistance, and have explained the simple spout shape of Fig. 23a by reference to the work of Gibson (G4), who showed that a conical diffuser with an angle of 5°35′ offers minimum resistance to fluid flow. Analogy with a diffuser is misleading since it disregards the important effect of solids movement on spout shape. It is therefore not surprising that spout shapes widely different from that postulated by these workers (e.g., as in Fig. 23c and d, where the spout tends to converge rather than diverge in the lower part of the bed) have been observed, though admittedly under experimental conditions quite different from those used by Mukhlenov and Gorshtein in their investigation.

B. Voidage Distribution

1. *Annulus*

The solids in the annulus of a spouted bed are essentially in a loose-packed bed condition (E2, M12). Any variation in the volumetric voidage in different parts of the annulus is minor, being entirely due to lack of homogeneity in the orientation of particles. A somewhat tighter packing in the lower part of the bed, compared to the upper part, has been visually observed by Thorley *et al.* (T1).

2. *Spout*

The spout is like a riser through which particles are being transported in a dilute phase, with the added features of a decreasing gas flow and an increasing solids flow along the height. The spout voids are therefore determined by interaction between the gas and solids flow patterns.

Mathur and Gishler (M10) estimated the vertical voidage profile in the spout by two different methods outlined below, both of which rely on certain observed data to describe the gas and solids flow.

In their first method, the downward solids flow in the annulus, calculated from the particle velocity at the wall, should equal the upward flow in the spout at any bed level. Knowing the upward linear velocity of particles in the spout (measured by high-speed cine-photography in a half-column) and the spout diameter, the authors calculated the bulk density of the gas–solids suspension as a function of bed level, expressing the results in terms of voidage.

In their second method, as in the case of a vertical riser, the total pressure drop along the spout height is composed of (i) a solids static head equivalent to the dispersed-solids bulk density, (ii) an acceleration pressure drop, and (iii) a solids friction loss due to relative motion of the particles with respect to the gas and to the spout wall. Thus,

$$dP_{total} = dP_{weight} + dP_{acceleration} + dP_{friction}$$

or

$$dP_T = dP_w + dP_{(a+f)} \qquad (70)$$

From an energy balance over an increment of spout height dz

$$dP_{(a+f)} = -(1/2g_c \bar{U}_s A_s) \, d(m_s v^2) \dagger \qquad (71)$$

Now

$$dP_w = d(\rho_{bs} z) \qquad (72)$$

and

$$m_s = v A_s \rho_{bs} \qquad (73)$$

where ρ_{bs} is the bulk density in the spout. Dividing Eq. (71) by Eq. (72) and combining the result with Eq. (73) we get

$$dP_{(a+f)}/dP_w = -[d(v^3 \rho_{bs})/d(\rho_{bs} z)][1/2g_c \bar{U}_s] \qquad (74)$$

Using the vertical profile of v and $\bar{U}_s$ (estimated from the measured pressure drop profile as explained in Section IV,A), the authors evaluated the right-hand side of Eq. (74) for each 2-in. increment of the spout, assuming constant ρ_{bs} over this increment. This value, on substitution in Eq. (70) along with the measured total pressure drop for the increment, gave the pressure drop per foot due to the solids bulk density in the spout, from which the voidage profile was calculated. The results agreed well with those from method (a).

The theoretical analysis of Lefroy and Davidson (L2) discussed in Section IV,B is essentially a further development of the second method above; it enables the momentum balance, Eqs. (43) and (44), to be solved for voidage with the aid of equations describing the gas and solids flow pattern, instead of relying on actual measurements of pressure drop and particle velocities. The system of Eqs. (41)–(53), therefore, provides a

† In early work at the National Research Council of Canada, radial variations in the spout particle velocity were ignored.

more generalized method of computing the voidage profile from the independent variables of the system, within the limits of the simplifying assumptions made.

Direct measurements of spout voidage have been made by Soviet workers, using the piezoelectric technique mentioned in Section IV,B. Simultaneously with particle-velocity measurements, they recorded the frequency with which the solid particles collided with the piezo-crystal from the number of peaks observed on the oscilloscope per unit time. The local voidage at the probe tip was calculated from these data using the equation

$$\epsilon_s = 1 - \tfrac{1}{6}\pi d_p^3 N/vA_e \tag{75}$$

where N is the number of collisions per second and A_e is the cross-sectional area of the sensing element.

The experimental results obtained by different investigators in cylindrical columns are shown in Fig. 25, as vertical voidage profiles in the spout.

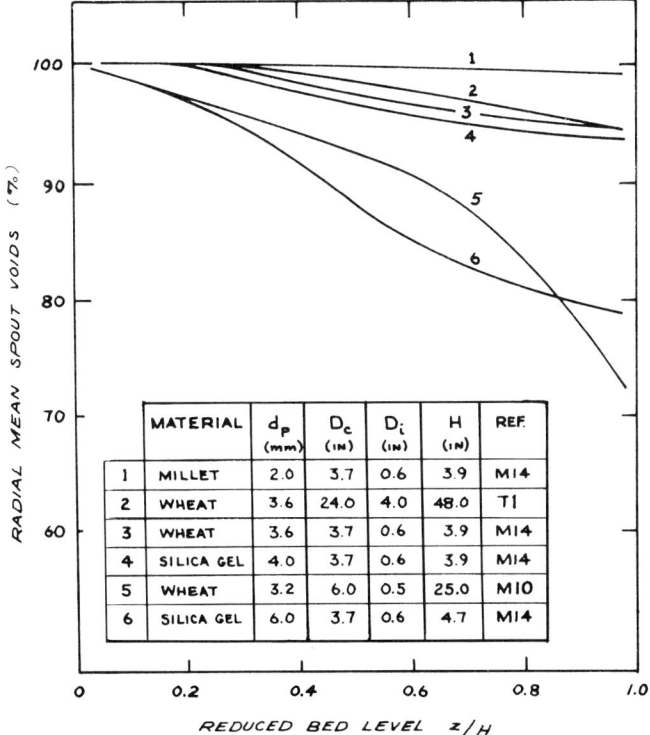

FIG. 25. Measured vertical voidage profiles in the spout.

The voidage value at a particular level represents the average voids over the spout cross section at that level. The upper limit of spout level covered by the data is the surface of the bed proper, no measurements having been made in the fountain above the bed. In all cases, the voidage is seen to decrease with increasing distance from the inlet orifice, from 100% at the orifice to its lowest value at the top, but the gradients and the lowest values of ϵ_s vary widely for the different systems studied, the latter ranging from 0.70 to 0.99 in Fig. 25. For beds in which $H = H_m$, it would be expected that the void fraction at the top of the spout would be somewhat closer to that of a loosely packed bed than 0.70.

On the basis of experimental results obtained in conical vessels using several solid materials, Mukhlenov and Gorshtein (M16) reached the conclusion that in a given bed, the spout voidage is substantially constant over the height of the spout. Their explanation of this observed behavior is based on the assumptions that:

(a) the weight of solids traveling up the spout remains substantially constant since cross-flow of solids from the annulus into the spout occurs primarily near the bottom;
(b) the upward solids velocity decreases along the spout height; and
(c) the cross-sectional area of the spout increases along the spout height.

Since (a) and (b) would not apply to the lowermost part of the spout, the constant voidage values reported by these workers were presumably reached after some initial distance from the inlet orifice. Spout-voidage data obtained with different cone angles, solid materials, and H/D_i ratios were correlated by the following empirical equation based on dimensional analysis:

$$\epsilon_s = \frac{2.17 \text{Re}_i^{0.33}}{\text{Ar}^{0.33}(H/D_i)^{0.5}(\tan \theta/2)^{0.6}} \quad (76)$$

The Reynolds number, which is based on the air velocity through the orifice and the diameter of the particle, varied from 50 to 1100; Archimedes numbers from 6.27×10^4 to 21.25×10^4; H/D_i from 1 to 9; and included cone angle from 20° to 60°.

Goltsiker (G7), who measured longitudinal as well as radial profiles in a conical vessel of 40° angle using a capacitance probe, also reports a substantially constant voidage over the spout height, except in the lower 4 cm (of an 11-cm-deep bed) which constituted a dilute zone. His radial profiles, measured at three different levels, exhibited a characteristic shape (see Fig. 26), and showed the voidage over a narrow region at the spout periphery to be noticeably lower (45–50%) than in the core of the spout

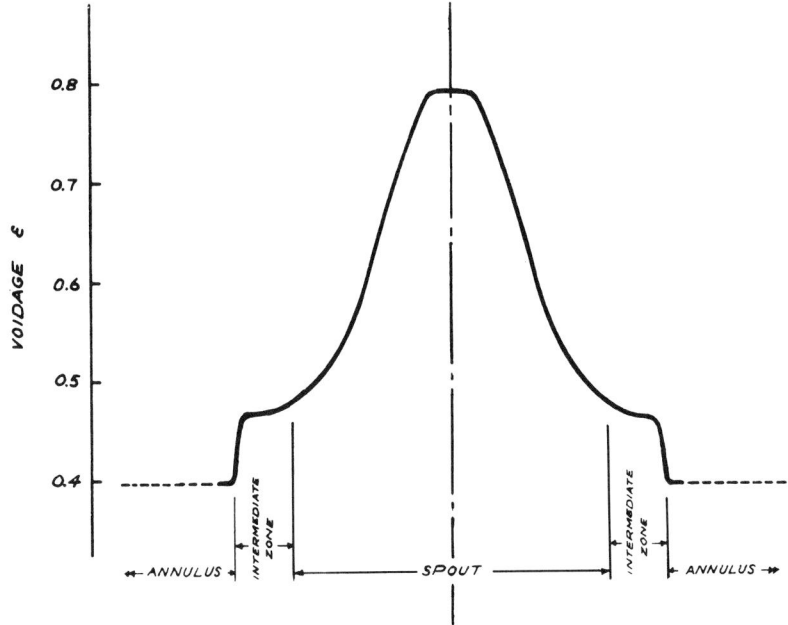

FIG. 26. Radial voidage profile in a conical spouted bed (Goltsiker's results, G7, quoted in R4).

(70–80%). This narrow region, which has a voidage intermediate between the dense-phase annulus and the dilute-phase spout core, has been identified by Romankov and Rashkovskaya (R4, p. 59) as the third zone (in addition to the annulus and the spout core), comprising "particles descending with a quick vortex-like movement," and can be seen in Goltsiker's photograph (Fig. 2b). According to these authors, particle velocities in the intermediate zone are much lower than in the spout core, but are still several times higher than, though in the same net direction as, in the slowly descending annular region. The existence of a distinct interface zone of finite width, noted by the Russian workers, appears to be a feature of spouting in conical vessels, and has not been observed in cylindrical spouted beds.

VI. Spouting Stability

It was explained in Section I that the regime of stable spouting is critically dependent on certain conditions; unless these are satisfied, the movement of solids becomes random, leading to a state of aggregative fluidization, and with increase in gas flow, to slugging. Spouting can be achieved only within certain limits of solids properties, while whether or not a material having properties within these limits will spout depends upon the geometry of the column, including to some extent the design of the gas inlet. A further overriding restriction on spouting stability is imposed by bed depth since spouting action for any given solids properties and column geometry would terminate beyond a certain maximum depth. The maximum spoutable depth can therefore be looked upon as an index of spouting stability, although a stably spouting bed of a depth smaller than the maximum would become unstable at excessively high gas-flow rates.

The first part of this section briefly summarizes the main findings concerning the effect of the various factors on spouting stability, while the second part deals with methods proposed for calculating the maximum spoutable bed depth.

A. Effect of Various Parameters

1. *Column Geometry*

   a. Orifice-to-Column-Diameter Ratio. In a given column, the maximum spoutable bed depth decreases with increasing orifice size until a limiting value of D_i is reached, beyond which spouting no longer occurs (see Fig. 27). On the basis of his data for spouting of several materials in cylindrical columns, Becker (B3) suggested that the critical value of the ratio D_i/D_c is 0.35. While this value is approximately in line with the data for wheat shown in Fig. 27, the critical value for finer materials appears to be considerably smaller; for example, 0.1 for 0.6-mm diameter particles in a 6-in. diameter column (M10). For conical vessels, the existence of an upper limit to the ratio of orifice to bed-surface diameter has been established by Romankov and Rashkovskaya (R4, p. 47), who have shown this ratio to be dependent on the Archimedes number for materials of 0.36–9-mm diameter.

   b. Cone Angle. The lower conical section of the bed facilitates the flow of solids from the annulus into the gas-jet region. With a flat instead of a conical base, a zone of stagnant solids with a conelike inner boundary is formed at the base, but this does not affect spouting stability. If the cone is too steep, on the other hand, spouting becomes unstable since the entire bed tends to be lifted up by the gas jet. This applies equally to cylindrical vessels with a conical base and to entirely conical vessels. The limiting cone angle depends to some extent on the internal friction characteristics of the solids, but for most materials its value appears to be in the region of 40° (E1, H3; R4, p. 60).

c. Inlet Design. In the early work at the National Research Council of Canada, it was found by trial and error that spouting is more stable when the orifice is somewhat smaller than the narrow end of the cone. This was subsequently rationalized by Manurung's (M7) demonstration that maximum stability is obtained with a design which does not permit the gas jet to be deflected from the vertical path before it enters the bed of particles. In his own experiments, he achieved this end by using the design shown in Fig. 28a, the main feature of which is that the gas inlet pipe protrudes a short distance above the flange surface. With this inlet, Manurung obtained somewhat higher maximum spoutable depths for several materials and was able to achieve stable spouting for coal beds containing a high proportion of fines, which would not spout with other gas inlets in which the gas pipe did not protrude.

The stabilizing effect of a slightly protruding inlet pipe has been confirmed by Reddy *et al.* (R1), who obtained better results with a converging nozzle inlet (Fig. 28b) than with a straight pipe. These workers consider

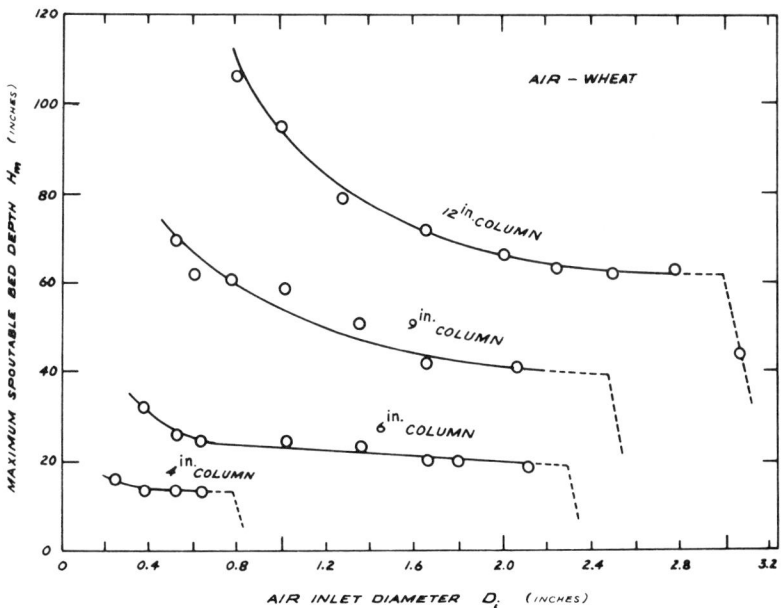

FIG. 27. Effect of orifice diameter on maximum spoutable bed depth; air–wheat (M10).

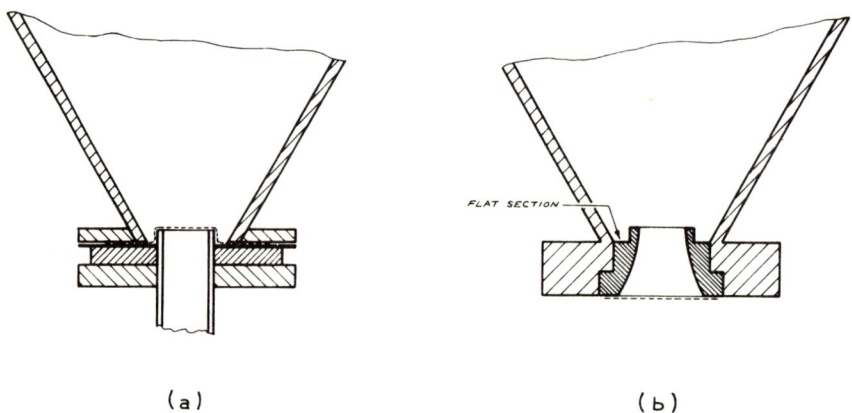

FIG. 28. Gas inlet designs for improved stability: (a) Manuring (M7); (b) Reddy et al. (R1).

that the flat section between the inlet nozzle and the lower end of the cone plays an important role in stabilizing the spouting flow pattern. Although the inlet of Fig. 28b gave maximum spoutable depth only slightly higher than those with the other types of inlets tried, the spouting action at any bed depth was more stable with this particular inlet. Malek and Lu (M3) did not observe any effect on H_m of their two inlet designs, but did find that the exact positioning of the screen had a marked effect on spouting stability. If the screen was loosely fitted over the orifice plate, spouting at any depth was unstable, but when the screen was placed below the orifice plate, satisfactory spouting resulted. The use of a converging–diverging gas inlet pipe has been mentioned by Berquin (B6), although it is not clear how spouting stability would be affected by such a design.

From the few observations cited above, it is clear that the exact design of the gas inlet has an important effect on spouting stability. However, the question of inlet construction has received insufficient attention. Indeed, it is even possible that some of the discrepancies in other aspects of spouting behavior observed by different investigators may be due to unspecified differences in inlet designs.

2. Solids Properties

a. Particle Size. Although the minimum limit of particle size for spouted bed operation has been quoted as 1–2 mm in Section I, Ghosh (G3) has suggested that spouting action can be achieved for much finer materials, as long as the gas inlet size does not exceed 30 times the particle diameter.

Using a very small air orifice, he was able to obtain a miniature spouted bed with glass beads as fine as 80–100 mesh (mean diameter of 0.16 mm). Spouting of such fine material, however, is of little practical interest since it cannot be achieved on a larger scale, except perhaps by the use of a bed with multiple spouts in parallel, such as that developed by Peterson (P3).

The maximum spoutable bed depth was found to decrease with increasing particle size by Malek and Lu (M3), who experimented with four different sizes of wheat (1.2–3.7 mm) in a 6-in. column. On the other hand, Reddy et al. (R1), who worked with mixed-size materials (alundum, glass spheres, and polystyrene), also in a 6-in. column, reported that H_m first increases with particle size and then decreases, a peak value being attained at a mean particle size of 1.0–1.5 mm. The observed variation of H_m, correlated by Reddy et al. with mean particle size, is likely to be also influenced by size distribution, which cannot be fully characterized by any particular mean diameter. Nevertheless, the existence of a peak H_m with respect to particle size alone is theoretically predictable from a comparison of the effect of particle size on the gas velocities required for spouting and for fluidizing a given material (R1). From Eq. (3), the effect of particle size and bed depth on spouting velocity, with all other variables held constant, is as follows:

$$U_{ms} = kd_p H^{1/2} \qquad (77)$$

while the general dependence of fluidization velocity on particle size can be expressed in the form

$$U_{mf} = Kd_p^n \qquad (78)$$

Since at H_m, $U_{ms} = U_{mf}$, it follows from Eq. (77) and (78) that

$$H_m = K'd_p^{(2n-2)} \qquad (79)$$

The value of n in Eq. (79) depends on the flow regime. Using either the generalized equation of Wen and Yu (W1) cited by Reddy et al. and analyzed by Kunii and Levenspiel (K3, p. 73), or the Richardson–Zaki equation (R3), in conjunction with the terminal velocity laws of Stokes and Newton, it can be shown that n changes from 2 for laminar flow to 0.5 for turbulent flow. Hence the exponent on d_p in Eq. (79) would change with increasing d_p from a value of 2 to -1, depending upon the flow regime, causing a peak value of H_m to occur with respect to particle size—at $n = 1$.

For the mixed-size materials used by Reddy et al., the peak value of H_m was found to occur at a mean particle size of 1.0–1.5 mm, the corresponding value of Reynolds number being about 70 in all cases, regardless of the particle density or the orifice size used (see Fig. 29). The experimental

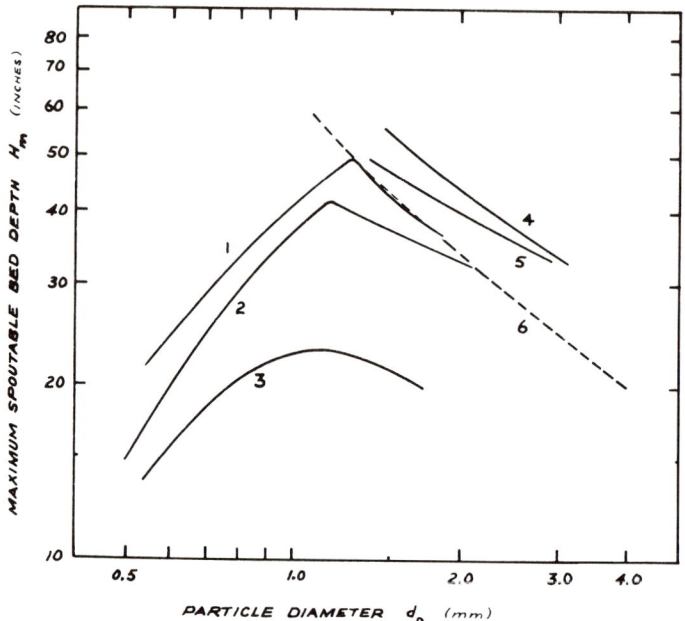

Fig. 29. Effect of particle size on maximum spoutable bed depth in a 6-in. diameter column. 1, Alundum, $D_i = 0.375$ in.; 2, alundum, $D_i = 0.50$ in.; 3, glass spheres, $D_i = 0.375$ in.; 4, polystyrene, $D_i = 0.375$ in.; 5, polystyrene, $D_i = 0.75$ in.; 6, wheat, $D_i = 1.0$ in.; 1–5 Reddy et al. (R1); 6 Malek and Lu (M3).

data in Fig. 29 follow slopes of 1.0–2.0 for the ascending section and of -0.4 to -0.8 for the descending section, which includes the data of Malek and Lu for different sizes of wheat particles. Thus the range of experimental slopes are similar to those predicted, despite the possible influence of size distribution in the data of Reddy et al., mentioned above.

b. Size Distribution. Uniformity of particle size favors spouting stability, since the lower permeability of a bed containing a range of sizes would tend to more effectively distribute the gas rather than produce a jet action. The presence of a small proportion of fines in a closely sized bed can seriously impair spoutability (K1, $-40 + 60$ mesh sand in a bed of $-20 + 30$ mesh sand), while the addition of a small proportion of coarse particles to a bed of finer material can also have the same effect (S2, $-9 + 14$ mesh alundum added to a bed of $-35 + 48$ mesh alundum). Nevertheless the limits of particle-size spread beyond which spouting would no longer occur are fairly wide, the latitude being greater with large particles than with small. Thus, beds of wood chips and cellulose acetate containing up to an eightfold size spread with particles of up to 3 cm in size could be satisfactorily spouted (C4); while, for coal and alundum with maximum size in the 2–4-mm range, limiting size spreads were seldom more than fivefold (M7, S2).

c. Density. Solids with widely differing densities, ranging from wood chips to iron pellets, have been spouted without any indication that any limits of particle density exist beyond which spouting action would not be achieved. Nor is there any clear evidence to show whether spouting stability is affected by particle density or not. The empirical correlation for calculating H_m proposed by Malek and Lu (M3) [Eq. (88)] implies that spouting stability is adversely affected by particle density; on the other hand, Fleming (F2) was able to spout deeper beds of alundum particles ($\rho_s = 2.46$ gm/cm^3) than of glass beads ($\rho_s = 1.55$–1.84 gm/cm^3) of the same size.

d. Particle Shape and Surface Characteristics. The above observations are confused by the effects of particle shape and surface characteristics, which were certainly different for the different materials used. Their effect has proved difficult to evaluate, partly because they are not easy to define. Using angle of repose as a combined criterion for both shape and surface (irregular and rough particles correspond to a high angle of repose), Fleming (F2) noted a direct dependence between angle of repose and spouting stability in his results for alundum, polystyrene, and glass beads. The empirical equation of Malek and Lu (M3) [Eq. (88)] also suggests that deeper beds of nonspherical particles can be spouted than of spherical particles. However, in beds of certain particles which deviate very widely from the spherical, true spouting action seems to terminate altogether. Thus, in the case of strongly ellipsoidal particles such as flaxseed and barley, Becker (B3) observed that although a through-channel resembling a spout was formed by the air jet in the bed, the resulting agitation of the solids was much more feeble than in true spouting. The pseudo spout formed was insensitive to changes in inlet-to-column diameter ratio and bed depth, and behaved like a solids-free channel, which probably owed its stability to the interlocking tendency of the particles. A similar phenomenon was observed by Reddy et al. (R1) in their experiments using deep beds of polystyrene.

3. *Gas Flow*

In relatively shallow beds ($H/D_c < 3$), an increase in gas flow much above that required for minimum spouting causes the spout above the bed surface to lose its well-defined shape, and though the movement of solids in the region above the bed becomes chaotic, the regular downward motion of particles in the annulus remains intact (M10). In deeper beds, on the other hand, the solids movement in the bed itself is disrupted at high flow rates. This disruption, in the case of coarse particles, takes the form of slugging, while with fine materials the bed first passes into the fluidized state, and with further increase in flow, to slugging. Phase diagram data similar to that of Fig. 1, reported for beds of various materials in 6-in. diameter (M7, M10) and 9-in. diameter (D3) columns, suggest that, in general, spouting stability with respect to gas flow rate increases with increasing particle size, increasing column diameter, decreasing orifice-to-column-diameter ratio, and increasing bed depth. For a given material and column geometry, the range of permissible gas flow for stable spouting becomes narrower as the bed depth approaches the maximum spoutable under those conditions. No data on the upper gas flow limits with larger columns have been reported, but since bed depths used in large columns would normally be well below the maximum spoutable, the tolerance for excess gas in such columns should be large.

B MAXIMUM SPOUTABLE BED DEPTH

Since at a bed depth just above the maximum spoutable, a spouting bed passes into the fluidized state, the limiting value of minimum spouting velocity for a given material and column geometry is similar to the minimum fluidization velocity for that material. Therefore, a simultaneous solution of an equation for minimum spouting velocity with an equation for minimum fluidization velocity should yield the value of maximum spoutable bed depth for a given system. Several correlations for prediction of maximum spoutable bed depth based on this approach have been proposed, with variation on the particular spouting and fluidization velocity equations used.

1. *Thorley, Saunby, Mathur, and Osberg* (T2)

These workers, who originally suggested the above approach, made use of the Ergun (E3) packed-bed equation for relating U_{mf} to gas and solids properties by putting $\Delta P/H$ in Ergun's equation equal to $(\rho_s - \rho_f)(1 - \epsilon_0)g/g_c$:

$$1.75\rho_f d_p U_{mf}^2 + 150\mu(1 - \epsilon_0)U_{mf} - d_p^2\epsilon_0^3(\rho_s - \rho_f)g = 0 \qquad (80)$$

For spouting velocity, they used Eq. (3). The values of H_m obtained by substituting U_{mf} calculated from Eq. (80) for U_{ms} in Eq. (3) agreed well with experimental results for several closely sized materials.

2. Becker (B3)

Instead of relying on a packed-bed equation for obtaining the fluidization velocity, Becker experimentally determined the limiting values of minimum spouting velocity for several materials. Although he found these to be similar to calculated fluidization velocities ($U_m \simeq 1.25 U_{mf}$), he nevertheless derived a separate empirical equation involving U_m which correlated his own data as well as the previous data of Mathur and Gishler (M10) in terms of Reynolds number and a drag coefficient at H_m [Eq. (8)]. Since U_m is independent of column geometry, it can be calculated by Eq. (8) from a knowledge of the gas and solids properties only. A further empirical equation which includes column-geometry variables, Eq. (10), enables the maximum spoutable bed depth to be calculated, using the value of U_m obtained from Eq. (8).

3. Manurung (M7)

Manurung has pointed out a basic similarity between the equations of Thorley et al. and of Becker, quoted above. Equation (80), used by the former workers for calculating spouting velocity at H_m ($U_m \simeq U_{mf}$), can be rearranged to the form

$$C_D = \frac{200(1 - \epsilon_0)}{\text{Re}_m \epsilon_0^3} + \frac{7.0}{3\epsilon_0^3} \qquad (81)$$

With a constant value of voidage substituted in the above equation, it becomes similar to Becker's empirical relation, Eq. (8). A similarity also exists in the second equation used for relating U_m to H_m [i.e., between Eqs. (3) and (10)]. In the limiting case of $H = H_m$, Eq. (3) for spouting velocity (used by Thorley et al.) becomes

$$U_m = (d_p/D_c)(D_i/D_c)^{1/3}[2gH_m(\rho_s - \rho_f)/\rho_f]^{1/2} \qquad (82)$$

which, on rearranging, takes a form similar to Becker's equation (Eq. 10), namely

$$(H_m/d_p)(d_p/D_c)^2(D_i/D_c)^{4/3}(C_D)(63) = 42 \qquad (83)$$

where the definition of C_D is the same as that used by Becker (see Eq. 8).

On testing the two sets of equations against his own experimental data for different solid materials (rapeseed, millet, coal, and polyethylene cubes), Manurung concluded that the assumption of constant voidage implied in Eq. (8) causes it to be less reliable than Eq. (81), since ϵ_0 in fact does vary depending on particle shape, size, and size distribution. Values of H_m calculated by Becker's method, therefore, showed poorer agreement with

Manurung's observed results than those from the equation of Thorley et al., [Eqs. (80) or (81), and (3) or (83)]. However, for coal particles which were rough and irregular, Manurung found it necessary to introduce a surface-roughness factor and a shape factor in Eq. (81) in order to obtain agreement between predicted and observed results:

$$C_D \bar{\psi} = \frac{200(1 - \epsilon_0)}{\bar{\psi} \operatorname{Re}_m \epsilon_0{}^3} + \frac{7.0}{3\epsilon_0{}^3} f\left(\frac{\delta}{d_p}\right) \qquad (84)$$

where δ represents depth of surface irregularities and $\bar{\psi}$ is the same shape factor as in Eq. (28). From experimental data for crushed coals, Manurung evaluated $f(\delta/d_p)$ in Eq. (84) to be $0.0146/d_v$ or $0.0136/d_p$, d_v and d_p being expressed in feet.

4. Reddy, Fleming, and Smith (R1)

Yet another method for calculating H_m based on the same approach is given by Reddy et al. Their experimental data were mainly for beds of mixed-size particles. For this reason, these workers chose the Smith and Reddy equation for spouting velocity, Eq. (13), using it in conjunction with the minimum fluidization velocity equation of Wen and Yu (W1),

$$U_{mf} = \frac{\mu}{d_p \rho_f} \left\{ \left[\frac{(33.7)^2 + 0.0408 d_p{}^3 (\rho_s - \rho_f) g}{\mu^2} \right]^{1/2} - 33.7 \right\} \qquad (85)$$

The experimental results for H_m obtained by Reddy et al., showed only limited agreement with values calculated from Eqs. (13) and (85). Although the observed trend of the data with varying particle Reynolds number, shown in Fig. 29, was correctly predicted, agreement between the calculated and experimental results for H_m was obtained only for those systems where the Reynolds numbers were close to or greater than the critical value of 70. For lower Reynolds numbers, the predicted values were found to be generally too high.

The following empirical equations, additionally proposed by Reddy et al., gave a closer correlation of their particular experimental data:

For $\operatorname{Re} \leq 70$,

$$H_m = 110.6 \, d_v^{1.26} \, D_i^{-0.33} \rho_s^{0.5} \qquad (86)$$

and for $\operatorname{Re} \geq 70$,

$$H_m = 20.4 \, d_v^{-0.57} \, D_i^{-0.125} \rho_s^{-0.2} \qquad (87)$$

H_m, D_i and d_v in the above equations are expressed in inches and ρ_s in pounds per cubic foot. The experiments were done with mixed-size beds of alundum, polystyrene, glass spheres, and glass shot, the range of particle size covered being 0.25–3.3 mm and of particle density, 62.5–247 lb/ft³. The term d_v in Eqs. (86) and (87) was determined by counting a large number of particles randomly sampled from the material, weighing the sample, and then converting the average particle weight thus obtained to the diameter of an equivolume sphere (F2). Only one column was used, of 6-in. diameter, the air inlet size varying between 0.25 and 0.75 in.

5. *Malek and Lu* (M3)

An empirical correlation was also proposed by Malek and Lu, whose experiments were restricted to particles of uniform size, in a range (d_p = 0.8–3.7 mm) somewhat coarser than that used by Reddy *et al.* Their own results for wheat, brucite, sand, polyethylene, polystyrene, millet, and timothy seed (ρ_s = 57–166 lb/ft³), obtained in 4-, 6-, and 9-in. columns, as well as previous data (B3, M10) for several other materials of size and density roughly within the same range, were all correlated within ±11% by the equation

$$H_m/D_c = 15(D_c/d_p)^{0.75}(D_c/D_i)^{0.4}\lambda^2/\rho_s^{1.2} \qquad (88)$$

where ρ_s is the absolute density of solids in pounds per cubic foot, and

$$\lambda = \frac{\text{surface area of particle}}{\text{surface area of equivolume sphere}} = \frac{0.205 s_p}{\bar{V}_p^{2/3}}$$

Values of λ ranged from 1.0 for particles such as millet, sand, and timothy seed to 1.65 for gravel. The above equation, based as it is on data for relatively coarse particles, should be comparable with Eq. (87). While the effects of the individual variables on H_m in the two equations are in qualitative agreement, the equations show wide differences in their respective values of the exponents on d_p, D_i, and ρ_s. Thus each equation must be regarded as having limited applicability.

6. *Lefroy and Davidson* (L2)

Semiempirical expressions for H_m have been developed by Lefroy and Davidson from their theoretical analysis discussed in Section IV,B and V,A. From measurements of spout diameter made in beds of kale seeds, polyethylene, and peas at maximum spoutable depths, they determined the value of the left-hand side of their force-balance equation [Eq. (66)] and found it to be about 0.36 in all cases. Therefore

$$[\pi(D_c^2 - D_s^2)/8D_s H_m]\tan\phi = 0.36 \qquad (89)$$

Assuming that $D_c \gg D_s$, and taking a constant value of $\phi = 33°$ for all three materials, Eq. (89) simplifies to

$$H_m = 0.72 D_c^2/D_s \qquad (90)$$

The above equation can be combined with Eq. (67) to yield the following expression for H_m in terms of primary parameters:

$$H_m = 0.67 D_c^{4/3}/d_p^{1/3} \qquad (91)$$

Although Eq. (91) was found to be in approximate agreement with the experimental data on which it was based, it does not encompass the observation of previous workers that H_m first increases before it starts to decrease with increasing particle size. This limitation of Eq. (91) has been attributed by Lefroy and Davidson in the case of very fine particles to breakdown of the assumption of uniform pressure across a horizontal section of the annulus, which is essential to the formulation of Eq. (90).

To take into account the effect of inlet diameter, which is missing in Eq. (91), another semiempirical equation for H_m was derived implicitly (by combining equations 31 and 32 of their paper) on the basis that the momentum transfer rate at the spout inlet is observed to be roughly one-half of the total upward force necessary to support the bed at minimum spouting:

$$H_m = 0.192 d_p D_c^4/D_i^2 D_s^2 \qquad (92)$$

However, when the above equation is combined with Eq. (67) to eliminate the dependent variable D_s, the expression obtained,

$$H_m = 0.168 d_p^{1/3} D_c^{8/3}/D_i^2 \qquad (93)$$

shows the effect of d_p on H_m to be contradictory to that given by Eq. (91) and of D_c to be considerably different, Eq. (91) being the more realistic inasmuch as it is more consistent with experimental results. The conflict between Eqs. (91) and (93) probably arises from mutually incompatible assumptions made by Lefroy and Davidson in developing the various models on which their equations are based.

7. Comparison of Methods

An attempt to assess the general applicability of the different calculation methods discussed above has been made by the present authors. The observed data for this assessment were selected so as to represent a wide range of the independent variables involved. In calculating H_m for closely sized solids, the mean screen aperture was used as d_p in all the equations, while for mixed sizes the reciprocal mean and the equivolume (sphere) mean diameters were also tried, wherever these were available. The shape

TABLE VI
COMPARISON BETWEEN OBSERVED AND CALCULATED VALUES OF H_m

Source of data	Solids	d_p (mm)	ρ_s (lb/ft³)	D_i (in.)	H_m, (observed, in.)	Becker, Eqs. (8) and (10) (in.)	ψ	Malek and Lu, Eq. (88) (in.)	λ	Thorley et al., Eqs. (3) and (80) (in.)	ϵ_0
				Different solids, 6-in. diameter column							
M9	Brucite	0.6	156.5	0.49	27.5	27.2	0.8	44.8	1.1	35.7	0.42
	Coffee bean	7.6	39.5		20.0	18.0	0.5	40.5	1.2	10.7	0.40
	Lima bean	12.7	83.0		11.4	11.3	0.4	11.3	1.2	6.8	0.40
	Mustard seed	2.2	75.7		34.0	29.5	1.0	32.9	1.0	33.2	0.42
	Rapeseed	1.7	68.9		30.0	31.0	1.0	43.5	1.0	35.3	0.42
	Peas	6.3	86.6		12.0	11.9	1.0	12.6	1.0	15.2	0.42
	Ottawa sand	0.6	145.0		27.0	25.5	1.0	40.6	1.0	33.9	0.42
	Shale	1.0	128.8		36.0	36.7	0.8	38.1	1.1	44.0	0.42
	Gravel	3.6	166.6		25.0	24.3	0.8	24.3	1.65	22.1	0.40
	Wheat	3.2	85.9		30.0	29.7	0.7	29.2	1.17	22.7	0.40
M7	Polystyrene	2.6	65.5	0.50	22.9	25.9	0.9	42.4	1.16	22.2	0.40
	Polyethylene	3.7	57.4		20.0	22.4	0.8	44.0	1.24	17.7	0.40
	Millet	1.3	80.8		32.0	30.7	0.8	62.8	1.23	27.5	0.40
	Nylon	2.8	68.6		35.8	29.1	0.7	56.5	1.41	21.6	0.40
	Coal, rounded	2.5	89.0		27.2	25.3	1.0	24.8	1.05	23.8	0.40
				Wheat, various column diameters							
	D_c—in.										
M4	4	3.8	82.0	0.75	8.8	9.2	0.7	9.4	1.16	6.0	0.42
	4	3.8	82.0	1.0	7.8	8.3	0.7	8.4	1.16	4.9	0.42
	6	3.8	82.0	0.375	28.0	28.6	0.7	29.5	1.16	27.8	0.42
	6	3.8	82.0	2.0	15.0	15.1	0.7	15.2	1.16	13.8	0.42
	9	3.8	82.0	0.5	70.0	61.4	0.7	63.2	1.16	68.2	0.42
	9	3.8	82.0	2.0	53.3	47.0	0.7	47.9	1.16	43.0	0.42
M9	12	3.2	85.9	1.0	95.0	99.6	0.7	97.6	1.17	89.6	0.40
	12	3.2	85.9	2.0	62.0	71.0	0.7	73.9	1.17	56.4	0.40

factors required in Eq. (10) and in Eq. (88), and the packed-bed voidage for use in Eq. (80), had to be estimated in the case of solids for which these properties were not specifically reported. Data involving particles for which no reasonable basis could be found for estimating these properties were excluded. From a comparison of the calculated and observed values of H_m, the following conclusions concerning the applicability of each equation can be drawn:

(a) Becker's equation [Eq. (10)] gives good agreement with observed data over a wider range of experimental conditions than any of the other equations (see Table VI).

(b) The equations of Thorley et al. [Eqs. (3) and (80)] and of Malek and Lu [Eq. (88)] also give fairly good predictions, though neither one is as consistently reliable as Becker's (Table VI). It should be noted that calculation by the method of Thorley et al. is very sensitive to the value of packed-bed voidage, which often has been assumed. This possibly explains why Manurung, who used measured packed-bed voidage values, obtained better agreement of his particular data for which ϵ_0 varied as widely as from 0.37 to 0.53 (as against 0.37 to 0.41 for Becker's solids) with the equation of Thorley et al. than with Becker's equation (The latter, as already pointed out, does not allow for variation of ϵ_0.)

(c) The general applicability of the equations of Reddy et al. [Eqs. (13) and (85), and Eq. (87)] and of Lefroy and Davidson [Eq. (91)]—tested only for D_i/D_c ratios similar to those used in the experiments of these authors—was found to be poor, the disagreement between the observed and calculated values of H_m being often in excess of 50%.

(d) None of the equations gave satisfactory predictions for beds of mixed-size particles, regardless of which mean diameter was used. In view of the strong influence of the presence of small proportions of fines noted previously, it would seem that a much closer size analysis than normally reported is necessary to enable a more relevant characterization of size distribution as it affects H_m.

Acknowledgments

The authors are indebted to the National Research Council of Canada for financial support and to W. A. Smith for technical help.

I. P. Mukhlenov and A. E. Gorshtein of the Lensovet Technical Institute, Leningrad, took the trouble to translate the manuscript of this monograph into Russian and offered many helpful criticisms. Useful comments on the manuscript were also received from G. L. Osberg and W. S. Peterson of the National Research Council of Canada.

Nomenclature

A_a Cross-sectional area of annulus
A_e Cross-sectional area of sensing element
A_s Cross-sectional area of spout
A_r Archimedes number, $gd_p^3 \rho_f (\rho_s - \rho_f)/\mu^2$
B Empirical constant in Eq. (50)
C_1, C_2 Constants in Eq. (31)
C_D Drag coefficient, $4d_p (\rho_s - \rho_f) g / 3\rho_f U^2$
c Concentration of bed particles in discharge at any instant
c_0 c at time zero
D_b Diameter of upper surface of bed
D_c Column diameter
D_i Fluid inlet diameter
D_s Spout diameter
d_p Particle diameter, or characteristic dimension of particle
$(d_p)_i$ Particle diameter of size fraction x_i
d_v Diameter of sphere of same volume as the particles
e Coefficient of restitution
F Volumetric feed rate of solids
F_r, F_z Hydrodynamic forces on an element of particles in the spout wall (Fig. 24)
G Fluid mass flow rate for unit of column cross-section
G_{ms} G at minimum spouting
g Acceleration of gravity
g_c Gravitational constant
H Bed depth
H_m Maximum spoutable bed depth
H_s Spout height (Fig. 14)
K, k Proportionality constants
M Momentum gained by particles or lost by fluid jet
m Mass of element of particles in the spout wall
m_s Mass flow rate of particles in the spout at any level
N Number of collisions per unit time
n Number of particles accelerated per unit time

P Pressure at any point
P_a Fluid pressure at the spout wall
P_b Downward force per unit cross-sectional area of the annulus
P_f Static pressure of fluid
P_s Fluid pressure at the column wall of a spouted bed
ΔP Pressure drop
ΔP_e Excess pressure drop ($\Delta P_m - \rho_b H$)
ΔP_m Maximum pressure drop prior to onset of spouting
ΔP_{mf} ΔP at minimum fluidization
ΔP_s Spouted-bed pressure drop
Re Superficial particle Reynolds number, $d_p U \rho_f / \mu$
Re_i Orifice Reynolds number, $d_p u_i \rho_f / \mu$
$(Re_i)_{ms}$ Re_i at minimum spouting
Re_m Re at $H = H_m$
r Radial distance from axis
r_s Spout radius
s Coefficient in Eq. (7)
s_p Surface area of a particle
T Dimensionless time, Ft/V
t Time
$\bar{t}$ Mean residence time of solids
$f(t)$ Fraction of feed particles in discharge after time t
U Superficial fluid velocity
U_a Upward superficial fluid velocity in the annulus
U_{aH} U_a at $z = H$
U_m U_{ms} at $H = H_m$
U_{ms} Minimum superficial fluid velocity for spouting
U_{mf} Minimum superficial fluid velocity for fluidization
U_r Volumetric rate of radial fluid percolation per unit area of spout-annulus interface
U_s Superficial spouting fluid velocity
$\overline{U_s}$ Volumetric fluid flow rate through the spout per unit of spout cross-sectional area
u Upward interstitial fluid velocity

u_i Fluid velocity through inlet orifice
u_s Upward interstitial fluid velocity in the spout
u_{sH} u_s at $z = H$
V Total volume of bed
V_b Perfectly mixed volume
V_d Deadwater volume
V_p Plug-flow volume
$\overline{V_p}$ Volume of a single particle
v Upward particle velocity in the spout
v_a Velocity of particle 1 prior to collision (Fig. 19)
v_c Velocity representing particle cross-flow across the spout wall
v_m Radial-mean upward particle velocity in the spout
v_0 v at $r = 0$
v_r v at radial position r
v_{3r} Entrainment velocity (Fig. 19)
v_t Free fall terminal velocity of a particle
W Overall bed circulation rate
W_a Downward solids mass flow rate at annulus top
w Solids mass cross-flow rate from annulus into spout per unit height of bed
w_n Weight of initially untraced region of bed
w_t Weight of initially traced region of bed
x Concentration of tracer in the initially traced bed region
x_i Mass fraction of particles of size $(d_p)_i$
x_n Concentration of nontracer in the bed after complete mixing
x_t Concentration of tracer in the bed after complete mixing
z Vertical distance from fluid inlet

GREEK LETTERS

α Angle of internal friction
β Coefficient in Eqs. (43) and (44)
δ Depth of surface irregularities
ϵ Voidage
ϵ_a Voidage in the spouted bed annulus
ϵ_0 Loosely packed bed voidage
ϵ_s Spout voidage
θ Included angle of cone
$\Psi, \overline{\Psi}, \lambda$ Particle shape factors, defined in text
μ Fluid viscosity
ρ_b Solids bulk density or specific weight
ρ_{bs} Solids bulk density in the spout
ρ_f Fluid density
ρ_s Particle density
τ Interparticle shear stress
ϕ Angle of repose

References

B1. Baerns, M., *Ind. Eng. Chem., Fundam.* **5,** 508 (1966).
B2. Barton, R. K., Rigby, G. R., and Ratcliffe, J. S., *Mech. & Chem. Eng. Trans.* **4,** 105 (1968).
B3. Becker, H. A., *Chem. Eng. Sci.* **13,** 245 (1961).
B4. Becker, H. A., and Sallans, H. R., *Chem. Eng. Sci.* **13,** 97 (1961).
B5. Berquin, Y. F., *Genie Chim.* **86,** 45 (1961).
B6. Berquin, Y. F., U.S. Patent 3,231,413 (1966).
B7. Berti, L., Operational criterion of a spouted bed oil shale retort. D.Sc. Thesis, Colorado School of Mines, Golden, 1968.
B8. Bowers, R. H., Stevens, J. W., and Suckling, R. D., British Patent 855,809 (1960).
B9. Buchanan, R. H., and Manurung, F., *Brit. Chem. Eng.* **6,** 402 (1961).

B10. Brown, R. L., and Richards, J. C., "Principles of Powder Mechanics," p. 70. Pergamon, Oxford, 1970.
C1. Charlton, B. G., Morris, J. B., and Williams, G. H., *U.K. At. Energy Auth., Rep.* AERE-R4852 (1965).
C2. Chatterjee, A., *Ind. Eng. Chem., Process Des. Develop.* **9**, 531 (1970).
C3. Cholette, A., and Cloutier, L., *Can. J. Chem. Eng.* **37**, 105 (1959).
C4. Cowan, C. B., Peterson, W. S., and Osberg, G. L., *Eng. J.* **41**, 60 (1958).
D1. Danckwerts, P. V., *Chem. Eng. Sci.* **2**, 1 (1953).
D2. Davidson, J. F., and Harrison, D., "Fluidized Particles," p. 8. Cambridge Univ. Press, London and New York, 1963.
D3. Dumitrescu, C., and Ionescu, D., *Rev. Chim. (Bucharest)* **18**, 552 (1967).
E1. Elperin, I. T., Zabrodsky, S. S., and Mikhailik, V. D., "Collected Papers on Intensification of Transfer of Heat and Mass in Drying and Thermal Processes," p. 232. Nauka i Tekhnika, BSSR, 1967.
E2. Epstein, N., *Ind. Eng. Chem., Process Des. Develop.* **7**, 158 (1968).
E3. Ergun, S., *Chem. Eng. Progr.* **48**, 89 (1952).
F1. Fisons Fertilizers Ltd., Levington, U.K., Personal communication (1969).
F2. Fleming, R. J., The spoutability of particulate solids in air. M.A.Sc. Thesis, University of Toronto, Toronto, Canada, 1966.
G1. Gelperin, N. I., Ainshtein, V. G., Gelperin, E. N., and L'vova, S. D., *Khim. Tekhnol. Topl. Masel* **5**, No. 8, 51 (1960).
G2. Gelperin, N. I., Ainshtein, V. G., and Timokhova, L. P., *Khim. Mashinostr.*, No. 4, p. 12 (1961).
(*Kiev*) No. 4, p. 12 (1961).
G3. Ghosh, B., *Indian Chem. Eng.* **7**, 16 (1965).
G4. Gibson, A., "Hydraulics and Its Applications," 5th ed., p. 93. Constable, London, 1952. Cited in ref. P1, p. 5–32.
G5. Gishler, P. E., and Mathur, K. B., U.S. Patent 2,786,280 (1957).
G6. Goltsiker, A. D., Rashkovskaya, N. B., and Romankov, P. G., *Zh. Prikl. Khim.* **37**, 1030 (1964).
G7. Goltsiker, A. D., Doctoral dissertation, Lensovet Technological Inst., Leningrad, 1967 (quoted in Romankov and Rashkovskaya, R4, Chapter 1).
G8. Gorshtein, A. E., and Mukhlenov, I. P., *Zh. Prikl. Khim.* **37**, 1887 (1964).
G9. Gorshtein, A. E., and Mukhlenov, I. P., *Zh. Prikl. Khim.* **40**, 2469 (1967).
G10. Gorshtein, A. E., and Soroko, V. E., *Izv. Vyssh. Ucheb. Zaved., Khim. Khim. Tekhnol.* **7**, No. 1, 137 (1964).
H1. Happel, J., *Ind. Eng. Chem.* **41**, 1161 (1949).
H2. Heiser, A. L., Lowenthal, W., and Singiser, R. E., U.S. Patent 3,112,220 (1963).
H3. Hunt, C. H., and Brennan, D., *Aust. Chem. Eng.* p. 9 (1965).
I1. I.C.I. Fibres Ltd., Harrogate, U.K., Personal communication (1969).
I2. Indian Explosives Ltd., Bihar, India, Personal communication (1968).
K1. Koyanagi, M., The design, construction and determination of the properties of a spouted bed. B.A. Sc. thesis, University of British Columbia, Canada, 1955.
K2. Kugo, M., Watanabe, N., Uemaki, O., and Shibata, T., *Bull. Hokkaido Univ., Jap.* **39**, 95 (1965).
K3. Kunii, D., and Levenspiel, O., "Fluidization Engineering." Wiley, New York, 1969.
L1. Lama, R. F., Pressure drop in spouted beds. M.Sc. thesis, University of Ottawa, Canada, 1957.

L2. Lefroy, G. A., and Davidson, J. F., *Trans. Inst. Chem. Eng.* **47**, T120 (1969).
L3. Leva, M., "Fluidization," p. 170. McGraw-Hill, New York, 1959.
L4. Levenspiel, O., *Can. J. Chem. Eng.* **40**, 135 (1962).
M1. Madonna, L. A., and Lama, R. F., *AIChE J.* **4**, 497 (1958).
M2. Madonna, L. A., and Lama, R. F., *Ind. Eng. Chem.* **52**, 169 (1960).
M3. Malek, M. A., and Lu, B. C. Y., *Ind. Eng. Chem., Process Des. Develop.* **4**, 123 (1965).
M4. Malek, M. A., Madonna, L. A., and Lu, B. C. Y., *Ind. Eng. Chem., Process Des. Develop.* **2**, 30 (1963).
M5. Malek, M. A., and Walsh, T. H., *Can. Dep. Mines Tech. Surv.*, Rep. FMP **66/54-SP** (1966).
M6. Mamuro, T., and Hattori, H., *J. Chem. Eng., Jap.* **1**, 1 (1968).
M7. Manurung, F., Studies in the spouted bed technique with particular reference to its application to low temperature coal carbonization. Ph.D. Thesis, University of New South Wales, Australia, 1964.
M8. Matheson, G. I., Herbst, W. A., and Holt, P. H., *Ind. Eng. Chem.* **41**, 1099 (1949).
M9. Mathur, K. B., *in* "Fluidization" (J. F. Davidson and D. Harrison, eds.), Chapter 17, Academic Press, New York, 1971.
M10. Mathur, K. B., and Gishler, P. E., *AIChE J.* **1**, 157 (1955).
M11. Mathur, K. B., and Gishler, P. E., *J. Appl. Chem.* **5**, 624 (1955).
M12. Matsen, J. M., *Ind. Eng. Chem., Process Des. Develop.* **7**, 159 (1968).
M13. Mikhailik, V. D., "Collected Works on Research on Heat and Mass Transfer in Technological Processes and Equipment," p. 37. Nauka i Tekhnika, BSSR, 1966.
M14. Mikhailik, V. D., and Antanishin, M. V., *Vesti Akad. Nauk BSSR, Ser. Fiz. Tekh. Nauk* No. 3, p. 81 (1967).
M15. Mukhlenov, I. P., and Gorshtein, A. E., *Zh. Prikl. Khim.* **37**, 609 (1964).
M16. Mukhlenov, I. P., and Gorshtein, A. E., *Khim. Prom. (Moscow)* **41**, 443 (1965).
M17. Mukhlenov, I. P., and Gorshtein, A. E., *Vses. Konf. Khim. Reactrom (Novosibirsk)* **3**, 553 (1965).
N1. Nikolaev, A. M., and Golubev, L. G., *Izv. Vyssh. Ucheb. Zaved., Khim. Khim. Tekhnol.* **7**, 855 (1964).
P1. Perry, J. H., "Chemical Engineers' Handbook," 4th ed., pp. 20–41. McGraw-Hill, New York, 1963.
P2. Peterson, W. S., *Can. J. Chem. Eng.* **40**, 226 (1962).
P3. Peterson, W. S., Canadian Patent 739,660 (1966).
Q1. Quinlan, M. J., and Ratcliffe, J. S., *Mech. Chem. Eng. Trans.* p. 19 (1970).
R1. Reddy, K. V. S., Fleming, R. J., and Smith, J. W., *Can. J. Chem. Eng.* **46**, 329 (1968).
R2. Reger, E. O., Romankov, P. G., and Rashkovskaya, N. B., *Zh. Prikl. Khim.* **40**, 2276 (1967).
R3. Richardson, J. F., and Zaki, W. N., *Trans. Inst. Chem. Eng.* **32**, 35 (1954).
R4. Romankov, P. G., and Rashkovskaya, N. B., "Drying in a Suspended State," 2nd ed. Chemistry Publishing House, Leningrad Branch, 1968 (in Russian).
S1. Singiser, R. E., Heiser, A. L., and Prillig, E. B., *Chem. Eng. Progr.* **62**, 107 (1966).
S2. Smith, J. W., and Reddy, K. V. S., *Can. J. Chem. Eng.* **42**, 206 (1964).
S3. Smith, W. A., B.A.Sc. Thesis, University of British Columbia, Canada, 1969.
T1. Thorley, B., Mathur, K. B., Klassen, J., and Gishler, P. E., Report. National Research Council of Canada, Ottawa, 1955.

T2. Thorley, B., Saunby, J. B., Mathur, K. B., and Osberg, G. L., *Can. J. Chem. Eng.* **37**, 184 (1959).
T3. Tsvik, M. Z., Nabiev, M. N., Rizaev, N. U., and Merenkov, K. V., *Uzb. Khim. Zh.* No. 4, 64 (1967).
T4. Tsvik, M. Z., Nabiev, M. N., Rizaev, N. U., Merenkov, K. V., and Vyzgo, V. S., *Uzb. Khim. Zh.* No. 2, 50 (1967).
V1. Volpicelli, G., and Raso, G., *Atti Accad. Naz. Lincei, Cl. i Sci. Fis., Mat. Natur., Rend.* (Rome) **35**, 331 (1963).
V2. Volpicelli, G., Raso, G., and Massimilla, L., *Proc. Eindhoven Fluidization Symp.,* p. 123. Netherlands Univ. Press, Amsterdam, 1967.
V3. Volpicelli, G., Raso, G., and Saccone, L., *Chim. Ind.* (*Milan*) **45**, 1362 (1963).
V4. Vyzgo, V. S., Pavlova, A. I., and Nabiev, M. N., *Uzb. Khim. Zh.* No. 4, p. 5 (1965).
W1. Wen, C. Y., and Yu, Y. H., *AIChE J.* **12**, 610 (1966).
Z1. Zabrodsky, S. S., "Hydrodynamics and Heat Transfer in Fluidized Beds," p. 111. MIT Press, Cambridge, Massachusetts, 1966.
Z2. Zenz, F. A., and Othmer, D. F., "Fluidization and Fluid-Particle Systems." Van Nostrand-Reinhold, Princeton, New Jersey, 1960.

RECENT ADVANCES IN THE COMPUTATION OF TURBULENT FLOWS*

W. C. Reynolds

Department of Mechanical Engineering
Stanford University
Stanford, California

I. Background and Overview 193
   A. The Stanford Conference 194
   B. Types of Closure 198
II. Mean-Velocity Field Closure 200
   A. Theory 200
   B. Examples 206
III. Mean Turbulent Energy Closure 216
   A. Theory 216
   B. Examples 223
IV. Mean Reynolds-Stress Closure 231
V. Opportunities and Outlook 236
   A. New Ideas for Homogeneous Flows 236
   B. Suggestions for the Future 242
   Nomenclature 244
   References 245

I. Background and Overview

The objective of this survey is to provide a brief but reasonably complete account of the state of the art of turbulent-flow computations, and to reflect the excitement of current debate on equation models in this field. The review has been written for readers with a basic background in turbulent transport, such as is given in a contemporary graduate course on this subject.

Our survey will be limited to methods that have a reasonable scientific

---

* This article was prepared for an American Institute of Chemical Engineers short course on turbulence given in November, 1970. Contributions that have appeared more recently are therefore not included.

basis, that show promise for extension to wider classes of flows, and that have been developed to the point where some amount of useful technical information can be obtained. Emphasis is placed on the physical assumptions rather than on the numerical techniques. The central ideas of contemporary methods are highlighted, and individual sources for more detailed descriptions are referenced. An effort has been made to both relate and critique the methods. The formulations have frequently been generalized to increase the range of their applicability. Also, certain new ideas on modeling of the pressure–strain term in the Reynolds-stress equations are included. It is hoped to acquaint newcomers with the techniques now available for computing turbulent flows, and to stimulate those with more experience to move forward in productive new directions.

Many schemes have been proposed for computation of turbulent flows. Regrettably, scant data are available for comparison with the predictions, with the exception of data for two-dimensional steady incompressible turbulent boundary layers. In other classes of flows (such as free shear layers, unsteady boundary-layer flows, flow with strong boundary layer–inviscid region interaction, rotation, or buoyancy, separated flows, or cavity flows) the data are very spotty, and therefore our ability to evaluate such computations is very limited. But it does seem clear that there is considerable opportunity for improving and extending the existing methods for all these flows.

A. THE STANFORD CONFERENCE

In 1968 a conference was held at Stanford on turbulent boundary layer prediction-method calibration (S3), where for the first time a large number of methods, totaling 29, were compared on a systematic basis. This comparison established the viability of prediction methods based on various closure models for the partial differential equations describing turbulent boundary layer flows, and has stimulated considerable more recent work on this approach.

The work leading up to the 1968 conference produced a volume of target boundary-layer data. A committee headed by D. Coles surveyed over 100 experiments and selected 33 flows for inclusion in this volume. The data of each experiment were carefully reanalyzed and recomputed for placement in a standard form; critiques were solicited from the experimenters; and all this was documented in a tidy manner by E. Hirst and D. Coles (S3). These data now stand as a classic base of comparison for turbulent boundary-layer prediction methods. Only the hydrodynamic aspects of these layers were considered; a corresponding standard for thermal behavior is still lacking, although the data of W. M. Kays and his associates are rapidly becoming such a collection. Moreover, the flows selected were relatively mild. Very strong pressure gradients, transpiration, roughness, rotation, and other interesting effects were not included.

Sixteen flows were selected as mandatory computations (most predictors did the others as well). Predictors were required to start these computations in a prescribed way, to use a prescribed set of free-stream conditions, to plot the results on a standard form, and to report all free-parameter adjustments. Most of the prediction programs were set up by graduate students for operation on the Stanford computer, so that by the time of the conference considerable experience with the various methods had been developed by the host group. This was very useful in preparing a paper on the morphology of the methods (Reynolds, in S3). The predictors sent their computations to Stanford shortly before the meeting, and they were compiled there for review at the conference.

Comparisons were limited to three integral parameters of the mean flow available from all computations: the momentum thickness θ, the shape factor H, and the friction factor C_f. Mean-profile comparisons were made by many authors, and a few even made comparison with turbulence data.

Figure 1 shows the common comparison for the easiest flow, a flat plate boundary layer. On shifted scales we show H, C_f (CF) and R_θ (RTH) vs. x for the 29 methods examined at the conference. The letters on the left identify the method. Note that all but one method is able to handle this flow reasonably adequately (see S3 for method code key).

Figure 2 shows the comparisons for a more difficult flow, an "adverse" pressure gradient (decelerating free-stream) flow. We note that some methods do reasonably well, while others do quite poorly. One predictor exemplified the integrity of the conference by producing a calculation that failed to fit his own data!

A small committee headed by H. Emmons studied the results and undertook to rank the methods. Figure 3 shows a comparison of the rankings of two evaluation committee members. Methods based on partial differential equations are shown as P, those including a turbulence differential equation are indicated by P+, and integral methods are shown as I. The committee noted that several different kinds of methods performed quite well, and that certain methods were consistently poor. They went on to recommend abandoning the poor methods, in view of the success of the better ones.

While there were a number of successful and attractive integral methods tested at the conference, one had to be impressed with the generality and

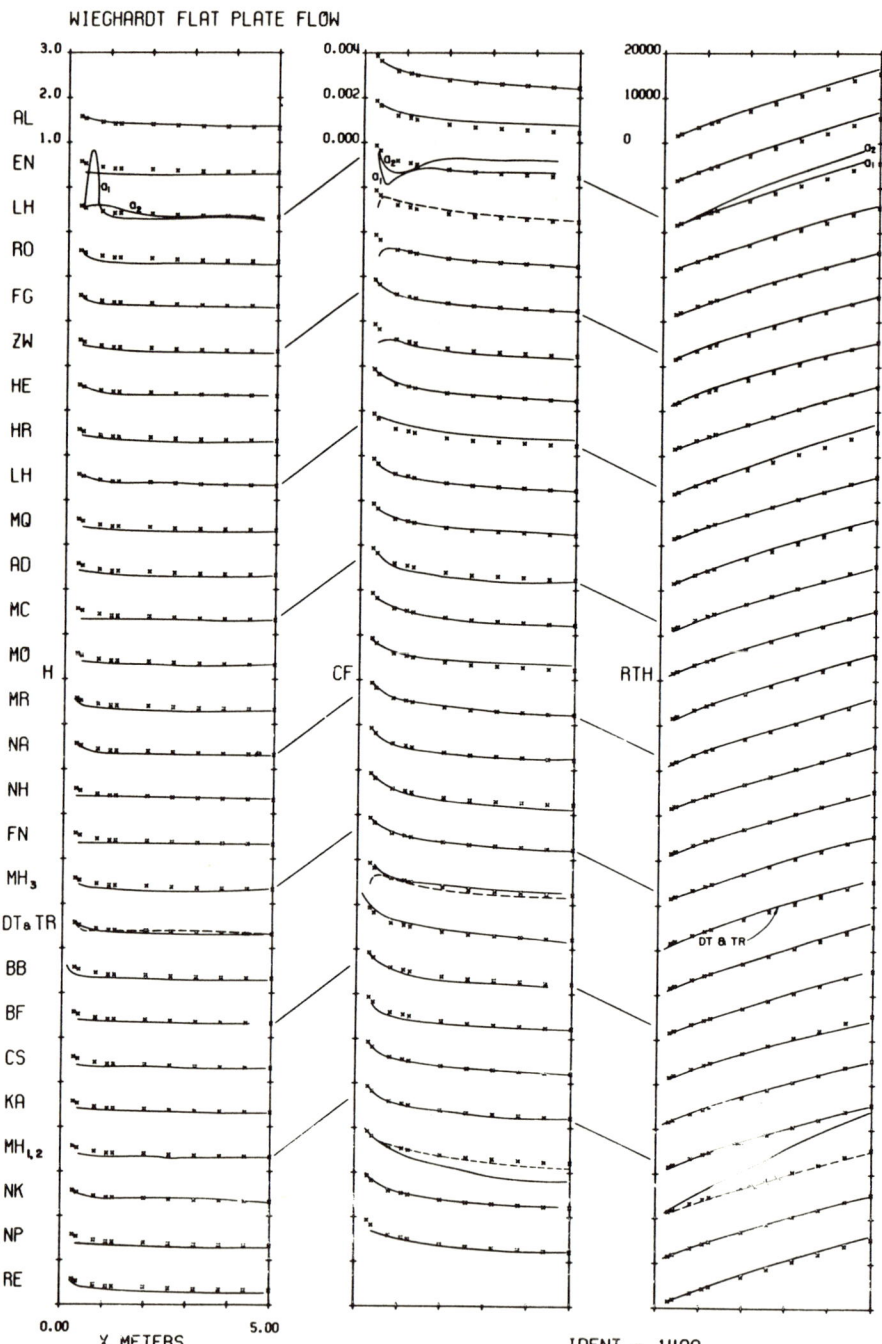

Fig. 1. TBLPC test flow no. 1400.

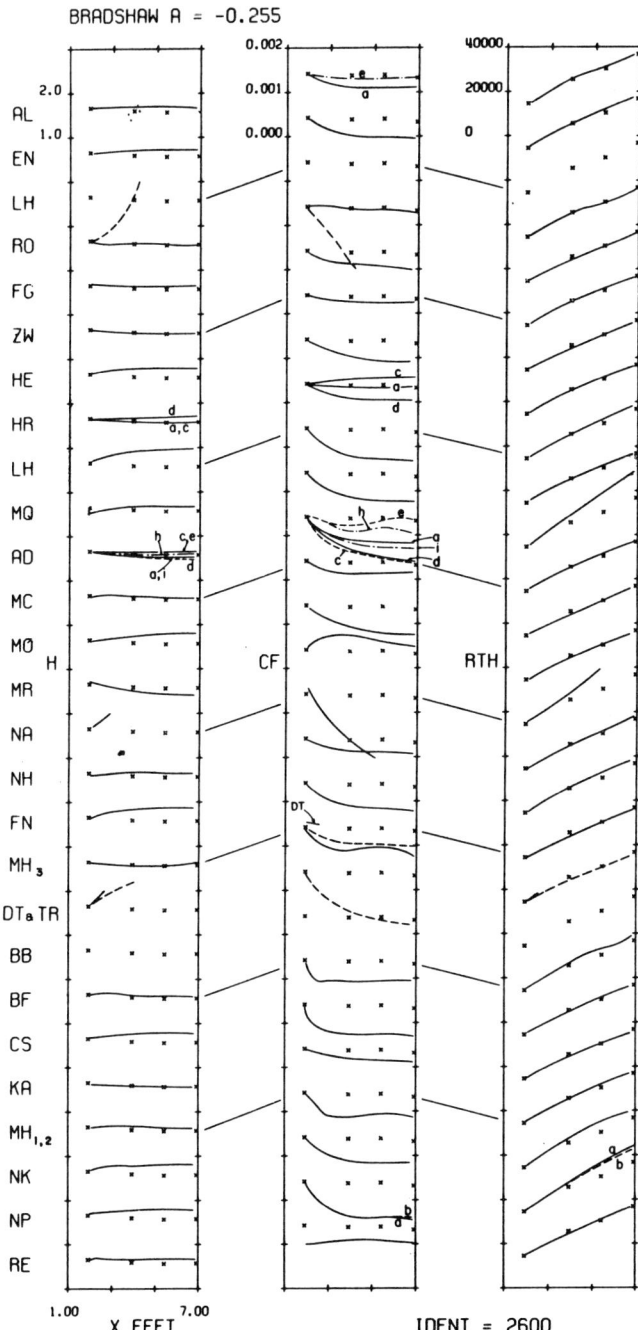

Fig. 2. TBLPC test flow no. 2600.

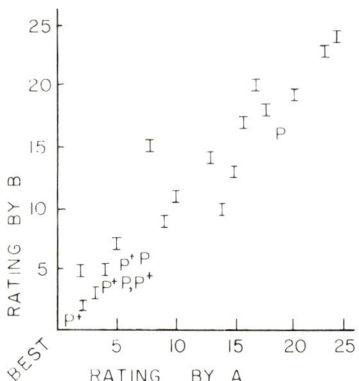

FIG. 3. TBLPC evaluation committee rankings of the methods. A lower ranking is better. I, integral method; P, partial differential equations; P+, turbulence partial differential equations.

speed of computations based on the partial differential equations. These schemes can be extended to new situations much more readily than integral methods. While integral methods are indeed useful in certain special cases, there is a definite interest in use of partial differential equation schemes. In view of the advantages of partial differential equation methods, integral methods are omitted from further consideration in this review. However, the development of adequate partial differential equations may well stimulate development of new integral methods based upon these equations. Two of the better integral methods were developed in this spirit (papers by McDonald and Camarata and by Hirst and Reynolds, in S3).

B. Types of Closure

The partial differential equations are obtained by time-averaging the Navier–Stokes equations. Unfortunately, in so doing information is lost to the point that the resulting equations are not closed. Additional equations may be derived by manipulations with the Navier–Stokes equations before the averaging process, but the number of independent unknowns increases more rapidly than the number of equations, and rigorous closure remains impossible. Apart from direct numerical solution of the unaveraged equations, about which we will comment later, the only hope lies in replacing some of the unknowns in the equations by terms involving other unknowns to bring the number of unknowns down to the number of describing equations. Such assumed relations are called "closure assumptions."

In turbulent shear flows it has seemed most convenient to work with the velocities as independent field variables, rather than with their Fourier transforms as is often done in isotropic turbulence. The simplest closure involves only the mean momentum equations; these contain unknown turbulent stresses for which a closure assumption must be made. This is the MVF[1] closure (mean velocity field). Models of this sort have been applied to a wide variety of flows, and work quite well for most boundary-layer flows of the Stanford conference. MVF methods are denoted by P in Fig. 3.

The next formal level of closure is at the level of the dynamical equations for the turbulent stresses, which we shall call mean Reynolds-stress closures (MRS). There have only been a few experimental calculations at this level, and such closures are not yet tools for practical analysis.

An intermediate closure level using the dynamical equation for the mean turbulent kinetic energy (MTE closure) has dominated more recent calculations, and has developed to the point of utility as an engineering tool. MTE closures are denoted by P$^+$ in Fig. 3. Since MTE closures permit calculation of at least one feature of the turbulence, such methods work better than MVF closures in problems where the turbulence behavior lags behind sudden changes in mean flow conditions. In addition, they give more useful information for only a little additional effort, indeed for considerably less effort than MRS closures. They do not give adequate detail on the turbulent structure and do not work well when the structure (but not the energy) depends explicitly upon some effect, such as rotation. MRS closures will be needed for these problems (although MRS closures have not yet been tested in cases where they are really needed to obtain accurate mean velocity predictions). It would appear, then, that MTE closures will remain important for some time, serving both as useful engineering tools and as guides to the development of more complex models.

Another approach that has promise for study of turbulence structure is the fluctuating velocity field (FVF) closure, adopted by Deardorff (D3). Using the analog of a MVF closure for turbulent motions of smaller scale than his computational mesh, Deardorff carried out a three-dimensional unsteady solution of Navier–Stokes equations, thereby calculating the structure of the larger-scale eddy motions. While it is likely that calculations of such complexity will remain beyond the reach of most for some time to come, results like Deardorff's should serve as guides for framing closure models.

Truly fresh alternative approaches to turbulence are few, and this review would not be complete without the mention of two that show promise for

[1] The acronyms for closure type used in this review are as follows: FVF, fluctuating velocity field; MVF, mean-velocity field; MVFN, Newtonian MVF; MTE, mean turbulent energy; MTEN, Newtonian MTE; MTES, structural MTE; MTEN/L, MTEN closure with dynamical length scale equation; MRS, mean Reynolds-stress; MRS/L, MRS closure with dynamical length scale equation.

future research. The first is Busse's (B4) and Howard's (H8) work in fixing bounds on the overall transport behavior of turbulent flows *without any closure approximations*. The second is the use of multipoint velocity probability densities (L4) with closure assumptions being made on the probability densities rather than on velocity moments. Neither of these schemes is presently developed as a general analytical tool, but either could spark a major revolution in turbulence theory.

In the sections that follow we will outline the theoretical framework of the MVF, MTE, and MRS closures, and give examples and commentary on applications of each. Readers unfamiliar with the differential equations should consult Hinze (H6) or Townsend (T1). In several instances we reformulate the constitutive models in an effort to extend their generality.

Following up on the concerns expressed about invariance at the Stanford conference, we have made certain extensions to put the basic equations in a properly invariant form. One must not read too much into this, however. P. Bradshaw (private communication) has cited Russell's (R2) wisdom: "A philosophy which is not self-consistent cannot be wholly true, but a philosophy which is self-consistent can very well be wholly false.... There is no reason to suspect that a self-consistent system contains more truth."

II. Mean-Velocity Field Closure

A. Theory

The equations for the mean-velocity field U_i and pressure P in an incompressible fluid with constant density and viscosity are

$$\partial U_i/\partial x_i = 0 \tag{1a}$$

$$\frac{\partial U_i}{\partial t} + U_j \frac{\partial U_i}{\partial x_j} = -\frac{1}{\rho}\frac{\partial P}{\partial x_i} + \nu \frac{\partial^2 U_i}{\partial x_j\, \partial x_j} - \frac{\partial R_{ij}}{\partial x_j} \tag{1b}$$

where $R_{ij} = \overline{u_i u_j}$. We will loosely call R_{ij} the Reynolds-stress tensor (actually $-\rho R_{ij}$ is the stress tensor). The over bar denotes a suitable average, and u_i is the instantaneous fluctuation field. Note that we employ the summation convention for cartesian tensors.

Closure is obtained through assumptions that relate the Reynolds stresses R_{ij} to properties of the mean velocity field U_i. The most productive approach has been to use a consitutive equation involving a turbulence length scale, usually called the "mixing length." A generalization of the usual assumption is

$$R_{ij} = \tfrac{1}{3}q^2\delta_{ij} - 2(2S_{mn}S_{mn})^{1/2}l^2 S_{ij} \tag{2a}$$

where

$$S_{ij} = \tfrac{1}{2}(\partial U_i/\partial x_j + \partial U_j/\partial x_i) \tag{2b}$$

is the strain-rate tensor, $q^2 = R_{ii} = \overline{u_i u_i}$, and l is a turbulence length scale. Throughout we shall denote such length scales by l, often subscripted. For the special case of simple shearing motion, where

$$S_{ij} = \begin{pmatrix} 0 & \tfrac{1}{2}\,dU/dy & 0 \\ \tfrac{1}{2}\,dU/dy & 0 & 0 \\ 0 & 0 & 0 \end{pmatrix} \tag{3}$$

Eq. (2) gives

$$R_{ij} = \begin{pmatrix} q^2/3 & -l^2\,|\,dU/dy\,|\,dU/dy & 0 \\ -l^2\,|\,dU/dy\,|\,dU/dy & q^2/3 & 0 \\ 0 & 0 & q^2/3 \end{pmatrix} \tag{4}$$

Now, if the spatial distribution of l is assumed, Eqs. (1) and (2) form a closed system for the variables U_i and $P + \rho q^2/3$. Note that the combination of $\rho q^2/3$ with P means that q^2 need not be evaluated.

Another closure approach used at this level is generalized as

$$R_{ij} = \tfrac{1}{3}q^2\delta_{ij} - 2\nu_T S_{ij} \tag{5}$$

where ν_T is the turbulent or eddy (kinematic) viscosity. An assumption of the spatial distribution of ν_T also suffices for closure. Occasionally these approaches are mixed. Comparison of Eqs. (2) and (5) gives

$$\nu_T = l^2(2S_{mn}S_{mn})^{1/2} \tag{6}$$

and consequently assumptions about l are often used to determine ν_T, or vice versa.

Mellor and Herring (M2), observing that Eqs. (2) or (5) imply that the Reynolds-stress deviations from $\tfrac{1}{3}q^2\delta_{ij}$ are proportional to the strain rates (and hence that the principal axes of the stress deviation and strain rate are aligned), call these closures "Newtonian." Accordingly, we denote them by MVFN. The success of the Newtonian model is remarkable, especially since for even the weakest of turbulent shear flows the principal axes are not aligned (C4).

In MVFN calculations the mixing length l is assumed in terms of the geometry of the flow. In a thin free shear layer, such as a jet or wake, the assumption that l is proportional to the local width of the layer seems to work quite well, with something like

$$l = 0.1\delta \tag{7}$$

This behavior is also used in the outer region of a turbulent boundary layer. Near a wall, l is experimentally found to be proportional to the distance from the wall, and the relations

$$l = \kappa y, \quad \kappa \approx 0.41 \tag{8}$$

seem to hold for smooth walls, rough walls, with modest compressibility, with transpiration, and in just about any axial pressure field.

In the viscous region immediately adjacent to a wall, the calculations are improved if l is reduced, with

$$l = \kappa y [1 - \exp\{-(yu^*/\nu)/A^+\}] \tag{9}$$

where $u^* = (\tau_w/\rho)^{1/2}$ is the friction velocity based on the local wall shearing stress τ_w, and A^+ is a parameter characterizing the thickness of the viscous region on the familiar $y^+ = yu^*/\nu$ scale. A^+ is known to depend upon both the streamwise pressure gradient and the transpiration velocity (for suction or blowing). Physical models of the wall layer can be used to suggest Eq. (9).

Kays and associates (K2) have correlated their turbulent boundary layer data to produce the A^+ correlation shown in Fig. 4. There P_0^+ is the streamwise pressure-gradient parameter

$$P_0^+ = (\nu/\rho u^{*3}) \, dP/dx \tag{10a}$$

and V_0^+ is the transpiration parameter

$$V_0^+ = V_0/u^* \tag{10b}$$

where V_0 is the injection velocity normal to the porous wall. Kays also modifies Eq. (9) by using the *local* shear stress $\tau(y)$ rather than τ_w in u^*.

In boundary-layer calculations, most workers simply use zonal models, with Eq. (9) in the inner region [which becomes Eq. (8) further from the wall] and something like Eq. (7) in the outer portion of the flow. Byrne and Hatton (B5) use a three-layer model as the basis for ν_T assumptions. Mellor and Herring (M2) have used concepts from the theory of matched asymptotic expansions to obtain composite representation for l valid across an entire turbulent boundary layer. A typical distribution of l in a boundary layer is shown in Fig. 5.

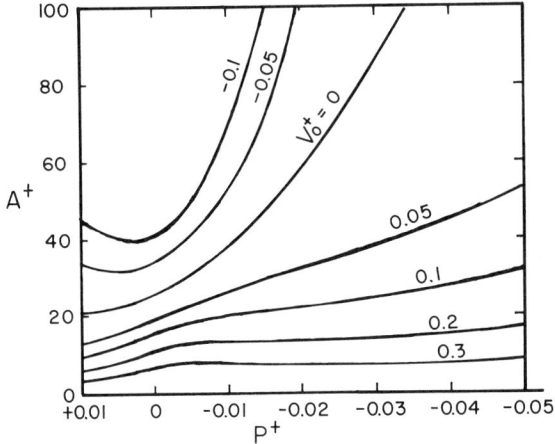

FIG. 4. Wall-layer thickness parameter as used by Kays.

For steady two-dimensional incompressible boundary layers, the MVFN equations reduce to $[U_i = (U, V, W), x_i = (x, y, z)]$

$$\frac{\partial U}{\partial x} + \frac{\partial V}{\partial y} = 0 \tag{11a}$$

$$U\frac{\partial U}{\partial x} + V\frac{\partial U}{\partial y} = -\frac{1}{\rho}\frac{dP_\infty}{dx} + \frac{\partial}{\partial y}(-\overline{uv}) \tag{11b}$$

and

$$-\overline{uv} = l^2 \mid dU/dy \mid \partial U/\partial y \tag{11c}$$

or

$$-\overline{uv} = \nu_T \, dU/\partial y \tag{11d}$$

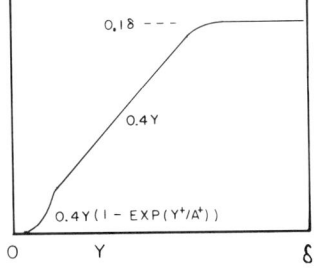

FIG. 5. Typical mixing-length distribution in boundary layers; not to scale.

These equations are of parabolic type, and may be solved by a forward marching technique. The upstream profile $U(x_0, y)$ must be specified, and the free-stream pressure distribution $P_\infty(x)$ must be known. $V(x_0, y)$ is then determined by Eq. (11a). The numerical problems are straightforward but not a trivial aspect of a successful method. Implicit schemes have been most successful, although explicit marching methods can be used if the wall region is treated separately.

In order to handle the rapid variations near a wall, one must either use a fine computational mesh in this region or else employ a special treatment. The variation in shear stress is, to a first approximation, small across this region, and the "law of the wall" is known to be followed by the mean-velocity profile very near the wall for most turbulent boundary layers. One simple approach is therefore to patch the numerical solution at the first computation point away from the wall to the empirical wall law,

$$U/u^* = (1/\kappa) \ln(yu^*/\nu) + B \qquad yu^*/\nu > 30, \qquad B \approx 5, \quad \kappa \approx 0.4 \quad (12)$$

This sets the value of U in terms of the wall shear stress (taken as the shear stress at the first mesh point) and y value at that point, and V may be taken as zero (or V_0) there. These conditions then provide boundary conditions for the numerical solution in the outer part of the flow, and a nearby uniform computational mesh in the outer region is usually feasible.

For transpired boundary layers or strong favorable pressure gradients, the shear stress variation in the wall region is significant and a better analysis is required. One approach is to use a solution to the governing equations obtained by assuming parallel flow (neglecting axial derivatives, except for pressure). This "Couette flow" solution is obtained by analytical or numerical solution of ordinary differential equations, and these solutions may often be precomputed in parametric form. A semitheoretical wall-layer treatment of this sort is very effective in permitting large computational steps in the streamwise direct on.

The Couette flow analysis uses the constitutive equation as its basis. The total shear stress in the boundary layer is written as

$$\tau/\rho = (\nu + \nu_T) \partial U/\partial y \quad (13)$$

Equation (13) may be integrated and expressed in dimensionless form,

$$\frac{U}{u^*} = \int_0^{u^+} \frac{\tau^+}{(1 + \nu_T/\nu)} dy^+ \quad (14)$$

where $y^+ = yu^*/\nu$ and $\tau^+ = \tau/\tau_w$. Thus, to develop the inner-region solutions, one needs to know the shear stress distribution $\tau(y)$. In the Couette flow approximation, the convective terms are deleted, and the shear stress emerges from the momentum equation as

$$\tau^+ = 1 + U^+V_0^+ + P^+y^+ \tag{15}$$

Loyd et al. (L2) have found this inadequate for strongly accelerated flows beyond $y^+ = 5$. Since the patching will take place at a much larger value of y^+ (perhaps around 30–50), a better shear stress distribution is needed. Loyd et al. noted that for fully asymptotic flow, where $U/U_\infty = f(y/\delta)$ throughout the entire layer (such flows can be realized with strong acceleration), the shear stress distribution is

$$\tau^+ = 1 + U^+V_0^+ + P^+y^+ \left[1 - y^{-1} \int_0^y (U/U_\infty)^2 \, dy\right] \tag{16}$$

and they use this expression to obtain a better shear stress distribution for use in the Couette analysis. These integrations are carried out at each streamwise step in the computation to patch the inner and outer solutions.

Recently W. M. Kays (unpublished work) has found that improvements in the prediction of flows with sudden changes in wall conditions are possible if empirical "lag equations" are used for the parameters P^+ and V_0^+ in determining A^+ from the correlation of Fig. 4.[2] Loyd et al. (L2) use

$$dP_e^+/dx^+ = (P^+ - P_e^+)/C_1; \quad dV_{0e}^+/dx^+ = (V_0^+ - V_e^+)/C_2 \tag{17a,b}$$

with C_1 and C_2 of approximately 3000. Here P^+ and V_0^+ are the actual values, and P_e^+ and V_{0e}^+ are the "effective" values used in reading A^+ from Fig. 4, and $x^+ = xu_*/\nu$.

Fine wall mesh schemes have been used to avoid this patching process. It is critical to use a good implicit-difference scheme in this case. Mellor (M1) developed a good linearized iteration technique which has since been adopted by others.

The approach to calculation of the temperature field and heat transfer follows closely the hydrodynamic calculation outlined above. For incompressible flow of a fluid with constant and uniform properties, neglecting the input to the thermal field by viscous dissipation, the thermal-energy equation (obtained by a combination of the energy and momentum equations) is

$$\frac{\partial \Theta}{\partial t} + U_j \frac{\partial \Theta}{\partial x_j} = \alpha \frac{\partial^2 \Theta}{\partial x_j \, \partial x_j} - \frac{\partial}{\partial x_j} (\overline{u_j \theta}) \tag{18}$$

Here Θ denotes the mean temperature and θ the local temperature fluctuation. The terms $\overline{u_j \theta}$ represent transports of internal energy by turbulent motions, and it is these terms that bring the closure problem.

[2] Kays has subsequently modified Fig. 4 to a slightly different form.

The common approach to the thermal problem is to assume

$$-\overline{u_j \theta} = \alpha_T \, \partial \Theta / \partial x_j \tag{19}$$

where α_T is the "turbulent diffusivity for heat," analogous to ν_T. With knowledge of α_T, and with U_j from solution of the hydrodynamic problem, the thermal problem is closed. It is usually assumed that

$$\nu_T / \alpha_T = \mathrm{Pr}_T \tag{20}$$

where Pr_T is a turbulent Prandtl number. For gases Pr_T is experimentally found to be approximately 0.7–1 in typical boundary-layer flows, and a constant value often suffices. More elaborate correlations of Pr_T with other properties of the flow have also been proposed (C1, S1). The choice of Pr_T is particularly important for liquid–metal heat transfer.

In examining the nature of α_T and ν_T in the viscous region of boundary layers, use has often been made of an unsteady two-dimensional parallel-flow Stokes model (C1). While such analysis may well yield the relevant dimensionless groupings, and possibly a fairly reasonable form for the α_T and ν_T distributions, failure to consider the now well-established strong three-dimensional unsteady features of the laminar sublayer (K5) would seem to render quantitative results questionable. Since the heat-transfer rate in boundary layers is strongly dependent on the assumptions made in this region, it would seem that at present the best results will be obtained with models having high empirical content, such as the A^+ correlation of Fig. 4 and the Pr_T correlations of Simpson (S1). New theories based on more accurate models of the wall layer will probably get considerable attention.

Though the concept of a turbulent viscosity has been displeasing to many, one cannot deny the success that its users have enjoyed. An interesting interpretation of ν_T is obtained by multiplying Eq. (5) by S_{ij}, which gives

$$\nu_T / \nu = -R_{ij} S_{ij} / (2 \nu S_{mn} S_{mn}) \tag{21}$$

The numerator is the rate of production of turbulence energy, and the denomenator is the rate of dissipation of mechanical energy by the mean field. The turbulent viscosity can therefore be described in terms of the rate of turbulence production.

B. Examples

Many examples of MVF calculations have now been published, and we shall now look at a small but representative collection. Readers should see the original papers for description of the details.

Most published computations have dealt with boundary layers. The numerical techniques employed have varied considerably, and hence the computational costs initially varied widely among programs. But now most workers have adopted implicit-difference schemes, with special wall-region treatment as outlined above, and/or a linearized iteration technique (M1), so that run times are now reasonably uniform. A typical two-dimensional compressible boundary layer can now be treated in under one minute on a typical large computer.

Among the pioneers and current advocates of the MVFN equations were A. M. O. Smith and his colleague T. Cebeci. They have elected to specify the eddy-viscosity distribution, using a form derived from the mixing-length model in the inner region and a uniform value reduced by multiplication by an intermittency factor in the outer region. The curves marked CS on Figs. 1 and 2 are by their method. Cebeci et al. (C2, C3) have extended their method to include heat transfer and compressibility.

Spalding has been an active explorer of turbulent boundary-layer computational methods. His early work with Patankar (P1) was based on the MVFN equations with mixing length specifications, and their complete program descriptions served as the seed for numerous computational efforts elsewhere. Figure 6 shows their computation of a wall jet flow as presented at the Stanford conference. This computation was among the few "more difficult" flows voluntarily presented by predictors to illustrate the range of their method. Spalding has now essentially abandoned this method in favor of MTE models.

G. Mellor and co-workers have used MVF closures for a variety of problems, and their unpublished work on the theoretical foundations of the theory has been both educational and useful in writing this review. Mellor and Herring startled the Stanford conference by presenting two methods, one based on MVF closure and a second based on MTE closure; except in

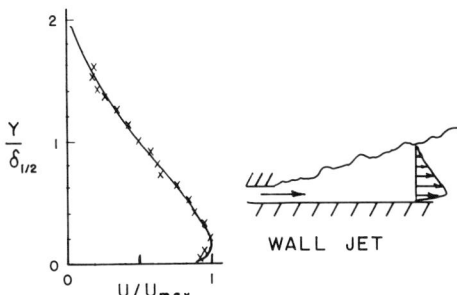

FIG. 6. Patankar-Spalding MVFN wall jet prediction using best-fit constant; crosses are data points.

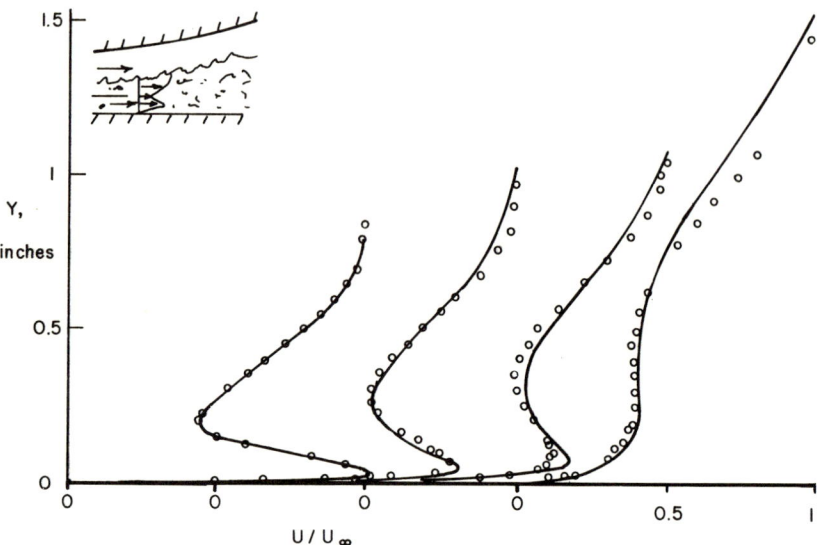

FIG. 7. Dvorak MVFN calculation for a wall jet in a boundary layer subjected to strong adverse pressure gradient.

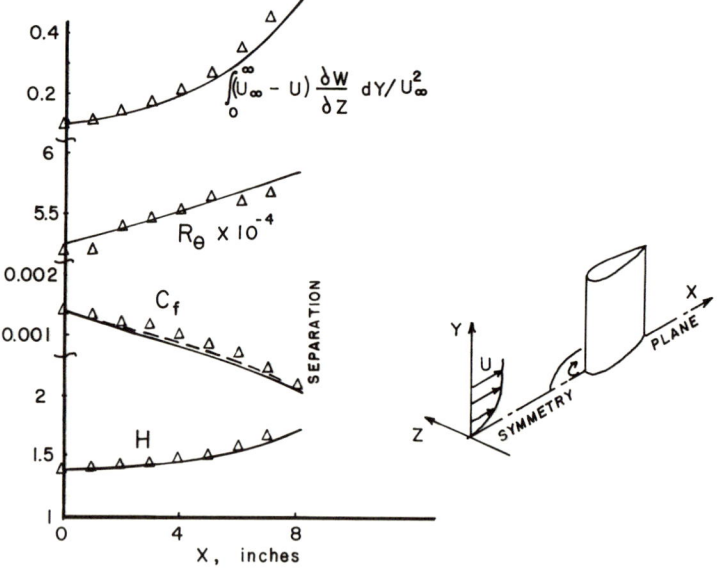

FIG. 8 Comparison of calculations on the symmetry plane in a three-dimensional flow: —— MVFN; ---- MTES, MTES/N. (Wheeler and Johnston, unpublished.)

one case the H, θ, and C_f predictions by the two methods were absolutely indistinguishable, both being judged among the best at the conference (shown as MH on Figs. 1 and 2). Mellor (M1) has also used a MVF method to study certain classes of three-dimensional boundary layers, and Herring and Mellor (H5) have extended the method to compressible boundary layers.

Since the Stanford conference, interest in the MVFN prediction methods has spread. F. Dvorak (private communication) has been looking at applications to more difficult flows of interest in aircraft design, and has kindly provided Fig. 7 as an illustration of his work. With some adjustment of the eddy-viscosity prescription, Dvorak is able to predict the growth of a boundary layer with tangential injection upstream and a strong adverse pressure gradient. This flow has two overlaid mixing layers, which suggests the variation in ν_T used by Dvorak, though it would seem difficult to make really accurate calculations if the downstream data were not available to guide the ν_T tailoring.

The MVFN equations have been used in the calculation of three-dimensional boundary layer flows by Mellor (M1) and currently by A. J. Wheeler and J. P. Johnston (unpublished). We remark that the MVFN model assumes that the shear stress is aligned with the strain rate. In spite of the strong experimental evidence (J1) that this does not hold, the MVFN equations work remarkably well in predicting the mean velocity field in three-dimensional boundary-layer flows where the pressure field (rather than the turbulent stress field) has the primary influence on the three-dimensionality; most boundary layers of engineering interest may be of this type. Figure 8 includes integral-parameter predictions using Mellor

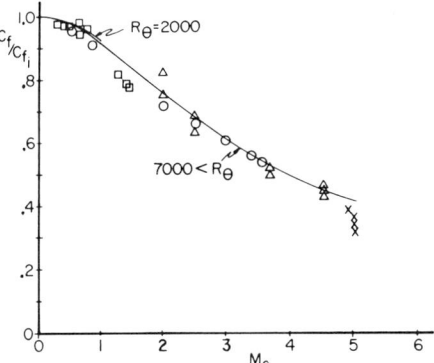

FIG. 9. Herring and Mellor MVFN calculation for the skin friction factor on a flat plate in compressible flow.

and Herring's MVFN method, by A. J. Wheeler and J. P. Johnston (private communication), of the flow along the symmetry plane in a boundary layer approaching an obstacle. Except very near the separation point, results are excellent. The MTE predictions on Fig. 8 will be discussed in Section III,B.

The prediction of turbulent boundary-layer separation by MVF methods has not been very successful. Indeed, it may be appropriate to identify turbulent separation in terms of the turbulence near the wall, and this will require use of a more sophisticated model (MTE or MRS), quite possibly in their full (rather than boundary-layer) form.

MVFN methods have been used with some success in compressible flows. Figure 9 shows a prediction of Herring and Mellor (H5) of the Mach number correction to the skin friction factor for a flat-plate boundary layer. Figure 10 shows their prediction for the boundary layer on a waisted body of revolution. We note that, while the momentum thickness is quite accurately predicted, the velocity-profile details are in considerable error.

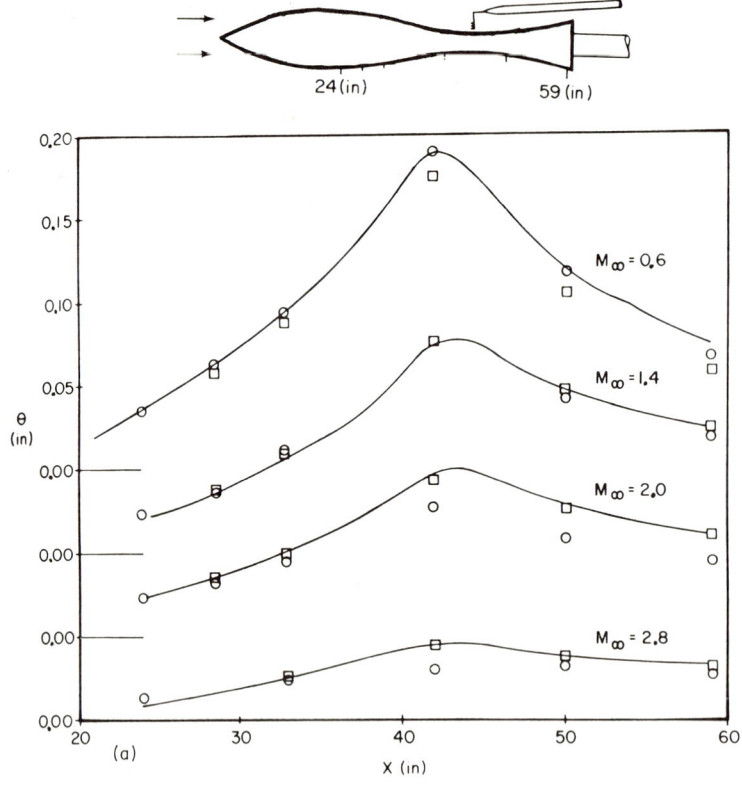

FIG. 10a.

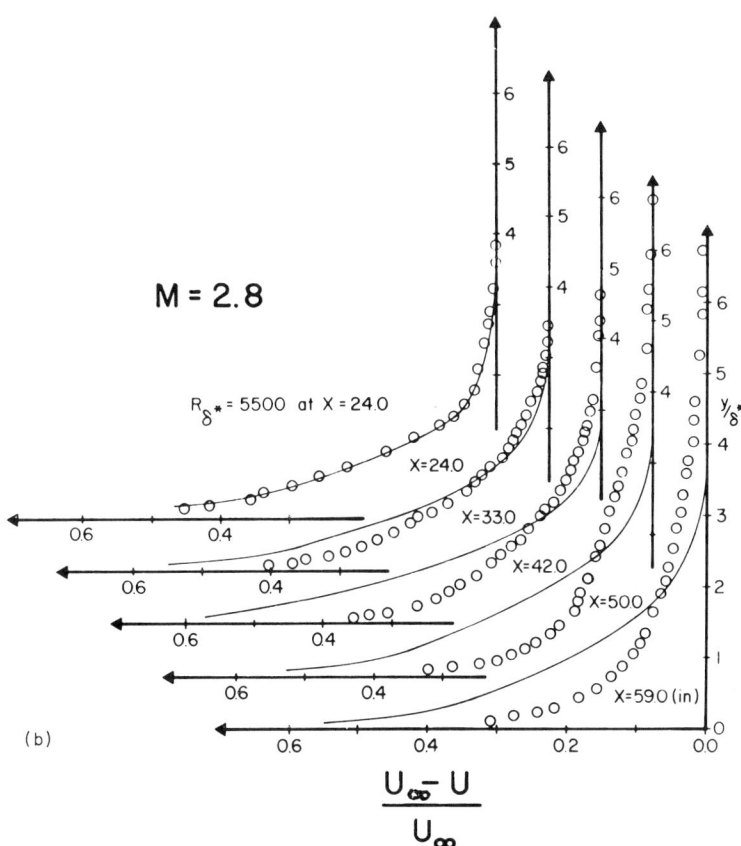

Fig. 10. Herring and Mellor MVFN calculation for a compressible boundary layer on a waisted body of revolution: (a) integral parameters, (b) profiles.

Indeed, MVFN methods are often much better in predicting integral properties of the flow than in predicting local details. Geometrical effects neglected in the analysis are the probable cause of much of the discrepancy.

Figure 11 shows a prediction by J. M. Healzer and W. M. Kays (private communication) of the heat-transfer coefficient (based on enthalpy difference) in an adiabatic rocket nozzle boundary-layer flow, made with an extended MVFN method, no chemical reactions being considered. The accuracy of this prediction attests to the value of such methods in contemporary engineering analysis.

MVFN methods have been used in contained and recirculating flows, where the boundary-layer approximations no longer apply. Spalding and

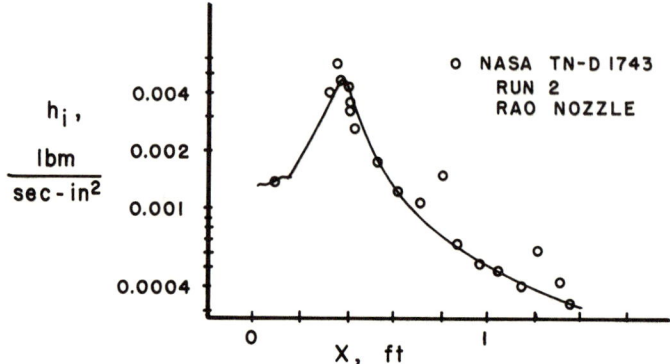

Fig. 11. J. M. Healzer and W. M. Kays (unpublished) MFVN calculation of the heat-transfer coefficient in a supersonic rocket nozzle.

his co-workers have led these efforts (G3). The numerical treatment is critical here, for the equation system is elliptic rather than parabolic, and the entire field must be solved simultaneously. Computational times are consequently considerably longer, with several minutes being required for a typical flow. Recently, Chin and R. A. Seban (private communication)

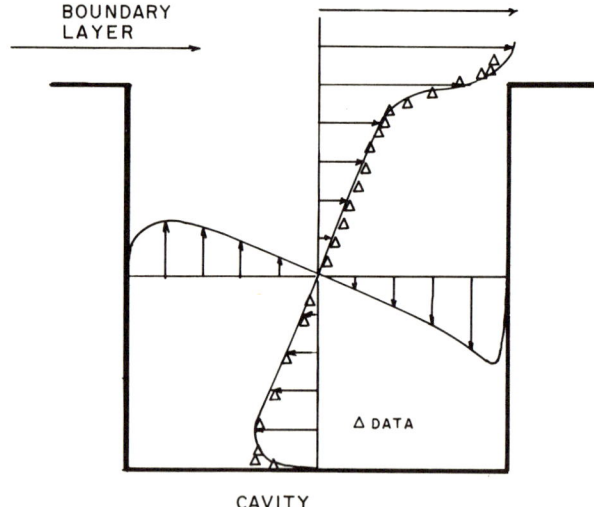

Fig. 12. Seban and Chin MVFN calculation of the recirculating flow in a square cavity; △ data points.

studied an improvement of Spalding's upwind difference treatment as applied to the flow in a cavity under a turbulent shear flow. The results of their computation are shown in Fig. 12. They used a simple wall-region patching treatment, with a linear mixing length near the walls, a uniform mixing length in the central region of the cavity, and a constant mixing length in the external shear layer. The computational mesh was 41 × 41 in the cavity, with closer spacing near the walls. In order to obtain convergence in the solution of the difference equations, over 1000 relaxation iterations were required, and the computation took 20 min on a CDC 6400 computer. While the velocity distribution in the central cavity is predicted very well, the heat transfer from the cavity bottom is not. R. A. Seban (private communication) states that an improved wall-region treatment is required, but that the relaxation iteration became nonconvergent when this was tried. He suggested that perhaps the time-dependent MVFN equations would have to be solved in order to compute the final steady-state flow.

MVFN equations have not been tested in very many time-dependent flows, for there are practically no data for comparison. Moreover, the computation costs skyrocket with every added dimension. However, if the time dependence is periodic, a Fourier analysis can be used to reduce the problem to a sequence of steady problems. If the flow is parallel and the periodic component takes the form of streamwise traveling waves of small amplitude, then the MVFN equations may be reduced to ordinary differ-

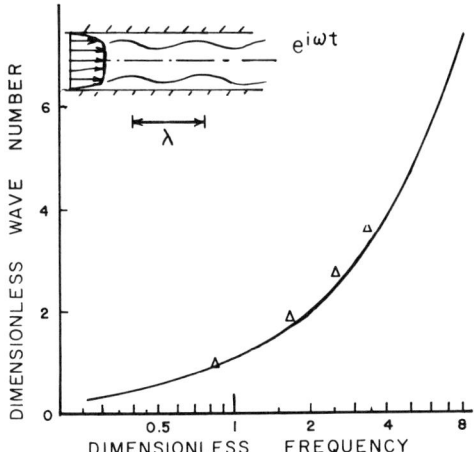

Fig. 13. Hussain and Reynolds MVFN calculation of the dispersion relationship for plane waves in turbulent channel flow.

ential equations for the periodic disturbance similar in structure to those used in analysis of the stability of laminar flows. We have been looking at the results of such computations for periodic disturbances in shear flows and for flows over waving boundaries. Our experimental observations of small periodic disturbances in turbulent channel flow (H9) indicate that a dispersion relation exists between the frequency and streamwise wave number of disturbance eigenfunctions. Figure 13 shows our predictions for this relationship as compared with our experimental data. The predictions were made using the eddy-viscosity distribution calculated from the mean

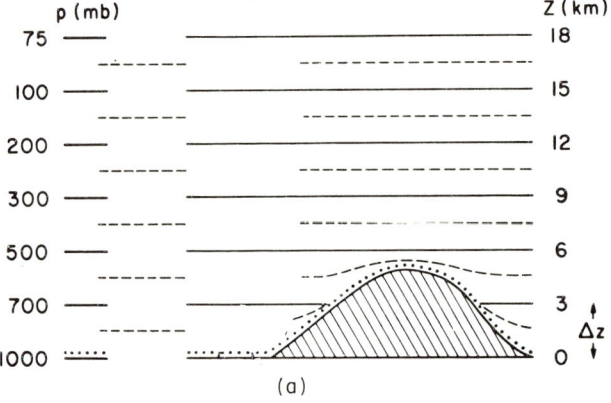

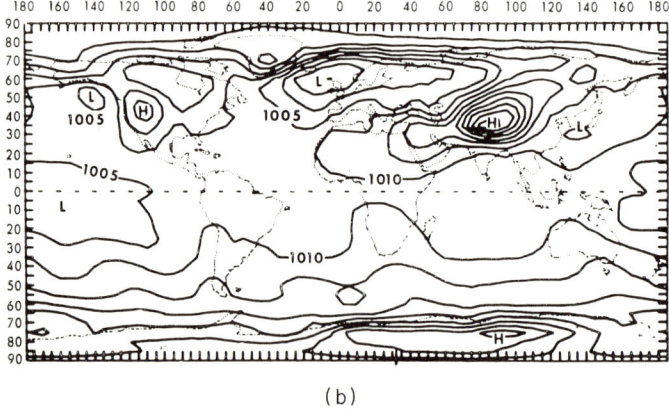

FIG. 14. NCAR 6-layer atmospheric-circulation model. (a) Model, (b) calculated sea-level isobars.

velocity profile, a fine wall mesh, and Eq. (5) in the time-dependent MVFN equations. Note that the MVFN model seems to work well in this unsteady flow.

We have also applied this approach to flows over waving boundaries, and in particular to Kendall's (K4) and Stewart's (S4) flows. In neither case did our predictions agree with the measurements; Davis (D2) used a similar MVF model with curvilinear coordinates, apparently with greater success. An experiment on turbulent channel flow with a waving wall has just been completed in our laboratory. The wave-induced wall-pressure oscillation is predicted fairly well by MVF theory for upstream-running waves, but not at all well for downstream-running waves. This suggests that the MVF model is weakest in flows with a "critical layer," i.e. a point where the mean velocity matches the wave speed. The ability to predict such flows by MVF methods would seem questionable, in view of the rapid changes in strain rate to which the turbulence is subjected. In all probability, a MTE or MRS method would work much better, and we intend to explore calculations along these lines.

MVF methods fail in any flow where the nature of the turbulence is altered by some parametric effect, such as rotation, which does not appear parametrically in the equations of mean motion. Such effects can be included in MVF methods only by alteration of the l or ν_T specification, and hence MTE or MRS methods are clearly to be preferred for such cases.

The most ambitious application of MVFN equations has been to atmospheric general circulations. The National Center for Atmospheric Research has developed an elaborate model in which the velocity components, temperature, and humidity are calculated over the entire earth. The goal is to obtain an accurate 14–30 day weather forecast. The computational mesh involves six vertical layers and 5° grid spacing at the equator, with fewer points near the poles. The horizontal grid therefore varies from about 500 to 100 km on a side. A turbulent viscosity model is used to handle sub-grid-scale turbulence. The effects of sun, snow, water, mountains, and precipitation are simulated. The main features of global weather patterns are reproduced. The dearth of field data makes quantitative comparison difficult, and initialization almost impossible. Kasahara (K1) reports that "better" results are obtained with a 2.5° mesh, but with present computers a 24-hr computation requires about 24 hr with this finer mesh. Figure 14 shows a computation from the 5° model. It seems quite possible that such calculations will someday become a routine part of our weekly weather forecasts, though refinements in the physical model may be required.

III. Mean Turbulent Energy Closure

A. Theory

The MVF equations assume that the turbulence adjusts immediately to changes in mean conditions, and that a universal relationship exists between the turbulent stresses and the mean strain rates. To avoid these assumptions, one must include differential equations for the Reynolds stresses (called "dynamical" or "transport" equations). MRS closures use these equations; MTE closures are somewhat simpler, and employ a single equation for the turbulent kinetic energy in conjunction with constitutive or structural equations relating the turbulent stresses to the turbulent kinetic energy. Thus MTE methods can to a degree handle the delayed response of turbulence structure to sudden changes in mean conditions, and they are now being studied by several groups for such use.

Equations for the Reynolds stresses R_{ij} may be developed from the Navier–Stokes equations (H6, T1). These are

$$\frac{\partial R_{ij}}{\partial t} + U_k \frac{\partial R_{ij}}{\partial x_k} = -R_{ik} \frac{\partial U_j}{\partial x_k} - R_{jk} \frac{\partial U_i}{\partial x_k}$$

$$- \frac{\partial}{\partial x_k} (\overline{u_i u_j u_k}) - \frac{1}{\rho} \left[\frac{\partial}{\partial x_j} (\overline{u_i p}) + \frac{\partial}{\partial x_i} (\overline{u_j p}) \right]$$

$$+ \overline{\frac{p}{\rho} \left(\frac{\partial u_i}{\partial x_j} + \frac{\partial u_j}{\partial x_i} \right)} + V_{ij} \qquad (22)$$

Here V_{ij} is the viscous term, to be discussed shortly. A contraction of these equations gives the equation for the turbulent kinetic energy. With $q^2 = \overline{u_i u_i}$, this may be written as

$$\frac{\partial q^2/2}{\partial t} + U_k \frac{\partial q^2/2}{\partial x_k} = -R_{ik} \frac{\partial U_i}{\partial x_k} - \frac{\partial}{\partial x_k} (\overline{u_k u_i u_i}/2)$$

$$- \frac{1}{\rho} \frac{\partial}{\partial x_k} (\overline{u_k p}) + V_{ii}/2 \qquad (23)$$

The first term on the right is the "turbulence production." The more common form of V_{ij} is

$$V_{ij} = \nu \frac{\partial^2 R_{ij}}{\partial x_k \partial x_k} - 2\mathfrak{D}_{ij} \qquad (24a)$$

where
$$\mathfrak{D}_{ij} = \nu \overline{\frac{\partial u_i}{\partial x_k} \frac{\partial u_j}{\partial x_k}} \tag{24b}$$

Then,
$$\frac{V_{ii}}{2} = \nu \frac{\partial^2 \overline{q^2/2}}{\partial x_k \partial x_k} - \mathfrak{D} \tag{24c}$$

where
$$\mathfrak{D} = \mathfrak{D}_{ii} \tag{24d}$$

This form is appealing because the first term in $V_{ii}/2$ can be interpreted as a "gradient diffusion" of turbulent kinetic energy, and the second is negative-definite (suggestive of "dissipation" of turbulence energy). However, the rate of entropy production is proportional to

$$\epsilon = \nu \overline{\left(\frac{\partial u_i}{\partial x_j} + \frac{\partial u_j}{\partial x_i}\right) \frac{\partial u_i}{\partial x_j}} \geq 0 \tag{25}$$

Properly ϵ is called the "dissipation," but not $\mathfrak{D}$. We might call $\mathfrak{D}$ the "isotropic dissipation."

A second form of V_{ij} is

$$V_{ij} = \nu \frac{\partial}{\partial x_k} \overline{[u_i \sigma_{jk} + u_j \sigma_{ik}]} - \nu \overline{\left[\sigma_{jk} \frac{\partial u_i}{\partial x_k} + \sigma_{ik} \frac{\partial u_j}{\partial x_k}\right]} \tag{26a}$$

where
$$\sigma_{ij} = \partial u_i/\partial x_j + \partial u_j/\partial x_i \tag{26b}$$

For Eq. (26a),
$$\frac{V_{ii}}{2} = \nu \frac{\partial}{\partial x_k} \overline{[u_i \sigma_{ik}]} - \epsilon \tag{26c}$$

The appearance of ϵ makes this form appealing, even though the first term can no longer be interpreted as "gradient diffusion."

MTE methods require closure assumptions for the last three terms in Eq. (23), and there has been heated debate on this point. There seems to be universal agreement that the dissipation term should be modeled by the constitutive equation

$$\epsilon = Cq^3/l_\epsilon \tag{27}$$

where l_ϵ is a "dissipation length scale", and C is a function of the dissipation

Reynolds number $R_\epsilon = ql_\epsilon/\nu$, C being constant for $R_\epsilon \gg 1$. We note that for $R_\epsilon \gg 1$, ϵ is independent of ν; this is a reflection of the belief that the small-scale eddies responsible for the final dissipation of mechanical energy can handle all the energy that is fed to them from and by larger scale motions, and hence that the larger eddies control the dissipation rate. The spectral transfer process for $R_\epsilon > 1$ results from the inertial nonlinearity, which suggests Eq. (27). The remainder of the viscous term is only important very near a wall; though it is strictly incorrect, reasonable results have been obtained by taking $\mathfrak{D} = \epsilon$, and writing

$$V_{ii}/2 = \nu \frac{\partial^2 q^2/2}{\partial x_j\, \partial x_j} - \epsilon \tag{28}$$

Mellor and Herring's model is more complicated [see Eqs. (58) and (43)].

The main MTE argument stems over the treatment of the pressure–velocity correlation term and the triple velocity correlation term. One widely used approach is the "gradient diffusion" model, where one sets

$$\overline{\left(u_k \frac{p}{\rho} + u_k \frac{u_i u_i}{2} \right)} = -N_Q \nu_T \frac{\partial q^2/2}{\partial x_k} \tag{29}$$

where N_Q is a constant (or specified function). There is strong feeling in some quarters that this model ignores the dominance of transport processes by large-scale eddy motions. A generalization of a "large-eddy transport" model (B3) is

$$\overline{\left(u_k \frac{p}{\rho} + u_k \frac{u_i u_i}{2} \right)} = G q^2 Q_k \tag{30}$$

where G is a constant (or specified function), and Q_k is a global vector velocity scale characteristic of the large eddy motions. The choice for this closure is of considerable importance; neglecting the viscous diffusion terms, the equation system based on Eq. (29) is of elliptic type, while with Eq. (30) the system is hyperbolic. This mathematical difference is suggestive of substantial physical differences in the model. Both approaches have been used quite successfully, however; and it is not easy to make a strong case for either, solely from testing against experiments.

Having closed the q^2 equations, one must relate the Reynolds stresses to q^2 in order to have a closed system. Again two approaches have been hotly debated. The more common approach uses the constitutive equation Eq. (26), together with an additional constitutive equation relating the turbulent viscosity to the turbulent kinetic energy,

$$\nu_T = ql F(R_T) \tag{31}$$

Here l is a turbulence-length scale, and F describes the dependence upon the turbulence Reynolds number $R_T = ql/\nu$ with $F = $ const for $R_T \gg 1$. The length scales l and l_ϵ must be specified (either algebraically, or through a differential equation) to close the equation system. Since use of (31) again implies Newtonian behavior, we shall refer to this MTE closure as MTEN.

Observing that the Newtonian structure is never observed in turbulent shear flows, but that persistently strained flows apparently develop an "equilibrium structure," Bradshaw (B3) prefers to relate the Reynolds stresses directly to q^2. A generalization of his constitutive equation is

$$R_{ij} = a_{ij}q^2 \tag{32}$$

where a_{ij} depends upon the type of strain. For the case of pure shear [Eq. (31)], a reasonable form of Eq. (32) is (C4, T1)

$$a_{ij} = \begin{pmatrix} 0.48 & -0.16 & 0 \\ -0.16 & 0.26 & 0 \\ 0 & 0 & 0.26 \end{pmatrix} \tag{33}$$

Lighthill (L1) suggested a general form which gives $a_{12} = -0.16$ but does not correctly represent the diagonal terms,

$$a_{ij} = \tfrac{1}{3}\delta_{ij} - 0.32 S_{ij}/(2S_{mn}S_{mn})^{1/2} \tag{34}$$

We will denote MTE closures involving an assumed turbulence structure [e.g., Eq. (33)] by MTES.

One would like to assume constant values for a_{ij} in thin shear layers. However, on a symmetry axis in a pipe or free jet flow, where $R_{12} = 0$, one has $a_{12} = 0$, and hence to use Eq. (32) in such flows one must specify a variation in a_{ij}. Hence, one must have a good "feel" for the flow to obtain a good prediction. This requirement for intuition is less important in simpler boundary-layer flows, where a uniform value of a_{ij} produces reasonable results.

There is a more fundamental objection to the MTES idea. Recently Lumley (L3) has argued that the homogeneous flows upon which Eqs. (32) and (33) are based do not really reach equilibrium, and that instead the turbulence time (and length) scales continually increase. Champagne et al. (C4) experiments confirm this expectation. Hence, a structural model cannot be fully correct in homogeneous flows.

MTEN and MTES closures both fail in the case of a sudden removal of the mean strain rate, where it is known that a very slow relaxation of the structure toward isotropy takes place. The MTEN model instantly becomes isotropic, while the MTES model retains a permanent structure (unless one twiddles with the a_{ij}). This may not be a serious objection as long as these methods are used in shear flows having reasonably persistent strain.

To summarize, the MTE closures commonly employed use one of the following two forms: Using Eqs. (5), (27)–(29), and (31), with $l = l_\epsilon$,

$$\frac{\partial q^2/2}{\partial t} + U_j \frac{\partial q^2/2}{\partial x_j} = 2\nu_T S_{ij} S_{ij} - C\frac{q^3}{l} + \frac{\partial}{\partial x_j}\left[(N_Q \nu_T + \nu)\frac{\partial q^2/2}{\partial x_j}\right] \quad (35a)$$

$$\nu_T = qlF \quad (35b)$$

Or, using Eqs. (27), (28), (30), and (32)

$$\frac{\partial q^2/2}{\partial t} + U_j \frac{\partial q^2/2}{\partial x_j} = -a_{ij} q^2 S_{ij} - C\frac{q^3}{l} - \frac{\partial}{\partial x_k}(Gq^2 Q_k) + \nu \frac{\partial q^2/2}{\partial x_j \partial x_j} \quad (36)$$

For Eq. (35), values or distributions for C, F, and N_Q must be assumed, while for Eq. (36) values or distributions for a_{ij}, C, G, and Q_k are needed. Both forms require an assumption for the spatial distribution of the length scale l. The terms with ν are not important except very near walls, and are often neglected in the outer flow.

Most computations have used length-scale distributions of the sort described in Section II. Recently there has been some interest in using a differential equation for l, and the most extensive test of this approach has been by Spalding and Rodi (S2) and Ng and Spalding (N2). Their length-scale equation, which is based on a spectral transport equation (R1), can be generalized with slight modification as

$$\left(\frac{\partial}{\partial t} + U_j \frac{\partial}{\partial x_j}\right)\left(\frac{lq^2}{2}\right) = C_1 \nu_T l S_{ij} S_{ij} - C_2 q^3 + C_4 \frac{\partial}{\partial x_j}\left[\nu_T \frac{\partial}{\partial x_j}(lq^2/2)\right] \quad (37)$$

Spalding and his co-workers are able to obtain very good predictions of a variety of boundary layer and free shear flows, using essentially Eq. (37) to determine l, provided some adjustments in C_2 are made near solid walls.

Gawain and Pritchard (G1) proposed a more complicated hueristic integrodifferential equation for turbulent length scales. In effect their local length scale is determined by the mean velocity field in the *region* of the local point. The two-point tensor

$$\mathcal{R}_{ij}(x, \xi) = \frac{\overline{u_j(x+\xi)u_j(x-\xi)}}{\overline{u_k u_k}} \quad (38)$$

is used to define the length scale,

$$l^2 = \int \mathcal{R}_{ii} \xi^2 \, dV \bigg/ \int \mathcal{R}_{ii} \, dV \quad (39)$$

where dV denotes a volume integration. A form for $\mathcal{R}_{ii}$ is in effect assumed in terms of the mean velocity field, and the integrations are performed to obtain l. This length scale is then used in a MTEN calculation method, where reasonably accurate results are reported for plane Poiseuille flow and for an axisymmetric jet flow.

Harlow and Nakayama (H2), noting that the length scale will be used to determine the dissipation, proposed a closure model for the exact differential equation for $\mathcal{D}$ derivable from the Navier–Stokes equations. They experimented with the use of this equation in MTEN closures. The Los Alamos group (private communication) has now abandoned the MTE closure in favor of MRS closures, which also use the $\mathcal{D}$ equation for inference of length scales [see Eq. (61)]. They refer to the $\mathcal{D}$ equation as a "dissipation" equation, which as we have noted is not strictly correct.

Hanjalic et al. (H1) have used a dissipation-model equation to study a variety of boundary-layer flows in an extended MTEN model. Their formulation is purported to work in the viscous region, eliminating the need for wall-solution patching [see Eq. (62)].

The interest in and activity with dynamical equations for the dissipation (or length scale) suggests that such equations will presently become an important and well-advertised feature of MTEN prediction methods, and probably of MRS methods as well. The dissipation equation is discussed in greater detail in Section IV.

The boundary-layer form of Eq. (36) is (neglecting viscous terms for the outer region)

$$\left(\frac{\partial}{\partial t} + U_j \frac{\partial}{\partial x_j}\right) q^2/2 = aq^2 \frac{\partial U}{\partial y} - C \frac{q^3}{l} - \frac{\partial}{\partial y}(Gq^2 Q_2) \quad (40a)$$

$$a = -a_{12}, \quad \tau/\rho = -\overline{uv} = aq^2 \quad (40b,c)$$

Bradshaw et al. (B3) use Eqs. (40) to derive a differential equation for the turbulent shear stress τ. The transport velocity Q_2 is taken as $(\tau_{\max}/\rho)^{1/2}$, where $\tau_{\max}$ is the maximum value of $\tau(y)$ in the boundary layer. G and l are prescribed as functions of the position across the boundary layer, and a is essentially taken as constant. Together with Eqs. (10a,b), Eq. (36) gives a closed set of equations for U, V, and τ; this system is of *hyperbolic* type, with three real characteristic lines. Bradshaw et al. construct a numerical solution using the method of characteristics; it can also be done using small streamwise steps with an explicit difference scheme (N1; A. J. Wheeler and J. P. Johnston, private communications). There is a great physical appeal to the characteristics, especially since it is found that the solutions along the outward-going characteristic dominates the total solution. This may well be connected with physical observations on the nature of turbulent boundary layers (K5).

The boundary-layer form of Eq. (35) is (neglecting viscous terms for the outer flow)

$$\left(\frac{\partial}{\partial t} + U_i \frac{\partial}{\partial x_i}\right)\left(\frac{q^2}{2}\right) = \nu_T \left(\frac{\partial U}{\partial y}\right)^2 - C\frac{q^3}{l} + \frac{\partial}{\partial y}\left[N_Q \nu_T \frac{\partial q^2/2}{\partial y}\right] \quad (41a)$$

$$\nu_T = qlF \quad (41b)$$

Then, together with Eqs. (5) and (1a,b), this gives a closed system of equations for U, V, and q, provided C, N_Q, and l are specified. This system is of *parabolic* type.

Equations (40) and (41) do not hold in the viscous region near the wall. One must either modify these equations to include viscous effects, or else use special solutions, as discussed in Section II, in this region. Experiments reveal a nearly uniform distribution of q in the wall region, except very close to the wall ($yu^*/\nu < 20$). Moreover, the value of q/u^* seems to be nearly universal, with

$$q \approx 2.5u^* \approx (1/\kappa)u^* \quad (42)$$

This has been used as a "wall" boundary condition for the solution of Eqs. (36) or (41).

Mellor and Herring (M2) prefer to use equations containing the viscous terms to calculate the inner region directly. Now the manner in which V_{ii} is written and modeled becomes important, and their version of the MTEN equation can be written as

$$\left(\frac{\partial}{\partial t} + U_j \frac{\partial}{\partial x_j}\right)\left(\frac{q^2}{2}\right) = 2\nu_T S_{ij} S_{ij} - \frac{q^3}{l_\epsilon} + \frac{\partial}{\partial x_j}\left[(\tfrac{5}{3}\nu + \nu_T)\frac{\partial q^2/2}{\partial x_j}\right] \quad (43)$$

The factor $\tfrac{5}{3}$ yielded by their treatment of V_{ii} is a main point of the difference with others, and Mellor and Herring's rationale seems most cogent. They then use a fine mesh near the wall, with l and l_ϵ varying linearly in the wall region, and being uniform in the outer flow. Some of their predictions are discussed in Section III,B.

Mellor and Herring also examine MRS closures, and show how the MTEN closure results from the MRS equations with the additional assumption of small departures from isotropy. While this approach is academically interesting, even the most weakly strained flows are far from isotropic (C4), and hence the main selling point for MTE methods is that they work very well for predicting a wide class of turbulent shear flows. Examples are given in the following section.

MTE boundary-layer equations require the same upstream information as for MVF computations, plus the upstream distribution and free-stream distribution of q. Normally the free-stream turbulence is set at zero, but the effect of nonzero free-stream turbulence can be incorporated in a MTE calculation. If a dynamical equation for the length scale is used, then the upstream, free-stream, and wall-boundary conditions for l must be given. The upstream l distribution can be drawn using the ideas in Section I. At the boundary-layer edge $\partial l/\partial y = 0$ seems appropriate. In the wall region $l = 0$ if the calculation is carried to the wall, and $l = \kappa y$ if the mesh computation is patched to a wall region solution at the innermost mesh point. The need for this turbulence information makes MTE methods somewhat more difficult to use, but the ability of a good MTE method to predict more severe test flows may make the extra effort worthwhile.

In the so-called "log" region of turbulent boundary layers, the turbulence energy is essentially determined by a delicate balance between the production and dissipation terms in Eq. (23). With $q = u^*/\kappa$ and $l = \kappa y$ [see Eqs. (8) and (42)] a balance between the first two terms on the right in Eq. (40) gives

$$dU/dy = [C/(a\kappa)](u^*/\kappa y) \quad (44)$$

while for Eq. (41a,b) one has

$$dU/dy = [C/(F\kappa^2)]^{1/2}(u^*/\kappa y) \quad (45)$$

Hence, both models will give the proper logarithmic velocity profile [Eq. (12)], provided the coefficients in brackets are unity in each case.

Heat-transfer predictions made using MTEN closures have employed the models described in Section I. The hydrodynamic calculation yields ν_T, and Eq. (20) is then used in (18) to construct the temperature field.

B. EXAMPLES

The MTES approach has been advocated by Bradshaw and his coworkers (B2, B3). Their predictions for the Stanford conference must be judged among the very best. The ability of Bradshaw's MTES method to predict severe flows was demonstrated by their results for the Tillmann ledge flow, a boundary-layer flow immediately downstream of a turbulent reattachment point (judged the most difficult conference flow). Figure 15 shows their predictions at a point downstream in this flow, including results for two drastically different initial shear stress distributions. We note that the predictions are very insensitive to the initial (upstream) shear stress distribution. In spite of the modest disparity between the measured and

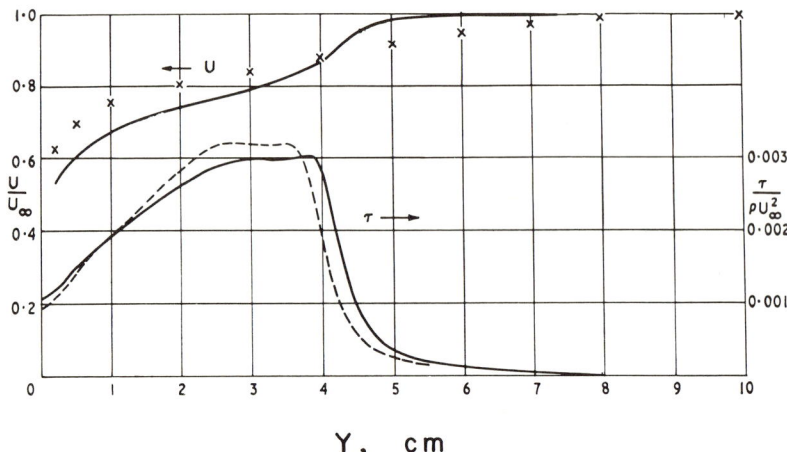

Fig. 15. Bradshaw's MTES calculation for the Tillmann Ledge flow: × experimental; — calculated; – – – – calculated with half initial shear stress; flow 1500, $X = 2.9$ m.

predicted velocity profiles, the predicted momentum thickness and skin friction were in considerably better agreement with the data than were the MTEN and MVF predictions. Indeed, many left the Stanford conference with the feeling that Bradshaw and Ferriss's MTES method was likely to be the wave of the future.

Nash (N1) has used a combination of Bradshaw's MTES ideas and a Newtonian assumption to treat three-dimensional turbulent boundary layers. Nash takes Bradshaw's structural assumption for the *total* shear stress vector,

$$[(\overline{uv})^2 + (\overline{wv})^2]^{1/2} = aq^2 \qquad (46)$$

but then uses the Newtonian approximation

$$\overline{uv}/\overline{uw} = (\partial U/\partial y)/(\partial W/\partial z) \qquad (47)$$

which assumes alignment of the shear stress and strain-rate vectors. The evidence is clear that (47) does not hold, yet it fortunately is worst in flows with strong spanwise pressure gradients, where the pressure gradients and not the shear stress control the mean velocity field. In Fig. 8 we show A. J. Wheeler and J. P. Johnston's (unpublished) calculation for the boundary layer along a plane of symmetry approaching an obstacle. The integral parameter predictions by Mellor and Hening's MVFN method, Nash's MTES/N method, and Bradshaw's MTES method are almost identical. Nash's (N1) own calculation for a point off the symmetry plane in a similar flow is shown in Fig. 16. Mellor (M1) made a similar calculation with his MVFN method with comparable results.

Bradshaw (B1) extended his MTES method to three-dimensional boundary layers, using the basic ideas to propose model equations for the vector sum and ratios of the two primary stresses, $\overline{-uv}$ and $\overline{-wv}$. Johnston (J1) has compared the result of predictions by this method with his own data for an infinite swept flow. In particular, data show that the stress vector does not align with the strain-rate vector as the Newtonian closures assume. It was hoped Bradshaw's structural model would work better on this flow; but Johnston's calculation shows that the angle of the shear stress vector is predicted quite poorly, although the mean velocity is predicted quite well. It is unlikely that MTE methods will ever predict this structural difference well, and one might hope that MRS methods will do considerably better.

A. McDonald and his associates (unpublished) have developed a MTES method following the lines of their integral method presented at the Stanford conference and are using this method in a variety of boundary-layer flows. They are also treating boundary layers using the full equations in order to study boundary layers near separation.

There has been considerable activity with MTEN computations. At the Stanford conference, Beckwith and Bushnell presented partial results from their MTEN method, and have since continued with its development. Spalding and his associates pushed ahead with MTES program development. Mellor and Herring (M2) have added to the theoretical framework through their application of the method of matched expansions to the selection of the length-scale distribution functions, and by showing how

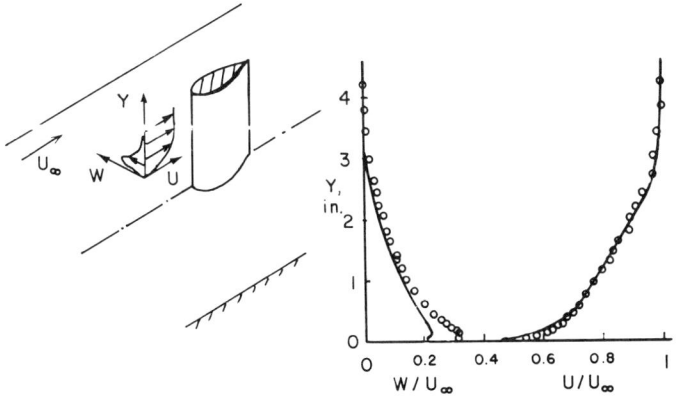

Fig. 16. Nash's MTES calculation of a three-dimensional turbulent boundary layer.

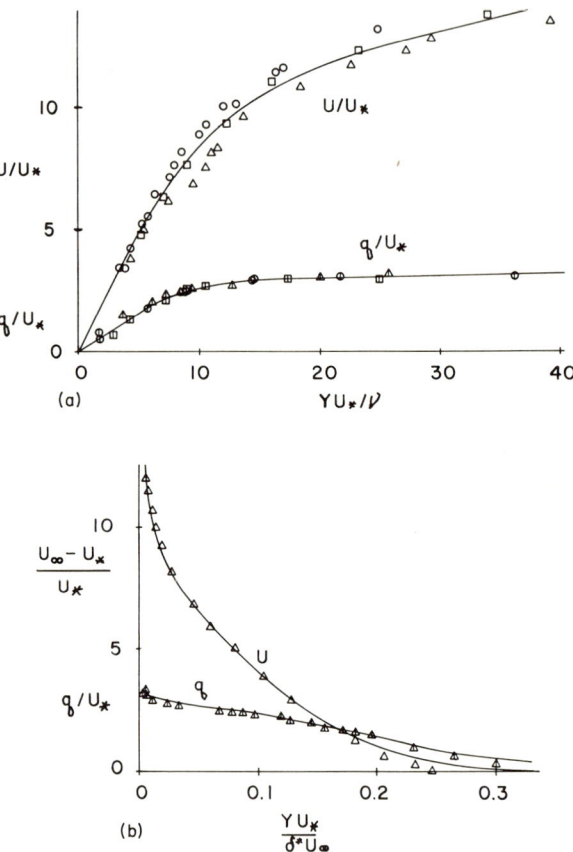

Fig. 17. Mellor and Herring's MTEN calculation for a flat plate boundary layer: (a) inner region; (b) outer region; ○, □, △ data.

the MTEN equations arise as a limiting case of MRS equations for nearly isotropic turbulence.

The ability of MTEN calculations to predict accurately the mean velocity field and turbulence kinetic-energy distribution is demonstrated by Fig. 17 from Mellor and Herring's contribution to the Stanford conference. Their use of a fine computational mesh near the wall is reflected in their accurate prediction of the inner regions.

The Mellor and Herring MTEN predictions were among the best at The Stanford Conference. We again note that these predictions were identical with those of their MVF method, for all but one test flow. Hence, for flows not too rapidly shocked by changes in free-stream or wall conditions, consistent MVF and MTE treatments may be expected to yield nearly the same results for the mean velocity and integral parameters. Of course, only the MTE calculation yields the turbulence-energy distribution directly. Figure 18 shows a Mellor–Herring MTEN calculation for a boundary layer responding to a sudden removal of adverse pressure gra-

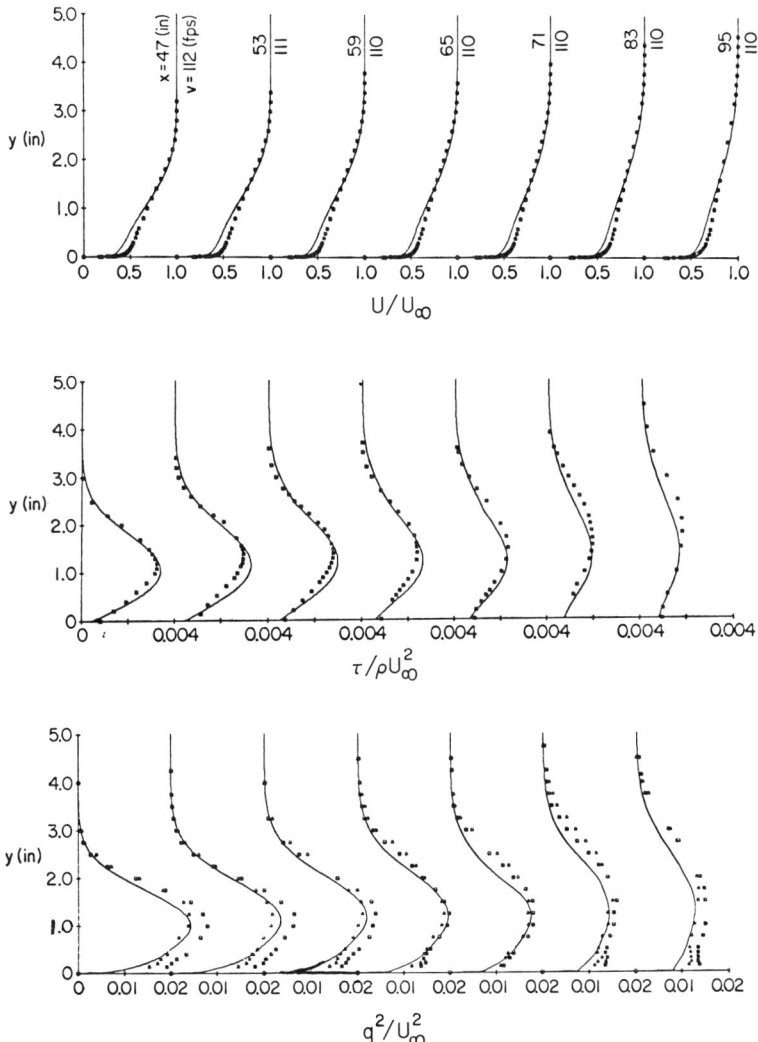

FIG. 18. Mellor and Herring's MTEN calculation for a relaxing boundary layer.

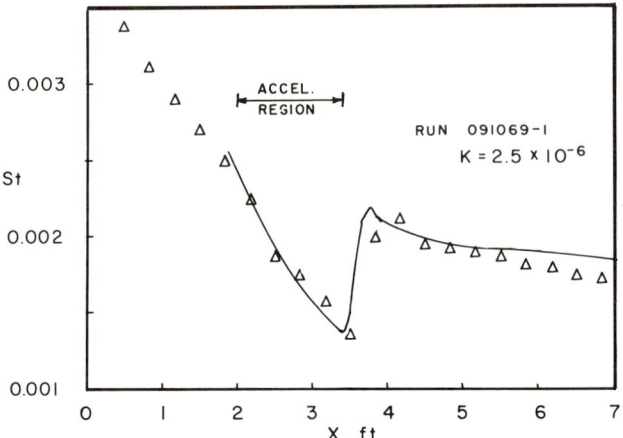

Fig. 19. Kays' MTEN calculation for the heat transfer to an accelerating boundary layer (St = Stanton number).

dient. Five years ago this would have been regarded as a "difficult" test flow, but we see that MTEN methods now handle it reasonably well.

The ability of MTEN methods to handle sudden changes in boundary conditions is evidenced by recent unpublished calculations by Kays and

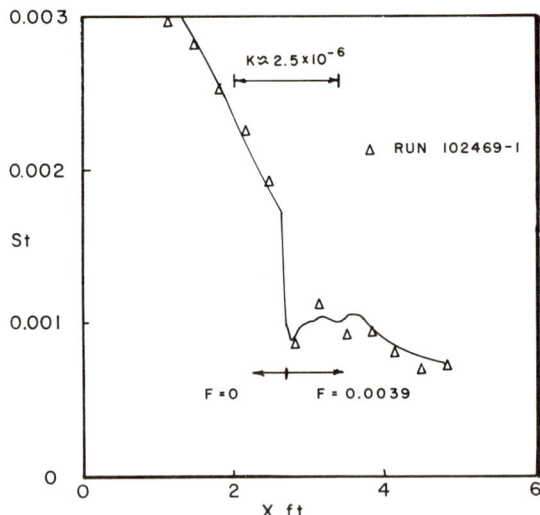

Fig. 20. Kays' MTEN calculation for the heat transfer to an accelerating boundary layer with transpiration.

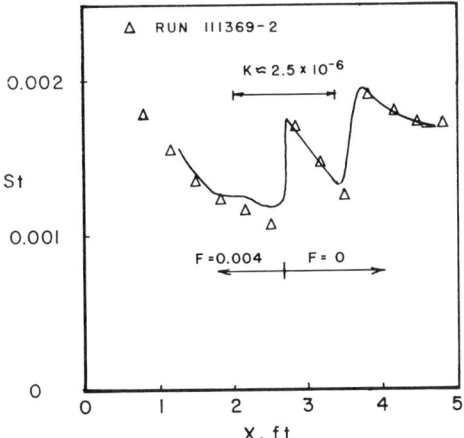

FIG. 21. Kays' MTEN calculation for the heat transfer to an accelerating boundary layer with changes in transpiration.

his co-workers (see Loyd et al., L2). They have modified an early Spalding MTEN program to the point where it successfully predicts the heat-transfer behavior of incompressible turbulent boundary layers with strong pressure gradients and with wall suction or blowing. With sudden changes in pressure gradient or blowing, the heat-transfer coefficient (or the Stanton number containing it) changes rapidly, and such calculations are more difficult for MVF methods.

For boundary layers the pressure gradient is conveniently represented

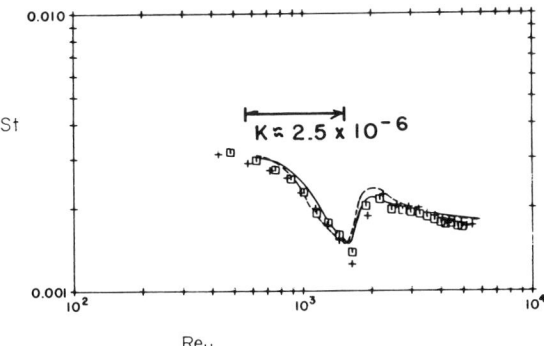

FIG. 22. Kearney's MTEN calculation of the effects of free-stream turbulence on an accelerating boundary layer. Initial free-stream turbulence intensity: ⊡ ——— 0.7%; + - - - - 3.9%.

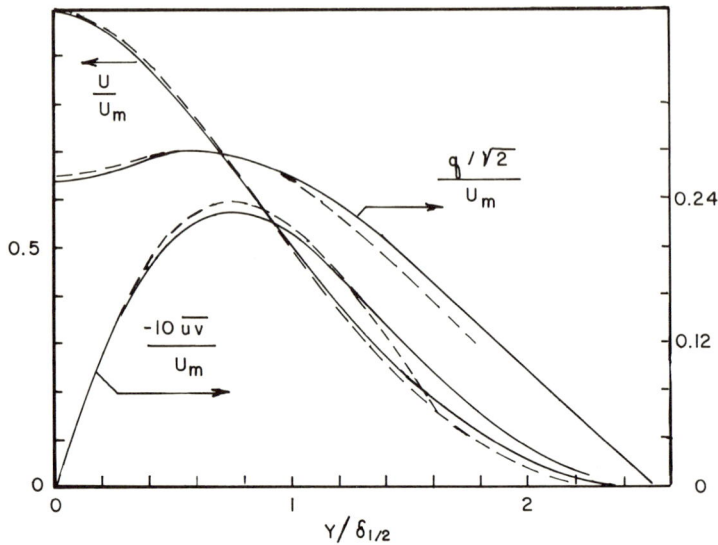

FIG. 23. Spalding and Rodi's MTEN calculation, incorporating their dynamical equation for the length scale, for a plane jet: — predictions MTEN-L; ---- experiment.

by the parameter

$$K = (\nu/U_\infty^2)(dU_\infty/dx) \tag{48}$$

Figure 19 shows an unpublished prediction by W. M. Kays of the heat transfer to a boundary layer undergoing strong acceleration followed by a relaxation to zero pressure gradient. Note that the sudden jump in Stanton number, as acceleration is removed, is predicted quite well. Figure 20 shows another Kays calculation for an accelerated boundary layer, with blowing beginning midway through the accelerated region and continuing through the relaxation to zero pressure gradient. Figure 21 shows a prediction for an accelerated boundary layer with blowing, with transpiration terminated upstream of the removal of acceleration. The remarkable success of these calculations suggests that MTEN methods are now developed to the point of utility as tools for engineering analysis.

The MTE methods include a calculation of the turbulence energy, and hence one may study the effects of variable free stream turbulence. Kearney et al. (K3) have compared such predictions with their data, and Fig. 22 shows a typical result for strongly accelerated turbulent boundary layer.

The MTEN methods have been applied to free shear flows to a limited degree by Spalding and co-workers. Figure 23 shows predictions by Spalding and Rodi (S2) for the asymptotic plane jet, using their model equation for the turbulence length scale. Gosman *et al.* (G3) have documented the Spalding MTEN program in detail, and advocate its application to heat and mass transfer in recirculating flows. Readers should be aware that such programs are under continual development, but this should not prevent their use in engineering analysis.

IV. Mean Reynolds-Stress Closure

In order to compute the structure of the turbulence (i.e., the R_{ij}), one must employ the dynamical equations for the R_{ij} [Eq. (22)]. This has been the subject of considerable recent interest, though only a few computational experiments have been carried out, and a truly "universal" general theory has yet to be established. We can expect considerable future activity on this front.

The problem is again to set up a satisfactory closure structure for the unknown terms in the dynamical equations, here the equations for R_{ij}. Some variation in approach is already evident, and interesting debate on the choices is likely over the next several years.

Examination of Eq. (22) shows that the R_{ij} equations contain a pressure–strain-rate correlation term that vanishes in the contraction [Eq. (23)]. The effect of this term must therefore be to transfer energy conservatively between the three components R_{11}, R_{22}, and R_{33}, and it is generally believed that this transfer tends to produce isotropy in the turbulent motions. Modelings of this term should incorporate this feature. A plausible model of this term, supported somewhat by the data of Champagne *et al.* (C4) is

$$P_{ij} = \overline{\frac{p}{\rho}\left(\frac{\partial u_i}{\partial x_j} + \frac{\partial u_j}{\partial x_i}\right)} = C\frac{q}{l}\left(\frac{q^2}{3}\delta_{ij} - R_{ij}\right) \quad (49)$$

An objection to this model rests on the observation that the fluctuating pressure field is given by a Poisson equation

$$\frac{1}{\rho}\frac{\partial^2 p}{\partial x_i\, \partial x_i} = \frac{\partial^2}{\partial x_j\, \partial x_j}[\overline{u_i u_j} - u_i u_j - U_i u_j - U_j u_i] \quad (50)$$

This suggests that the P_{ij} model should contain terms arising from interactions between the mean and fluctuating velocities, and should somehow reflect the dependence of the pressure fluctuations on distant velocity fluctuations.

Rotta (R1) studied the Poisson equation in some detail, and proposed Eq. (49) for the portion of P_{ij} independent of the mean-fluctuation interaction. He also proposed the form of additional terms that would take these interactions into account. More recently, Daly and Harlow (D1) have attempted to include these effects in a complex closure approximation still in an experimental stage [see Eq. (77a)]. Other MRS closure calculations have all used Eq. (49). Some new suggestions are explored in Section V.

The pressure–velocity terms have been modeled in all MRS computations of which I am aware by extensions of the gradient diffusion model (6.8). Donaldson and Rosenbaum (D4) use

$$(1/\rho)\overline{pu_i} = -ql_\epsilon \, \partial R_{ik}/\partial x_k \tag{51}$$

Daly and Harlow (D1) use a similar expression with a more complex coefficient. Mellor and Herring (M2) suggest

$$(1/\rho)\overline{pu_i} = -\tfrac{1}{3}ql_\mathrm{P} \, \partial q^2/\partial x_i \tag{52}$$

Various forms of gradient diffusion models have been suggested for the triple velocity term. Donaldson and Rosenbaum use and Mellor and Herring accept

$$\overline{u_i u_j u_k} = -ql_\mathrm{d}(\partial R_{ij}/\partial x_k + \partial R_{jk}/\partial x_i + \partial R_{ki}/\partial x_j) \tag{53}$$

The Daly–Harlow representation may be cast as

$$\overline{u_i u_j u_k} = -C(l_\epsilon/q)(\partial R_{ij}/\partial x_m)R_{km} \tag{54}$$

If the objections to a gradient-diffusion approach are valid, one should presumably use an extension of Eq. (30). A possible large-eddy transport model is

$$\overline{u_i u_j u_k} = R_{i.}Q_k + R_{ik}Q_j + R_{jk}Q_i \tag{55}$$

The viscous terms have also been handled in different ways. Donaldson and Rosenbaum (D4) and Daly and Harlow (D1) use V_{ij} in the form Eq. (25). Following Glushko (G2), Donaldson and Rosenbaum take

$$D_{ij} \equiv \frac{1}{2}\overline{\frac{\partial u_i}{\partial x_k}\frac{\partial u_j}{\partial x_k}} = \frac{1}{2}\frac{R_{ij}}{l_\mathrm{D}^2} \tag{56}$$

Daly and Harlow put

$$D_{ij} = \frac{2D}{q^2}R_{ij} \tag{57}$$

and use another differential equation for $D \equiv D_{ii} = \mathfrak{D}/(2\nu)$. Mellor and

Herring, invoking arguments of local isotropy and using kinetic theory as a guide, propose using Eq. (26) with

$$\overline{u_i \sigma_{kj}} + \overline{u_j \tau_{ki}} = \frac{\partial R_{jk}}{\partial x_i} + \frac{\partial R_{ik}}{\partial x_j} + \frac{\partial R_{ij}}{\partial x_k} \tag{58}$$

and

$$\nu \overline{[\sigma_{jk}\, \partial u_i/\partial x_k + \sigma_{ik}\, \partial u_j/\partial x_k]} = \tfrac{2}{3}(q^3/l_e)\delta_{ij} \tag{59}$$

In order to complete the closure, the various length scales in the models above must be prescribed or related to the other independent variables through a differential equation. Daly and Harlow use a dynamical equation for $\mathfrak{D}$, derived exactly from the Navier–Stokes equations and then closed by assumptions. The $\mathfrak{D}$ equation will be discussed presently. Daly and Harlow are now considering the use of two length-scale equations for the dissipating and energy-containing eddies.

Mellor and Herring (M2) have given considerable thought to the MRS

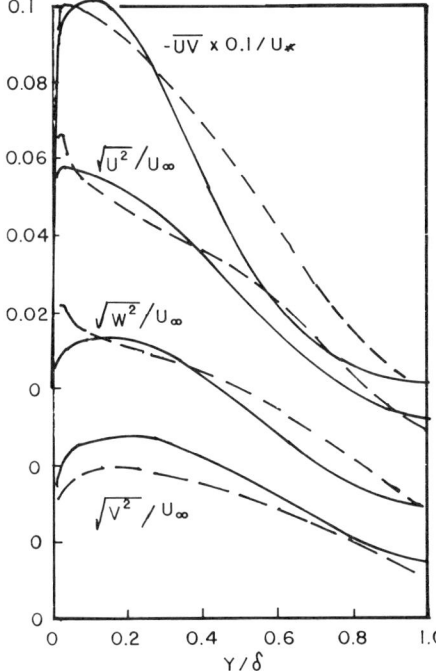

FIG. 24. Donaldson and Rosenbaum's MRS calculation for a flat plate boundary layer: — MRS; - - - - data.

closure, and show how the MTEN equations emerge from MRS equations if the turbulence structure is assumed to be nearly isotropic.

Calculations with MRS closure models have been carried out by Donaldson and Rosenbaum, by Daly and Harlow, and by Harlow and Romero (H4). Harlow and Romero used the model with moderate success to study the distortion of isotropic turbulent (see Section V). Donaldson and Rosenbaum considered plane turbulent boundary-layer flow in zero pressure gradient, and specified reasonable length-scale distributions for this calculation. Their prediction of the mean velocity profile is good (but not better than a good MVF or MTE calculation); their predicted turbulent stress distributions, shown in Fig. 24, are in substantial agreement with experiments. Daly and Harlow studied plane Poiseuille flow with a more complex model, obtaining less satisfactory results. Their model is not accurate near the wall, and is currently undergoing further extension and adjustment. Hirt (H7) gives a useful summary of the thinking behind ongoing developments in the Los Alamos group.

We now consider the "dissipation" transport equation. The dynamical equation for $\mathfrak{D}$ is derived by differentiating the momentum equation for u_i with respect to x_j, multiplying by $2\nu\, \partial u_i/\partial x_j$ and averaging. The result is

$$\frac{\partial \mathfrak{D}}{\partial t} + U_j \frac{\partial \mathfrak{D}}{\partial x_j} = \nu \frac{\partial^2 \mathfrak{D}}{\partial x_j\, \partial x_j} - 2\nu \left\{ \frac{\partial^2 U_j}{\partial x_i\, \partial x_k} \overline{\left(u_k \frac{\partial u_j}{\partial x_i} \right)} \right.$$

$$+ \frac{\partial U_j}{\partial x_k} \overline{\left(\frac{\partial u_j}{\partial x_i} \frac{\partial u_k}{\partial x_i} \right)} + \frac{\partial U_k}{\partial x_i} \overline{\left(\frac{\partial u_j}{\partial x_i} \frac{\partial u_j}{\partial x_k} \right)}$$

$$+ \overline{\frac{\partial u_j}{\partial x_i} \frac{\partial u_k}{\partial x_i} \frac{\partial u_j}{\partial x_k}} + \frac{1}{2} \frac{\partial}{\partial x_k} \overline{\left(u_k \frac{\partial u_j}{\partial x_i} \frac{\partial u_j}{\partial x_i} \right)}$$

$$\left. + \frac{\partial}{\partial x_j} \overline{\left(\frac{\partial u_j}{\partial x_i} \frac{\partial p}{\partial x_i} \right)} + \nu \overline{\left(\frac{\partial^2 u_j}{\partial x_k\, \partial x_i} \frac{\partial^2 u_j}{\partial x_k\, \partial x_i} \right)} \right\} \quad (60)$$

To obtain closure, one must propose models for all the terms between the braces on the right-hand side of Eq. (60). This requires a considerable amount of courage as well as insight; there is no direct experimental evidence about any of the terms, and one can really only conjecture as to their effect. Lumley (L3) has used a reasoned approach for the special case of homogeneous flows (Section V). Daly and Harlow, using qualitative ideas about the effect of each term, proposed as a model of Eq. (60):

$$\frac{\partial \mathfrak{D}}{\partial t} + U_j \frac{\partial \mathfrak{D}}{\partial X_j} = \frac{\partial}{\partial x_j} \left(\nu + \frac{2q^2}{\mathfrak{D}} R_{jk} \right) \frac{\partial \mathfrak{D}}{\partial x_k} - b_1 \frac{\mathfrak{D}}{q^2} R_{jk} \frac{\partial U_j}{\partial x_k} +$$

$$\frac{\mathfrak{D}}{q^2}\frac{\partial}{\partial x_j}\left(\frac{2q^2}{\mathfrak{D}}R_{jk}\frac{\partial q^2/2}{\partial x_k}\right) + b_2\frac{\partial}{\partial x_j}\left(q^2\frac{\partial R_{jk}}{\partial x_k}\right) - F\frac{\mathfrak{D}^2}{q^2} \quad (61)$$

Here b_1 and b_2 are "universal parameters, all with values near unity (or possibly equal to zero)," and F is a function of the turbulence Reynolds number. In effect they assume that $\mathfrak{D} = \epsilon$, through their treatment of the R_{ii} equation.

Hanjalic *et al.* (H1) propose a model of Eq. (60) which can be generalized as

$$\frac{\partial \mathfrak{D}}{\partial t} + U_j\frac{\partial \mathfrak{D}}{\partial x_j} = \frac{\partial}{\partial x_i}\left[(\nu + N_S\nu_T)\frac{\partial \mathfrak{D}}{\partial x_j}\right] + c_1\frac{\mathfrak{D}}{q^2}\nu_T S_{ij}S_{ij}$$
$$- c_2 f_2\frac{\mathfrak{D}^2}{q^2} + f_3\nu\nu_T\left(\frac{\partial^2 U_i}{\partial x_j \partial x_j}\frac{\partial^2 U_i}{\partial x_k \partial x_k}\right) \quad (62)$$

Here c_1 and c_2 are constants, and f_1, f_2, and f_3 are functions of the turbulence Reynolds number. They report "encouraging" results when this equation is used in an MTEN computational scheme. Clearly the use of such equations is presently quite experimental, and Eq. (62) is given here to illustrate the rather substantial differences in ideas as to how best to model Eq. (60).

For the special case of homogeneous flows at large turbulence Reynolds numbers, Eqs. (61) and (62) do have a common form:

$$\partial \mathfrak{D}/\partial t + U_j\, \partial \mathfrak{D}/\partial x_j = -C_1\mathfrak{D}^2/q^2 + C_2 \mathfrak{D}\mathcal{P}/q^2 \quad (63)$$

where $\mathcal{P}$ is the rate of production of turbulence energy. This form is probably quite adequate for homogeneous flows (Section V).

Lumley (L3) has studied the distortion of homogeneous turbulence by uniform strain using a limited MRS closure. In homogeneous flow, the R_{ij} equations become

$$\frac{\partial R_{ij}}{dt} = -R_{ik}\frac{\partial U_j}{\partial x_k} - R_{jk}\frac{\partial U_i}{\partial x_k} - \frac{1}{\rho}\overline{\left(u_j\frac{\partial p}{\partial x_i} + u_i\frac{\partial p}{\partial x_j}\right)} - 2\nu\overline{\left(\frac{\partial u_i}{\partial x_j}\frac{\partial u_i}{\partial x_j}\right)}$$
(64)

Lumley closes by taking

$$\nu\overline{\frac{\partial u_i}{\partial x_k}\frac{\partial u_j}{\partial x_k}} = \epsilon\frac{\delta_{ij}}{3} \quad (65)$$

$$-\frac{1}{\rho}\overline{\left(u_j\frac{\partial p}{\partial x_i} + u_i\frac{\partial p}{\partial x_j}\right)} = \frac{1}{T}\left(q^2\frac{\delta_{ij}}{3} - R_{ij}\right) \quad (66)$$

where T is a time scale of the turbulence [compare Eq. (49)]. He further assumes that the time scale is related to the dissipation rate by

$$T = C_1 q^2 / 2\epsilon \qquad (67)$$

which is equivalent to Eq. (27). The dissipation rate is in turn described by

$$d\epsilon/dt = -4\epsilon^2/q^2 \qquad (68)$$

as deduced by Lumley from scaling arguments based on Eq. (60) [compare Eq. (63)].

Lumley has solved the equation system for homogeneous shear, and compared the results with homogeneous strain and homogeneous shear experiments. Lumley's model predicts that the time scale T grows without bound, *so that homogeneous flows can never attain an equilibrium structure.* Champagne *et al.* (C4) experiments are consistent with Lumley's notion, but Lumley's model does not predict the observed structure very well. Some improvements on Lumley's model based on Eq. (63) are suggested in Section V.

It does seem clear that equilibrium is never obtained in homogeneous flows. In inhomogeneous flows the transport features apparently act to set the equilibrium structure. MTES methods really should not work in homogeneous flows, and we may well be suspicious of methods when "universal constants" are obtained by tests against such flows.

V. Opportunities and Outlook

A. New Ideas for Homogeneous Flows

It has become apparent, in preparing this review, that too little attention is being given to systematic development of the closure model. The approach has been to construct a comprehensive model, with numerous universal constants, and then to select these constants by optimizing the average fit to a number of selected flows. A more systematic approach would be to develop the closure model in a step-by-step approach, working gradually through a heirarchy of experimental flows.

In order to develop some feeling for what might be accomplished, the writer examined the following approximations to homogeneous flow:

(1) decay of isotropic turbulence (T1);
(2) return to isotropy in the absence of strain or shear (T2);
(3) development of structure under pure strain (T2);
(4) development of structure under pure shear (C4).

The starting point of this analysis was the dynamical equations for the turbulent kinetic energy and the dissipation. For homogeneous flows, these equations are

$$\frac{dq^2/2}{dt} = \mathcal{P} - \epsilon \tag{69}$$

and

$$\frac{d\epsilon}{dt} = -C_1 \frac{\epsilon^2}{q^2} + C_2 \frac{\mathcal{P}\epsilon}{q^2} \tag{70}$$

Here $\mathcal{P} = -R_{ij}S_{ij}$ is the rate of turbulence-energy production. Equation (69) is exact, and Eq. (70) is the form suggested by both the dissipation equations of Daly and Harlow (D1) and Hanjalic et al. (H1) and by the length-scale equation of Ng and Spalding (N2); [see Eq. (63)]. For his model of very weakly strained flows, Lumley (L3) developed the C_1 term with $C_1 = 4$ from first principles, and neglected the C_2 terms, in Eq. (68). Hanjalic et al. suggest $C_1 = 4$ and $C_2 = 3.2$. Ng and Spalding's empirical flow fitting is equivalent to $C_1 = 3.9$ and $C_2 = 3.3$. Daly and Harlow use $C_1 = 4$ and $C_2 = 2$.

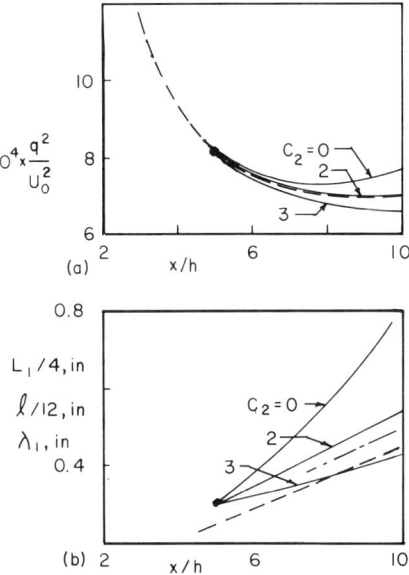

Fig. 25a. Turbulence energy in the (C4) flow-determination of C_2: ---- data.
Fig. 25b. Length scales in the (C4) flow. Note that $C_2 = 2$ models the observed length-scale percentage changes: ---- λ_1 data; ———— L_1 data.

If one considers the decay of homogeneous isotropic turbulence with zero strain for large turbulence Reynolds numbers, while modeling the dissipation by

$$\epsilon = q^3/l \qquad (71)$$

Eqs. (69) and (70) are found to produce

$$d(q^2/2)/dt = -q^3 l \qquad (72)$$

$$dl/dt = (C_1 - 3)q \qquad (73)$$

Now, experiments indicate that, for large time, $q^2 \sim t^{-1}$ and $l \sim t$. This requires $C_1 = 4$, which thus seems a clear choice.

To investigate C_2, we calculated the distribution of $\mathcal{P}$ from the data of (C4), and carefully determined an initial value for ϵ from the experimental q^2 distribution (taking the starting point at $x = 5$ ft in their experiments).

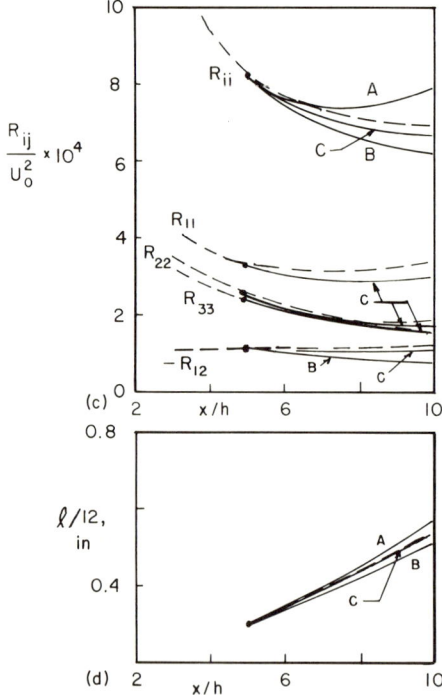

FIG. 25c. R_{ij} in the (C4) flow; all use (69) and (74). A-(77a) with $C_3 = 5$. B- (77c) with $C_4 = \frac{1}{2}$. C- (77d) with $C_4 = \frac{1}{2}$, $C_5 = \frac{1}{4}$: ----- data.

FIG. 25d. Dissipation-length scales in the (C4) flow. A, B, C as for Fig. 25c. Note that C reproduces the behavior deduced using (69) and (74): ----- from ϵ calculation.

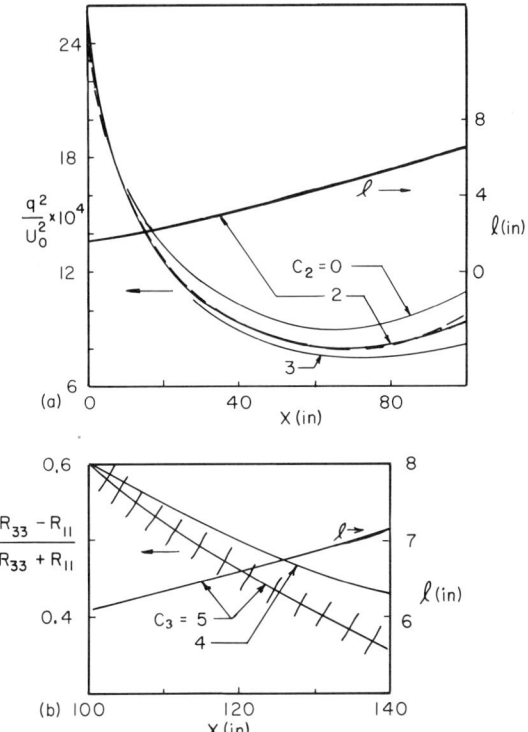

FIG. 26a. Turbulence energy in the (T2) flow-determination of C_2. Initial shape from exponential fit to Tucker and Reynolds' Fig. 6; ---- experiment.

FIG. 26b. Structure in the unstrained return-to-isotropy portion of the (T2) flow; determination of C_3. //// data range.

The differential equations (69) and (70) were then solved numerically for different values of C_2; $C_2 = 2$ is clearly preferred (Fig. 25a,b). A similar calculation was carried out for the (T3) flow (Fig. 26a), where $C_2 = 2$ also gives excellent agreement. Note that the predicted length-scale variations do model the integral-scale changes as measured (C4) (Fig. 25b). It therefore appears that a satisfactory model equation for the dissipation history in homogeneous flows is

$$\frac{d\epsilon}{dt} = -4\frac{\epsilon^2}{q^2} + 2\frac{\epsilon \mathcal{P}}{q^2} \tag{74}$$

Further, we considered the R_{ij} equations with the objective of obtaining a model that, with Eqs. (69) and (74), correctly predicts the measured

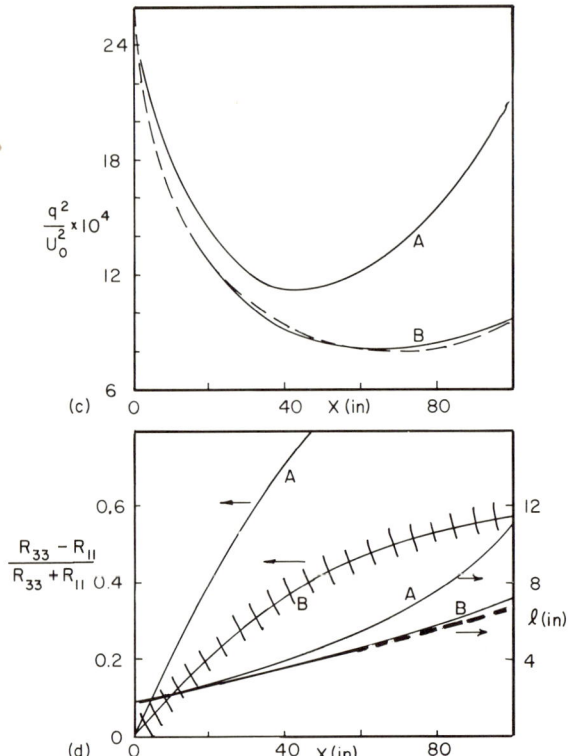

Fig. 26c. Turbulence energy in the (T2) flow. A- (77a) with $C_3 = 5$. B- (77c) with $C_4 = \frac{1}{4}$. - - - - experiment.

Fig. 26d. Structure in the straining region of the (T2) flow; A, B as for Fig. 26c. Note that B reproduces the length scale changes calculated using (69) and (74). - - - - l from ϵ equation; //// data range.

R_{ij}. Closure assumptions are required for the pressure–strain and dissipation terms. In all calculations we took Eqs. (26) and (59) in Eq. (22), and hence wrote

$$\frac{dR_{ij}}{dt} = -R_{ik}\frac{\partial U_j}{\partial x_k} - R_{jk}\frac{\partial U_i}{\partial x_k} + P_{ij} - \tfrac{2}{3}\epsilon\delta_{ij} \qquad (75a)$$

where the pressure–strain term is

$$P_{ij} = \overline{\frac{p}{\rho}\left(\frac{\partial u_i}{\partial x_j} + \frac{\partial u_j}{\partial x_i}\right)} \qquad (75b)$$

First

$$P_{ij} = C_3\epsilon(\tfrac{1}{3}\delta_{ij} - R_{ij}/q^2) \qquad (76)$$

was used. The first test was for the strain-free portion of the (T3) flow, where the structure is relaxing toward isotropy. Calculations showed that $C_3 = 5$ gives a good representation of the structure, energy, and production in this flow (Fig. 26b).

We then proceeded to try Eq. (76) with $C_3 = 5$ in the straining regions of the (T2) and (C4) flows, but were not satisfied with the energy predictions (see Figs. 25c,d, 26c,d). It appears that some alteration in either the P_{ij} or the dissipation terms is required, and we chose to experiment with the P_{ij}. A ground rule was that any proposed modification could not alter what has already been systematically established. The forms investigated were

$$P_{ij} = (5\epsilon + C_4 \mathcal{P}) \left(\tfrac{1}{3}\delta_{ij} - \frac{R_{ij}}{q^2} \right) \tag{77a}$$

$$P_{ij} = 5\epsilon \left(\tfrac{1}{3}\delta_{ij} - \frac{R_{ij}}{q^2} \right) + C_4 q^2 S_{ij} \tag{77b}$$

$$P_{ij} = 5\epsilon \left(\tfrac{1}{3}\delta_{ij} - \frac{R_{ij}}{q^2} \right) + C_4 (R_{ik}S_{kj} + R_{jk}S_{ki} + \tfrac{2}{3}\mathcal{P}\delta_{ij}) \equiv P_{ij}^* \tag{77c}$$

$$P_{ij} = P_{ij}^* + C_5 \left[R_{ik}\left(\frac{\partial U_j}{\partial x_k} - \frac{\partial U_k}{\partial x_j}\right) + R_{jk}\left(\frac{\partial U_i}{\partial x_k} - \frac{\partial U_k}{\partial x_i}\right) \right] \tag{77d}$$

We note that $P_{ii} = 0$ in each case. Equation (77a) is Daly and Harlow's form, with slightly different constants. Equations (77b)–(77d) are suggested by the notion that interactions between the mean strain rate and fluctuation fields contribute to the pressure fluctuations. The closures Eqs. (77a) and (77b) were unsatisfactory. For the (T2) flow, Eq. (77c) with $C_4 = \tfrac{1}{2}$ works very well (Fig. 26c,d), but it is not adequate for the (C4) flow (Fig. 25c,d). Equation (77d) reduces to Eq. (77c) for irrotational mean flow, i.e. the (T2) flow, and with $C_4 = \tfrac{1}{2}$ and $C_5 = \tfrac{1}{4}$, Eq. (77d) predicts the (C4) flow reasonably well (Fig. 25c,d).

It does seem clear that Eq. (76) is not adequate in flows with strain or shear. With the constants indicated, Eq. (77d) is

$$P_{ij} = 5\epsilon \left(\tfrac{1}{3}\delta_{ij} - \frac{R_{ij}}{q^2} \right) + \tfrac{1}{2}(R_{ik}S_{kj} + R_{jk}S_{ki} + \tfrac{2}{3}\mathcal{P}\delta_{ij})$$

$$+ \frac{1}{4}\left[R_{ik}\left(\frac{\partial U_j}{\partial x_k} - \frac{\partial U_k}{\partial x_j}\right) + R_{jk}\left(\frac{\partial U_i}{\partial x_k} - \frac{\partial U_k}{\partial x_i}\right) \right] \tag{78}$$

which is probably better. Further development is needed, and Eq. (78)

is offered here as an interim model.[3] However, it is not clear that Eq. (78) is a model of P_{ij}; it could just as well be a model for its complement [see Eq. (66)], as used by Lumley (L3) in Eq. (64)!

Rodi (private communication) pointed out that, with the constants as indicated above, Eq. (78) can be written as

$$P_{ij} = 5\epsilon \left(\tfrac{1}{3}\delta_{ij} - \frac{R_{ij}}{q^2} \right) - \tfrac{1}{2}(\mathcal{P}_{ij} - \tfrac{2}{3}\delta_{ij}\mathcal{P}) \qquad (79)$$

where $\mathcal{P}_{ij}$ is the production of the component ij,

$$\mathcal{P}_{ij} = -R_{ik}\, \partial U_j/\partial x_k - R_{jk}\, \partial U_i/\partial x_k \qquad (80)$$

This is appealing because the first part is proportional to the anisotropy of the structure and the second part to the anisotropy of the production.

B. Suggestions for the Future

It should not be long before simple boundary-layer flows are routinely handled in industry by MVF prediction methods. These methods are easy to use, require a minimum of input data, and give results which are usually adequate for engineering purposes. MTE methods will become increasingly important to both engineers and scientists, for they afford the possibility of including at least some important effects missed by MVF methods. The debate over the gradient-diffusion vs. large-eddy-transport closures will continue, and both methods will probably continue to be used with nearly equal success. MRS methods will be explored from the scientific side, but probably will not be used to any substantial degree in engineering work for some time to come.

Considerable effort is likely to be expended on the development of length-scale (or equivalent) equations, such as the dissipation equation discussed in Section IV. In this connection the two-point correlation

$$\mathcal{R}_{ij} = \overline{u_i(x)u_j(x+\xi)} \qquad (81)$$

could be used to advantage, either along the lines of Gawain and Pritchard (G1), or perhaps through a closure of its own dynamical equation (H6). This equation will of course involve six independent space variables, but by integration over the separations ξ these variables could be removed. Then, one might assume the form of $\mathcal{R}_{ij}$, say

$$\mathcal{R}_{ij} = R_{ij} \exp(-\xi_k/l_k) \qquad (82)$$

---

[3] J. L. Lumley (private communication) argues that the constants $C_1 - C_5$ should be functions of Reynolds number. The values suggested are probably most appropriate at large turbulence Reynolds numbers. See also Naot et al. (N3).

and carry out the integrations, thereby obtaining three additional differential equations of the transport type relating the integral scales l_i and the one-point correlations (turbulent stresses) R_{ij}. Experimental calculations along these lines would be most interesting.

The heavy computation approach (D3) might be used to test numerically the closure assumptions used in the simpler MRS and MTE models. It is hoped such computations will be documented in the future, with this use in mind.

The MRS closures will attract most interest for use wherever MTE methods fail. For example, in flows with rotation the Coriolis terms enter the R_{ij} equations, but drop out in the equation for $R_{ii} = q^2$. Therefore, an MRS method probably will be essential for including rotation effects, which are of considerable importance in many practical engineering and geophysical problems. Other effects that have not yet been adequately modeled and for which MRS methods may offer some hope include additive drag reduction, ultrahigh Reynolds numbers, separation, roughness, lateral and transverse curvature, and strong thermal processes that affect the hydrodynamic motions.

We might also see the complex closure models used as the basis for computationally simpler integral methods. The success of integral methods of this type at the Stanford conference should not be forgotten in the rush to use the full partial differential equations.

A disappointing aspect of the current status is that very little use has been made of the substantial advances made over the past decade in our understanding of the structure of turbulent shear flows. We know that large eddy structures dominate such flows; only MTES methods recognize this at all, and then not quantitatively; MTEN and MRS methods ignore it altogether. Indeed, the concept of a "transport theory" for turbulent correlations would seem antithetical to a large-eddy view. The wall region is known to be dominated by a particular correlatable structure. Also, the structure of the outer region of boundary layers has been extensively studied recently, and entrainment of nonturbulent fluid through the turbulent interface (superlayer) is known to be a critical process in turbulent shear flows. No real utilization of these two facts has been incorporated into any of the closure models. We know that the outer layer flow has a dual structure, intermittently consisting of turbulent and nonturbulent regions with considerably different character; yet all calculation methods are based on averages taken over long periods of time, averages that wash out this essential feature of the flow. Some believe a similar duality exists in the wall region; might not this also be incorporated?

In short, it seems that too much attention has been paid to the numerical aspects of the computations. Indeed, the difficulty of a first encounter with complex differencing schemes has made this necessary. But now we should begin a concerted effort to bring the new physical information into the turbulent flow computation methods, and we look for a better situation ten years hence.

Finally, for those who choose not to take up the computation game, some fresh thinking at the fundamental level may be fruitful. For example, what is it that we are working so hard to compute? What is the *operational definition of turbulence*?

ACKNOWLEDGMENTS

The Stanford TBLPC conference (T1) which contributed so substantially to this field had the following persons as its principal organizers. Executive Committee: M. Morkovin, G. Sovran, D. Coles; Host Committee: S. J. Kline, E. Hirst, W. C. Reynolds. Advisory Board: F. H. Clauser, H. W. Emmons, H. W. Liepmann, J. C. Rotta, I. Tani.

The author is indebted to several colleagues, cited at appropriate points in this report, who made their unpublished work available for reference. The author's study was supported in part by grants NASA-NgR-05-020-420 and NSF-GK-10034.

Nomenclature

A^+ Wall-layer thickness parameter [Eq. (9)]
a_{ij} Structure tensor [Eq. (32)]
$\mathfrak{D}$ Isotropic dissipation [Eq. (24d)]
K Pressure gradient parameter [Eq. (48)]
l Turbulence length scale
$\mathcal{P}$ Turbulence production $-R_{ij}S_{ij}$
P_{ij} Pressure-strain tensor [Eq. (49)]
P Mean pressure
p Fluctuation pressure
Pr_T Turbulence Prandtl number [Eq. (20)]
P_0^+ Pressure-gradient parameter [Eq. (10a)]
Q_i Large-eddy vector velocity scale
$q^2/2$ Turbulence kinetic energy density
Re Reynolds number
R_{ij} $\overline{u_i u_j}$, "Reynolds stress tensor"
S_{ij} Mean strain rate tensor [Eq. (2b)]
t Time

u_i Fluctuation velocity vector (u, v, w)
U_i Mean velocity vector (U, V, W)
u^* Friction velocity, $(\tau_w/\rho)^{1/2}$
V_0^+ Transpiration parameter [Eq. (10b)]
V_{ij} Viscous terms in R_{ij} equation [Eqs. (24a) and (25a)]
x_i Cartesian coordinate vector (x, y, z)
α Molecular thermal diffusivity
α_T Turbulent thermal diffusivity
ν Molecular kinematic viscosity
ν_T Turbulent kinematic viscosity
κ Karman constant
ρ Mass density
σ_{ij} See Eq. (26)
τ Shear stress
τ_w Wall shear stress
Θ Mean temperature
θ Fluctuation temperature
ϵ Dissipation of turbulence energy [Eq. (25)]

References

B1. Bradshaw, P., *J. Fluid Mech.* **46,** 417 (1971).
B2. Bradshaw, P. *et al.*, *Nat. Phys. Lab.*, *Aero Div. Rep.* **1182, 1217, 1271, 1286, 1287, 1288** (1966 et. seg.).
B3. Bradshaw, P., Ferriss, D. H., and Atwell, N. P., *J. Fluid Mech* **28,** 593 (1967).
B4. Busse, F. H., *J. Fluid Mech.* **41,** 219 (1970).
C1. Cebeci, T., "A model for Eddy-conductivity and Turbulent Prandtl Number," Rep. MDC-j0747/01. McDonnell Douglas Co., 1970.
C2. Cebeci, T., *Prepr., Int. Heat Transfer Conf., 4th* (1970).
C3. Cebeci, T., Smith, A. M. O., and Mosinskis, G., *AIAA Pap.* **69-687** (1969).
C4. Champagne, F. H., Harris, V. G., and Corrsin, S., *J. Fluid Mech.* **41,** 81 (1970).
D1. Daly, B. J., and Harlow, F. H., *Los Alamos Sci. Lab Prepr.* LA-DC-**11304** (1970).
D2. Davis, R. E., *J. Fluid Mech.* **42,** 721 (1970).
D3. Deardorff, J. W., *J. Fluid Mech.* **41,** 453 (1970).
D4. Donaldson, C. D., and Rosenbaum, H., "Calculation of Turbulent Shear Flows through Closure of the Reynolds Equations by Invariant Modeling," Rep. No. 127, Aero Res. Ass., Princeton University, Princeton New Jersey, 1968.
G1. Gawain, T. H., and Pritchard, J. W., *J. Comput. Phys.* **5,** 385 (1970).
G2. Glushko, G. S., *NASA Tech. Transl.* TTF-**10,080**; translation of *Izv. Akad. Nauk SSSR, Meckh.* **4,** 13 (1965).
G3. Gosman, A. D., Pun, W. M., Runchal, A. K., Spalding, D. B., and Wolfshtein, M., "Heat and Mass Transfer in Recirculating Flows." Academic Press, New York, 1969.
H1. Hanjalic, K., Jones, W. P., and Launder, B. E., "Some Notes on an Energy-dissipation Model of Turbulence" (internal report). Imperial Col., London, 1970.
H2. Harlow, F. H., and Nakayama, P. I., *Phys. Fluids* **10,** 2323 (1967).
H3. Harlow, F. H., and Nakayama, P. I., *Los Alamos Sci. Lab Prepr.* LA-DC-8635 (1969).
H4. Harlow, F. H., and Romero, N. C., "Turbulence Distortion in a Nonuniform Tunnel," Los Alamos Lab Rep. LA-4247. 1969.
H5. Herring, H. J., and Mellor, G. L., *NASA Contract. Rep.* NASA **CR-1444** (1968).
H6. Hinze, J. O., "Turbulence." McGraw-Hill, New York, 1959.
H7. Hirt, C. W., *Phys. Fluids* **12,** II-219 (1969).
H8. Howard, L. N., *J. Fluid Mech.* **17,** 405 (1963).
H9. Hussain, A. K. M. F., and Reynolds, W. C., "The Mechanics of Perturbation Wave in Turbulent Shear Flow," Rep. FM-6. Mech. Eng. Dept., Stanford University, Stanford, California. See, also, *J. Fluid Mech.* **41,** 241 (1970).
J1. Johnston, J. P., *J. Fluid Mech.* **42,** 823 (1970).
K1. Kasahara, A., "Simulation of the Earth's Atmosphere." Amer. Soc. Mech. Eng., New York, 1969.
K2. Kays, W. M. *et al.*, private communication; *ASME Pap.* **71-HT-44** (to appear in *J. Heat Transfer*).
K3. Kearney, D. *et al. Proc. Heat Transfer Fluid Mech. Inst.* (1970).
K4. Kendall, J. M., *J. Fluid Mech.* **41,** 259 (1970).
K5. Kline, S. J. *et al. J. Fluid Mech.* **30,** 741 (1967).
L1. Lighthill, M. J., *J. Fluid Mech.* **1,** 554 (1952).
L2. Loyd, R. J., Moffat, R. J., and Kays, W. M., "The Turbulent Boundary Layer on a Porous Plate; An Experimental Study of the Fluid Dynamics with Strong

Favorable Pressure Gradients and Blowing," Rep. HMT-13. Mech. Eng. Dept., Stanford University, Stanford, California, 1970.
L3. Lumley, J. L., *J. Fluid Mech.* **41,** 413 (1970).
L4. Lundgren, T. S., *Phys. Fluids* **10,** 969 (1967).
M1. Mellor, G. L., *AIAA J.* **5,** 1570 (1967).
M2. Mellor, G. L., and Herring, H. J., "A Study of Turbulent Boundary Layer Models," Parts I and II, Rep. SC-CR-70-6125. Sandia Lab., 1970.
N1. Nash, J. F., *J. Fluid Mech.* **37,** 625 (1970).
N2. Ng, K. H., and Spalding, D. B., "Some Applications of a Model of Turbulence for Boundary Layers Near Walls," Rep. B1/TN/14. Imperial Col., London, 1969.
N3. Naot, M.M., Shavit, M. M., and Wolfshtein, M., *Isr. J. Technol.* **8,** No. 3 (1970).
P1. Patankar, S. V., and Spalding, D. M., "Heat and Mass Transfer in Boundary Layers." Chem. Rubber Co., Cleveland, Ohio, 1967. See, also, *Int. J. Heat Mass Trans.* **10,** 1389 (1967).
R1. Rotta, J. C., *Phys.*, **129,** I; 547–752; **131,** II, 51–77 (1951).
R2. Russell, B., "A History of Western Philosophy." Allen & Unwin, London, 1961.
S1. Simpson, R. L., Whitten, D. G., and Moffat, R. J., *Int. J. Heat Mass Transfer* **13,** 125 (1970).
S2. Spalding, D. B., and Rodi, M. M., *Warme-Stuffubertragung* **3,** 85 (1970).
S3. Computation of Turbulent Boundary Layers-1968 AFOSR-IFP-Stanford Conference, Vols. 1 and 2. Thermosci. Div., Dept. Mech. Eng., Stanford University, Stanford, California.
S4. Stewart, R. H., *J. Fluid Mech.* **42,** 733 (1970).
T1. Townsend, A. A., "The Structure of Turbulent Shear Flow." Cambridge Univ. Press, London and New York, 1956.
T2. Tucker, H. J., and Reynolds, A. J., *J. Fluid Mech.* **32,** 657 (1968).

DRYING OF SOLID PARTICLES AND SHEETS

R. E. Peck and D. T. Wasan

Department of Chemical Engineering
Illinois Institute of Technology
Chicago, Illinois

I. Introduction	247
II. Estimation of Heat- and Mass-Transfer Coefficients.	248
A. Heat or Mass Transfer on the Surface of Drying Material	248
B. Analogy between Heat and Mass Transfer	250
C. Correlation for the Psychrometric Ratio	251
III. Moisture Movement through Porous Solids	252
A. Industrial Significance of Liquid Movement through Porous Solids	253
B. Theoretical Discussion	257
IV. Drying of Porous Solids—Batch Operations	258
A. General Introduction	258
B. Drier Design for Thin Materials	259
C. Drier Design for Packed-Bed Driers	273
V. Drying of Porous Solids—Continuous Operations	279
A. Rotary Driers	279
B. Tunnel Driers	288
VI. Summary	288
Nomenclature	289
References	290

I. Introduction

Drying, as the term is used in this chapter, is the unit operation of passing a gas over, or through the interstices of, a nonvolatile solid to remove adherent or loosely combined moisture by vaporizing it into the gas. The operation is sometimes called "air-drying of solids" because the carrier gas is frequently air. There is, however, no change in principle when some other gas is the carrier as, for example, safety requires if the "moisture" is a flammable solvent.

The chapter treats primarily the drying of porous materials in the form of sheets or slabs and of particles either in packed beds or freely flowing as in a rotary drier. Chief attention is given to information that has become available for the rational design of driers from fundamental principles. Those interested in more general aspects of the subject may be guided through the vast literature by the periodic reviews that have appeared in *Industrial and Engineering Chemistry* (M3, M4) and, in the case of the Russian literature, by Fulford's recent review (F3).

The treatment is divided into four sections. Section II deals with estimation of coefficients of heat transfer and of mass transfer. Because most, or all, of the latent heat of evaporation of the moisture is normally derived from the sensible heat of the carrier gas, our knowledge of the pertinent coefficients of heat transfer from the gas to the surface of the drying solid is summarized. A summary of the analogous mass-transfer coefficients records in condensed form gives our current knowledge of the means of estimating the rate of transport from the solid to the gas of the vapor evolved.

Section III is concerned with moisture movement through porous solids. The general theory of moisture distribution and the rate of moisture movement inside porous media is reviewed. The three theories of condensation—diffusion, capillarity, and vaporization—are discussed. The roles of various mechanisms causing liquid movement in solids are assessed.

The drying of porous solids in batch operations is discussed in Section IV. A general discussion of the drying of porous solids and sheets at ordinary temperatures is presented. A new model for the drying rate of porous solids in the falling-rate period is developed and tested with available data from the literature. Generalized charts are presented for sizing pan and packed-bed driers. Sample calculations are shown for evaluating drying schedules. Effects of humidity, air velocity, temperature, and shape on drying times are determined.

Section V deals with the drying of porous solids in continuous operations. The study of drying in rotary and tunnel dryers is presented based on the relationships derived from basic theory. The effect of the operating variables on drier performance is discussed. A suitable procedure is developed for sizing rotary and tunnel driers.

II. Estimation of Heat- and Mass-Transfer Coefficients

A. HEAT OR MASS TRANSFER ON THE SURFACE OF DRYING MATERIAL

1. *Forced Heat Convection in Laminar Flow of a Fluid Parallel to a Surface*

Pohlhausen (P4) in 1921 presented the direct solutions of the convection equations for the laminar boundary layer on the upstream portion of a flat plate placed edgewise in a stream of infinite extent. Only the case of steady-state heat flow was considered. His paper deals with two cases: (1) uniform surface temperature and (2) external perfect heat insulation of the plane plate.

For case (1) the following result is obtained:

$$N_{\text{Nu}_m} \equiv h_m L/k = 0.664 (LV_0/\nu)^{1/2} (\nu/\alpha_T)^{1/3} \qquad (1)$$

This equation is in good agreement with the result of direct measurements on air by Jakob and Dow (J2).

Pohlhausen's equation can be written in j-factor form:

$$j_{h_m} = (h_m L/k)(\nu/\alpha_T)^{-1/3}(Lv_0/\nu)^{-1} = 0.664(Lv_0/\nu)^{-1/2} = (h_m/c_p \rho v_0)(\nu/\alpha_T)^{2/3} \qquad (2)$$

$$j_{h_L} = (h_L L/k)(\nu/\alpha_T)^{-1/3}(Lv_0/\nu)^{-1} = 0.332(Lv_0/\nu)^{-1/2} \qquad (3)$$

where h_m is the average heat transfer coefficient for the whole plate and h_L is the local coefficient at distance L from the leading edge.

For air with $N_{\text{Pr}} = 0.71$, one finds

$$N_{\text{Nu}_m} = h_m L/k = 0.592 (Lv_0/\nu)^{1/2} \qquad (4)$$

Mass transfer j-factors j_M can be obtained by replacing the Prandtl number (ν/α_T) by the Schmidt number (ν/D_v), where D_v is the diffusivity of vapor.

Boundary-layer theory has been applied to solve the heat-transfer problem in forced convection laminar flow along a heated plate. The method is described in detail in numerous textbooks (E1, G5, S3). Some exact solutions and approximate solutions are also obtained (B2, S3).

2. *Forced Heat Convection in Turbulent Flow Parallel to a Plane Plate at Uniform Temperature*

This case is treated in the paper of Latzko (L1). Assuming that the plate is so thin that the leading edge does not affect the arriving stream, the following form is obtained:

$$h_m L/k = 0.0356 (v_0 L/\nu)^{0.8} (\nu/\alpha_T) \qquad (5)$$

Jakob and Dow (J1, J2) presented the following empirical equation from their experimental results:

$$h_m L_{\text{tot}}/k_f = 0.028 (v_0 L_{\text{tot}}/\nu_0)^{0.80} [1 + 0.40 (L_{\text{st}}/L_{\text{tot}})^{2.75}] \qquad (6)$$

where L_{tot} is the total length of the surface in the flow direction; L_{st} is the length of the hydrodynamic (unheated) starting section of the plate;

$N_{Re} = v_0 L_{tot}/\nu_0$; $N_{Num} = h_{mtot}/k_f$; k_f is the thermal conductivity of the air at the mean film temperature, $T_f = \frac{1}{2}(T_s + T_b)$, where T_s is the surface temperature and T_b is the bulk fluid temperature.

3. *Heat Transfer Coefficients for Forced Convection through Packed Beds*

Experimental data on heat and mass transfer in packed beds have been empirically correlated in *j*-factor form:

$$j_h = 0.91(G/a\mu_f\psi)^{-0.51}\psi \qquad (G/a\mu_f\psi < 50) \qquad (7)$$

$$j_h = 0.61(G/a\mu_f\psi)^{-0.41}\psi \qquad (G/a\mu_f\psi > 50) \qquad (8)$$

Here $j_h = (h_{loc}/C_{pb}G)(C_p\mu/k)_f^{2/3}$ and h_{loc} is the local heat-transfer coefficient; G is the superficial mass velocity; the subscript f denotes properties evaluated at the film temperature $T_f = \frac{1}{2}(T_s + T_b)$. Here T_s refers to the surface temperature and T_b to the bulk fluid temperature. The quantity ψ is an empirical coefficient that depends on the particle shape, e.g., $\psi = 1.0$ for spheres and $\psi = 0.91$ for cylinders. Values of ψ for other shapes are tabulated elsewhere (B2, B3).

B. ANALOGY BETWEEN HEAT AND MASS TRANSFER

The analogy between heat and mass transfer makes it possible to obtain the solutions of many mass-transfer problems at low mass-transfer rates from the results of heat transfer in similar situations.

The analogy among heat, mass, and momentum transfer was studied and a more generalized presentation of the data on heat- and mass-transfer coefficients was made (E1, G1, G5, S3, W3).

Wasan and Wilke (W3) developed the following expressions for the Sherwood number (i.e., the analog in mass transfer of the Nusselt number) from a cylindrical surface placed parallel to the stream in a turbulent flow:

$$\frac{h_m d}{k} = \frac{(f/2)(dv_0/\nu)(\nu/D_v)}{1 + f/2[13.0(\nu/D_v)^{0.80} - 13.0]} \qquad \text{for} \quad 0.2 \le \nu/D_v \le 2 \qquad (9)$$

and,

$$\frac{h_m d}{k} = \frac{(f/2)(dv_0/\nu)(\nu/D_v)}{1 + f/2[13.8(\nu/D_v)^{0.71} - 13.0]} \qquad \text{for} \quad 2 \le \nu/D_v \le 100 \qquad (10)$$

Heat-transfer Nusselt numbers can be obtained by replacing the Schmidt numbers by Prandtl numbers in the above expressions. These expressions for Nusselt numbers are based on the difference between wall and average concentration or temperature.

More recently, Kauh et al. (K3) calculated the local heat- and mass-transfer coefficients for a flat plate by extending the turbulent analogy theory treatment of Kestin and Persen (K4) and Gardner and Kestin (G1), and by employing the Spalding (S13) equation for the law of the wall.

Rai (R1) has presented experimental data for convective mass transfer from flat and cylindrical surfaces in axisymmetric flow. Based on the analogy between mass- and heat-transfer processes, Rai's experimental values may be employed in estimating heat-transfer coefficients.

C. Correlation for the Psychrometric Ratio

Based on the approximation that the effect of mass transfer on heat transfer is negligible, the correlation for the psychrometric ratio is obtained by several investigators (B1, C7, H3, L5, W4).

By modification of the analogy between transfer of momentum, heat,

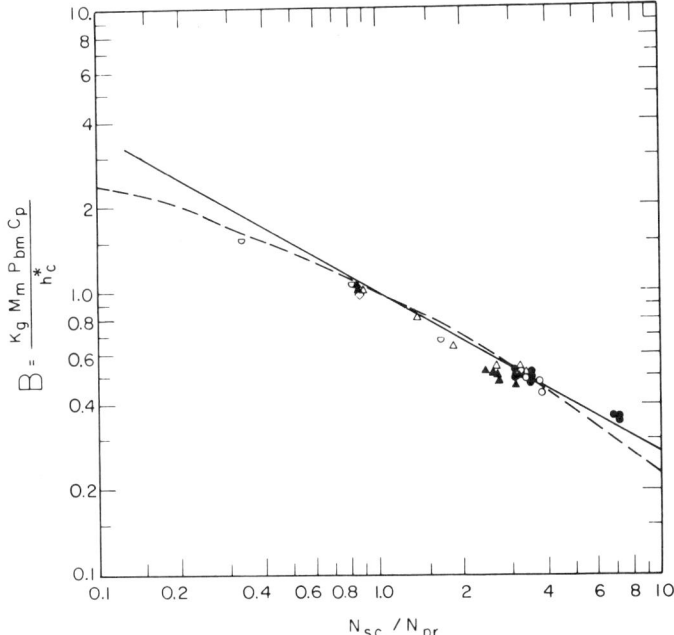

Fig. 1. Comparison of correlations for psychrometric ratios with the experimental data. — Henry and Epstein; – – – Wilke and Wasan; .. Arnold; ○ Bedington and Drew; .. Dropkin; □ Lynch and Wilke; ▲ Mark; ● Henry and Epstein.

and mass as developed for turbulent flow in pipes, Wasan and Wilke (W3) derived a new correlation for the psychrometric ratio for the wet cylinder in turbulent gas streams. Their proposed expression is

$$B = k_g P_{Bm} M_m C_p / h_c{}^* = [1 + 0.7(N_{Pr}^{0.77} - 1)]/[1 + 0.7(N_{Sc}^{0.77} - 1)] \tag{11}$$

More recently, Henry and Epstein (H3) reported data on psychrometric ratios for cylinders in cross-flow and spheres. Their experimental results, which covered the Lewis number range of 3.7 to 7.2, were identical for spheres and cylinders. Furthermore, their results could best be represented by an equation similar to that of Bedingfield and Drew (B1) as follows:

$$B = (N_{sc}/N_{pr})^{-0.567} \tag{12}$$

Figure 1 compares the experimental data of various investigators with Eqs. (11) and (12). Equation (11) compares more favorably with the experimental results at lower values of Schmidt to Prandtl number ratios, whereas Eq. (12) compares more favorably at higher values. It is evident that further work is needed to derive a theoretical relationship which encompasses the entire range of the experimental results. Furthermore, practically no data exist for the psychrometric ratio at high temperatures and high humidities.

III. Moisture Movement through Porous Solids

There are three general theories for interpretation of moisture distribution and rate of moisture movement inside porous solids. These theories can be listed as (1) diffusion theory, (2) capillary theory, and (3) vaporization–condensation theory.

Any model can be used to predict some drying data if enough parameters are used. However, no theory has been able to predict drying times where most of the resistance is in the solid phase.

To be valid, each theory has its specific requirements. The major factors that decide the mode of liquid movement through porous solids are the nature of the liquid, the structure of the solid, the concentration of liquid, and the temperature and pressure of the system.

The movement of liquid in a solid is caused by various forces. The possible mechanisms discussed in the literature are summarized as follows:

(1) liquid diffusion due to differences in moisture concentrations;
(2) liquid movement due to capillary forces;
(3) vapor diffusion in partly air-filled pores, due to differences in partial pressures;

(4) liquid or vapor flow due to differences in total pressure, generated by external pressure, capillarity, shrinkage, or high temperature inside the moist material;

(5) liquid moisture flow due to gravity.

A. Industrial Significance of Liquid Movement through Porous Solids

1. *The Evaporation–Condensation Theory*

The evaporation–condensation theory assumes that moisture migration occurs entirely in the gaseous phase (in the pores). The work of Gurr et al. (G6), Hutcheon (H7), and Kuzmak and Sereda (K9) showed that in a system subjected to a temperature gradient, this assumption is correct, even at relatively high pore saturation. The evaporation–condensation mechanism was utilized by Henry (H4), Cassie et al. (C3), Walker (W2), and others to describe the movement of moisture in beds of textile materials. On the basis of an examination of the liquid–vapor equilibria and of the mass- and energy-transfer processes in porous systems, Harmathy (H2) developed a theory for simultaneous mass and heat transfer during the pendular state of a drying material. He presented a set of differential equations which, when solved with appropriate initial and boundary conditions, yield the complete moisture content, temperature, and pressure history of the system. Further work would be needed to check the validity of this model.

2. *The Diffusion Theory*

The movement of moisture by diffusion was explicitly proposed as the principal flow mechanism by Lewis (L3), Tuttle (T7), Sherwood (S6), Newman (N1), Childs (C6), Kamei (K1), and many others. Sherwood (S6) assumed that the mechanism by which water travels from the interior to the surface is diffusion, and that the major resistance was in the solid for the falling-rate period. Newman (N1) applied the diffusion equations to the drying of solids of various shapes where the surface evaporation rates must be considered as well as the fluid flow within the solid. Sherwood (S7) recommended this procedure, but states its limitations, namely, that water movement is produced by capillarity and not by diffusion, and that the apparent success of the diffusional equations for calculating the drying time of such substances as wood and clay lies in the fact that these calculations were made by integration methods that compensate for the errors caused by assuming the wrong distributions obtained from diffusion equations. The diffusion equations apply only when the capillary tension produced in flow varies directly with the unsaturation of the solid, the body of the solid has a uniform composition, and the gravitational effect is negligible (C4). Possibly, such a situation is approximated in fine fibrous structure or even fine clays.

Buckingham (B4), Gardner (G2, G3), and Wilsdon (W5) attempted the difficult problem of applying variable diffusivities to the diffusion equations by employing a capillary potential instead of a concentration potential.

The evaluation of the parameters in terms of the concentrations has proved extremely difficult, and application of the resultant equations to decreasing drying rates has not been satisfactory.

Krischer (K7) gave a comprehensive treatise on the subject. The second falling-rate period starts when the moisture content at the surface reaches the equilibrium value. In both periods the vapor diffuses from the interior of the solid and the rate of vapor diffusion determines the rate of drying.

In Japan there seems to be a general trend to use diffusion theory to describe the moisture flow in a drying solid. Wakabayashi (W1) calculated the moisture distribution in clay during drying. He used the following diffusion equation to express the moisture movement:

$$\partial c'/\partial t = \partial(D_s \, \partial c'/\partial x)/\partial x$$

where c' is moisture concentration, t is drying time, x is the distance the moisture moves through clay, and D_s is moisture diffusivity. D_s varies with the internal moisture concentration. The above nonlinear differential equation was solved using the experimental values of the moisture diffusion coefficients of several kinds of clay. The validity of this treatment was attested by the good agreement between the calculated values and the experimental data for the moisture distribution in the clay. Any theory can be used to correlate data for limited materials if most of the drying variables are held constant.

In most cases, diffusion equations have been applied to calculate the moisture distribution without regard to the applicability or limitation of such equations. Hougen et al. (H6) pointed out the limitations of diffusion equations in accounting for the liquid movement in solids during drying, and by comparison with the experimental data they established that, for porous substances, flow is caused by capillarity rather than diffusion. Also, in cases where diffusion does play a role, integration of the diffusion equations is not available to account for the nonuniform initial moisture distribution, for cases where shrinkage occurs, or where the diffusivities are variable. By comparing the diffusion theory results with drying data obtained for clays, soap, paper pulp, and sand, they proved that the diffusion equations did not apply for these materials for various reasons.

3. The Capillary Theory

Ceaglske and Hougen (C4) showed that the movement of moisture in sand is controlled entirely by capillarity and gravity, and that diffusion is not involved. Diffusion equations cannot be made to apply by using variable diffusivity values.

Water held in the interstices of solids, as liquid covering the surface and as free water in cell cavities, is subject to movement by gravity and capillarity, provided passageways for the continuity of flow are present. Water flow due to a capillarity applies to water not held in solution and to all water above the fiber saturation point (as in textiles, paper, and leather) and to all water above the equilibrium moisture concentration at atmospheric saturation as in fine powers and granular solids, such as paint, pigments, minerals, clays, soil, and sand (H6).

Water vapor may be removed by vapor diffusion through the solid, provided an adequate temperature gradient is established by heating, which creates a vapor pressure gradient toward the surface. Vaporization and vapor diffusion may be applied to any solid where heating takes place at one surface and drying from the other, and also where water is isolated between granules of the solid.

Ceaglske and Hougen (C4) showed that the water may actually move in the sand toward a region of high concentration from a region of low concentration, provided the high-concentration region possesses a finer pore structure. For example, when a layer of fine sand is placed upon a layer of coarse sand with drying taking place from the surface of the fine sand, the course sand dries out more rapidly than the fine sand. The water concentration in the fine sand becomes higher than that in the coarse sand, and the water actually flows in the direction of the higher concentration. For the same water content, the capillary force in the fine sand is much greater than in the coarse sand. This behavior is entirely inconsistent with the principles of diffusion, but can easily be explained by the behavior of capillary forces.

The flow and distribution of water resulting from absorption, drainage, or evaporation was demonstrated by the extensive work of Haines (H1). He started with the behavior of small spherical particles of uniform diameter packed together as densely as possible. The intricate geometry of this situation was first solved by the classical work of Slitcher (S10) in 1898. Haines states that the moisture distribution in an unsaturated granular solid was determined by the suction produced by interfacial tension. He discussed both increasing and decreasing water contents and explained why, for a given suction, the water content was much greater during the decreasing stages.

Richards (R2, R3) and Klausner and Kraft (K6) developed mathematical relations for flow of liquids in a capillary model. Their equations have not been of much value in predicting internal resistance in drying.

Sherwood and Comings (S8) also believed that moisture movement in the drying of granular materials is caused by capillarity. Water evaporates from the small miniscuses exposed at the surface of granular materials. The small curvature of these miniscuses exerts sufficient capillary pull to draw water through any passage ending in air–water interfaces with larger curvatures. The water drawn to the surface is replaced by air. When the water from within has been drawn into passages affording small miniscuses, the capillary pull toward the surface ceases and a small amount of evaporation from the surface results in their retreat. When this condition is reached, drying is greatly retarded.

Leverett (L4) used the capillarity behavior of the components of a liquid mixture to explain the static vertical distribution of fluids of different densities in porous media.

Corben and Newitt (C8) showed that the drying characteristics of moist porous granular material are consistent with the capillary theory of moisture movement. The difference in the form of the drying curves are primarily due to the capillary action of the pores. The rate of drying during the constant-rate period is higher for porous than nonporous materials.

Akbar and Goerling (A2) presented a theoretical interpretation of shrinkage and moisture flow during the drying of gel and paste-type substances.

Ksenzhek (K8) developed a model for a porous body by representing it as a cube with regularly spaced intersecting capillaries. Liquid under pressure penetrates against the capillary forces. He showed that small pores block large ones, and this accounts for the fact that the liquid is not uniformly distributed throughout the solid. At low pressure, liquid penetrates only to a limited depth.

Markin (M1) made an extensive study of capillary equilibrium in porous solids, and formulated a model of a porous medium in which the

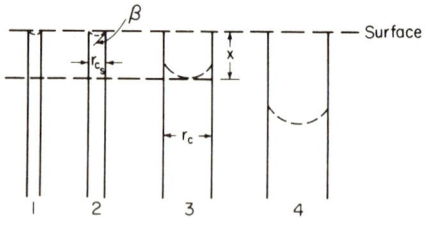

FIG. 2. Capillary model.

pore diameter changes along the length of a pore. This model also assumed Y-type pore intersections. A method of cycles was proposed for calculating the probability of filling a pore with a gas, and the equations for all the cycles were derived.

In a survey of recent Soviet research on the drying of solids, Fulford (F3) outlined the importance of internal moisture migration and suggested various approaches to determine the rate of moisture movement. The results of Soviet research showed that there seems to be a trend to applying the capillarity concept to describe moisture flow during the drying. The approach is very similar to that described in this section.

B. Theoretical Discussion

In drying a solid, the water may be retained as a hydrate or as liquid water in the capillaries. When water is removed, water will flow from the large capillaries into the smaller capillaries. This phenomenon can best be explained by use of a model.

Consider a system composed of cylindrical capillaries which can be considered as extending in the three directions. (See Fig. 2.)

Let r be the capillary radius. If the height of the meniscus is neglected, x is the distance from the surface of the solid to the liquid surface in a given capillary. There will be a certain radius (r_{cs}) of capillary which will just bring the water to the surface against gravity. The pressure below the surface at r_{cs} is $\pi - (2\sigma \cos \beta)/r_{cs}$ where π is atmospheric pressure and β is the contact angle between the liquid surface and the solid surface. Since all capillaries are assumed interconnected by capillaries in the other two directions, this pressure is constant under the surface at all capillaries smaller than r_{cs}. In the case of smaller capillaries the radius of curvature is greater than the radius of these capillaries. Also the pressure down a distance x is constant for all full capillaries. At the level x this pressure will be $\pi - (2\sigma \cos \beta)/r_c$.

The drop in pressure in the water from x to 0 would be

$$2\sigma \cos \beta [(1/r_{cs}) - (1/r_c)] \tag{13}$$

where r_c is the largest capillary that is full of water at the point x. A force balance would give

$$(x\rho g/g_c) \cos \alpha + \Delta p_f = 2\sigma \cos \beta [(1/r_{cs}) - (1/r_c)] \tag{14}$$

where ρ is the density, Δp_f is the friction drop in the capillaries due to flow of water, and α is the angle between the surface and the horizontal.

If ground or any solid is broken up so that r_c and r_{cs} are increased in size, the potential for flow as given by the right-hand side of the equation (sometimes called suction pressure) is reduced. A decrease in drying rate will result.

Most solids do not fit the above model, but all porous material have regions of high area-to-volume ratios. The net result is for water to move from regions of small area to regions of large areas. Mathematically, the system can be represented by capillaries of equivalent radii.

It is difficult to measure area-to-volume ratios inside a solid. With the present state of knowledge it is necessary to use empirical relations for friction drop or suction pressure inside the solid.

For high water concentrations in many materials, the friction term and gravity can be neglected. If gravity and friction are small enough to be neglected (all capillaries are either full or empty), $r_c = r_{cs}$ and there is very little moisture gradient in the solid.

IV. Drying of Porous Solids—Batch Operations

A. GENERAL INTRODUCTION

Toei *et al.* (T5, T6) believe that when the moisture content on the surface becomes less than the critical moisture content, the first falling-rate period starts and the evaporation occurs at the interior of the solid. The second falling-rate period starts when the moisture content at the surface reaches the equilibrium value. The evaporating plane retreats into the solid and dried-up zone begins to grow from the surface into the solid. The dried zone retains the equilibrium moisture content.

Kauh (K2, P2) assumed the applicability of the diffusion equation to describe the movement of moisture in the solid. He also took into consideration the surface temperature change from the time the solid is placed in the dryer until the time of completion of the drying process. He showed that the area for mass transfer at the surface was proportional to the $\frac{2}{3}$ power of the moisture concentration for balsa wood. The results show that the internal resistance in the case of thin solids is very small.

The analysis of Arzan *et al.* (A3) in the falling-rate period is based on a two-region moving-boundary model. At a given time there exists a submerged interface, parallel to the surface of the porous medium and gas stream, from which all evaporation takes place. As drying proceeds, the evaporative surface recedes into the porous medium, dividing the latter into two regions. The region above the surface contains solid and vapor, whereas the region below contains solid, liquid, and vapor. Each region is characterized by an effective thermal conductivity and effective molar diffusivity. One-dimensional heat and mass transfer is assumed.

Lester and Bartlett (L2) presented a theory of drying which helps to explain the form of the drying curve but it is of little use without a great deal of experimental data.

A survey of recent Soviet research on the drying of solids, made by Fulford (F3), shows that their approach to the drying process is more or less the same as discussed here.

Experimental data indicate that no model is of much value for most materials if there is a major resistance to drying in the solid phase. If the internal resistance is small, almost any theory is satisfactory. As an example, data on drying 10-cm thick material were correlated by Toei (T5, T6) and by Sheth (S9). Sheth correlated the data by assuming no resistance to flow in the solid. If no resistance to flow was assumed, the correlation was a little better than when using Toei's (T5, T6) model.

B. Drier Design for Thin Materials

When drying thin material, there is little effect on the drying time by friction due to flow in the capillaries. In many of these cases the effect of gravity is negligible and there will be no moisture gradient within the solid.

Heat transfer can be through the gas or by conduction from solid surfaces. The case that will be discussed here is when the solid resistance term can be neglected.

In the falling-rate period of drying, the area for mass transfer is smaller than the area for heat transfer since solid material will account for some of the area. The three basic relations are

$$dW/d\theta = -k_g A (P_s - P_R) \qquad (15)$$

where W gives pounds of water per pound of dry solid, k_g is the mass transfer coefficient, A is the area of water on surface per pound dry solid (in general it will be a function of W at the surface), P_s is vapor pressure of water on the wet surface, P_R is the partial pressure of water in air in the drier, and θ is time. If the sensible heat of the solid is neglected,

$$dW/d\theta = -(1/\lambda) h A_0 (T_R - T_s) \qquad (16)$$

where λ is the latent heat of evaporation, h is the total heat transfer coefficient, A_0 is the area for heat transfer per pound of solid, T_R is temperature in the drier, and T_s is temperature at the surface. Since there is no internal resistance to drying in this core, all capillaries of radius greater than r_{cs} are empty [see Eq. (13)]. As soon as dry capillaries start reducing the area for mass transfer, the rate of drying will start to decrease; this point will be called W_c, the critical water concentration.

If we take a basis of one square foot of surface area, there will be an area of these full capillaries which can be designated as some length squared (L^2). Capillaries in the other two directions are being emptied down to r_{cs} at the same time as those at the surface. The total volume of water is thus reduced to L^3. Let $L = 1$ when the total surface is available for mass transfer of water and we are at the critical water concentration (W_c). Let W be the water content at any time. Thus

$$W/W_c = L^3/1 \tag{17}$$

The area ratio exposed to drying at this time is given by

$$A/A_0 = L^2/1 \tag{18}$$

or,

$$A/A_0 = (W/W_c)^{2/3} \tag{19}$$

No material would fit the above capillary model, but most solids have regions where the surface-area-to-volume ratio inside the solid is greater than other regions. The water will migrate from regions of low area to regions of high area-to-volume ratios. This model can be approximated by equivalent capillaries of radius r_c and the water will be held in a restricted volume or cell.

Consider the case of drying a unit area and volume. If all the regions that have water were to be assembled in one place, the particle would have part wet area and part dry area. The wet area could stay wet after assembly and no new area need be created. The final area could be taken as L^2 per particle. In the assembly process water would also be contracted in the third direction so that the volume of water would be L^3. The area for mass transfer would be proportional to L^2 while the W/W_c would be proportional to L^3 where L was 1 at W_c.

When a material is in the falling-rate period, areas of dry surface and wet surface are assumed to be distributed over the complete area. With this assumption T_s is the surface temperature. In a case where the leading edge dries faster than the rest of the material there is a big variation in T_s. In this analysis T_s is an average surface temperature;

$$P_s = P_{R_s} \exp[-18\lambda(T_R - T_s)/RT_R T_s] \tag{20}$$

P_{R_s} is the vapor pressure of water at T_R and P_s is the vapor pressure of water at temperature T_s. Equation (15) becomes

$$\begin{aligned} dW/d\theta &= -k_g A (P_s - P_R) \\ &= -k_g A \{P_{R_s} \exp[-18\lambda(T_R - T_s)/RT_R T_s] - P_R\} \end{aligned} \tag{21}$$

Equation (19) can be substituted into Eq. (21):

$$dW/d\theta = -k_g A_0 (W/W_c)^{2/3} \{P_{R_s} \exp[-18\lambda(T_R - T_s)/RT_R T_s] - P_R\} \quad (22)$$

Equation (22) can be written in dimensionless form as

$$dC'/dF = -G'(C')^{2/3} \{\exp[-\phi(T_R/T_s - 1)] - S\} \quad (23)$$

and Eq. (16) becomes

$$dC'/dF = -(1 - T_s/T_R) \quad (24)$$

where

$$C' = W/W_c \quad (25)$$

$$F = \theta h T_R A_0 / \lambda W_c \quad (26)$$

$$G' = k_g P_{R_s} \lambda / h T_R \quad (27)$$

$$\phi = 18\lambda/RT_R \quad (28)$$

and

$$S = P_R/P_{R_s} \quad (29)$$

T_s/T_R can be eliminated from Eq. (23) with the use of Eq. (24) and C' can be evaluated as a function of F where G', ϕ, and S are parameters.

Approximate solutions can be obtained which will satisfy many drying problems. G' is a function of T_R and the psychrometric ratio. The psychrometric ratio can be assumed constant for an approximate solution.

1. *Generalized Plots*

Equation (23) contains G', S, and ϕ as parameters. $G'(k_g \lambda P_{R_s}/hT_R)$ is a function of T_R and the psychrometric ratio. The psychrometric ratio can be calculated and is fairly constant. Therefore G' is a function of T_R only (P_{r_s} being the vapor pressure at T_R). ϕ is a function of T_R and S is a function of T_R and T_W. In Figs. 3 and 4, moisture concentration and time, in dimensionless coordinates, have been plotted with T_R and $T_R - T_W$ as parameters. Given T_R, T_W, and $\theta h T_R A_0/\lambda W_c$, one can determine the drying schedule by using these plots. A_0/W_c is the area of heat transfer per pound of water at the critical water content.

The data from about 50 runs on balsa wood were fed into the computer and the best value of the exponent of W/W_c was calculated. The figure which resulted was 0.6, which checks the theoretical value.

The correlation that is presented used the following critical moisture concentrations for all drying conditions. For balsa wood, $W_c = 3.16$ lb water/lb dry solid. For clay, $W_c = 0.3$ gm/gm dry solid.

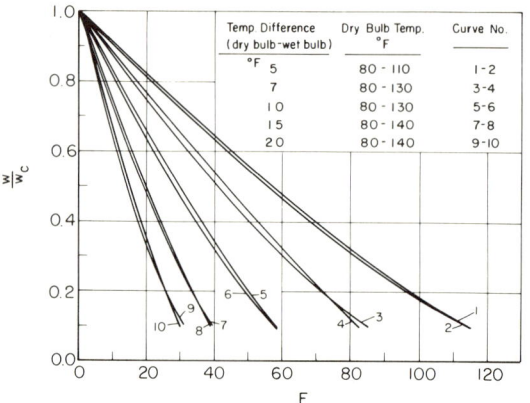

Fig. 3. Generalized drying curves.

Many runs were taken using various welding-rod fluxes. This flux had many compositions and densities. For the most dense (2.16 gm/cm³), W_c was taken as 0.055 gm/gm dry solid. For the least dense (1.35 gm/cm³), W_c was taken as 0.017. Most of the electrode samples were taken directly from the production line and there was almost no constant-rate data. In these cases the rate may remain constant due to an increase in solids temperature and the sample may not have a completely wet surface.

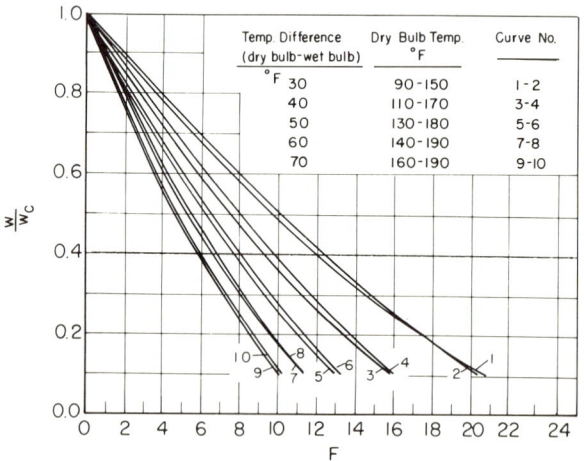

Fig. 4. Generalized drying curves.

Since no constant-rate data were available for the rods, the initial moisture was taken as the "critical water content." The theory deviated as much as 20% from experiment for this data. Because the errors involved were small, no attempt was made to obtain better values of the critical moisture content in this case.

Values of h are best obtained from the constant-rate data but may also be computed from any heat- or mass-transfer data.

The correlation presented in Figs. 3 and 4 represent the data from about 500 drying experiments. These experiments were on balsa wood, clay, and about 30 different welding rod coatings. The original data are available in theses (A1, C5, G4, K2, S16).

2. *Comparison of Theoretical and Experimental Drying Curves*

Experimental data from various sources (C5, K2, G4, S16) were taken for comparison. Kauh (K2) determined the drying schedules for balsa wood slabs of various thicknesses ($\frac{1}{8}, \frac{1}{4}, \frac{3}{8}$ in.) at different wind velocities (100–124 ft/min). It was not possible to apply boundary-layer theory to calculate heat- and mass-transfer coefficients because the length of the slabs was not recorded.

Figures 5–7, respectively, show the effect of humidity and slab thickness, in addition to the comparison of theoretical and experimental drying curves on the drying time. It is interesting to note that in the final stages of drying, the deviation is greater. The reason for this fact is that this stage of drying

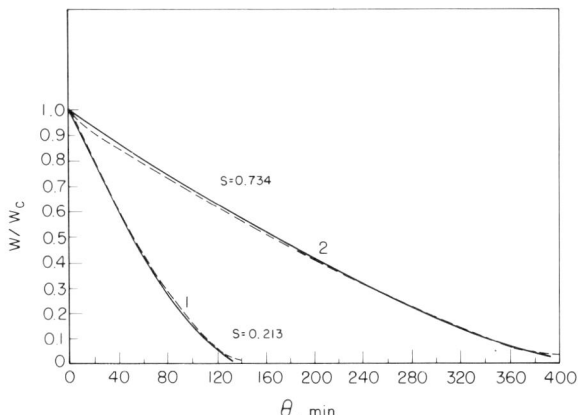

FIG. 5. Effect of humidity on drying time. - - - Experimental (K2); — theoretical; thickness $\frac{1}{2}$ in.; (1) run no. 11; (2) run no. 14.

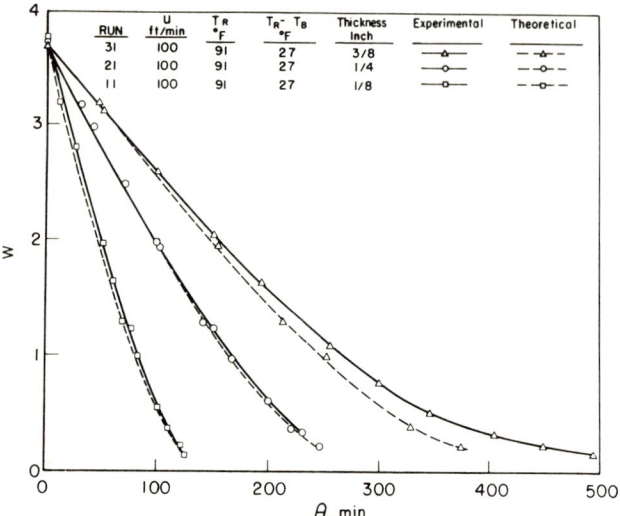

Fig. 6. Effect of thickness on drying time.

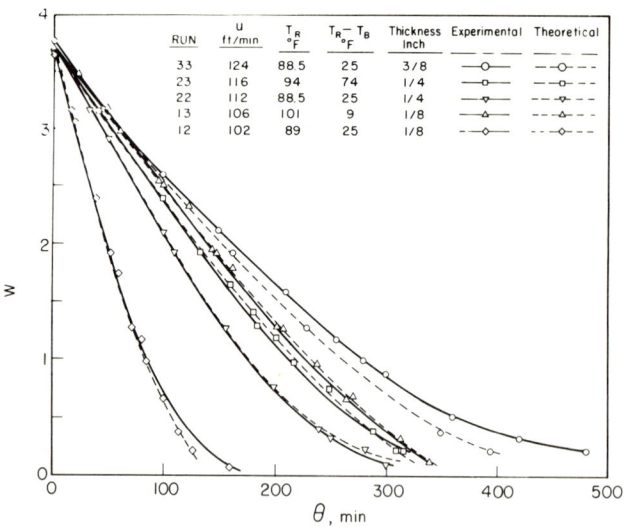

Fig. 7. Effect of humidity and thickness on drying time.

is in the second falling-rate period and there is appreciable resistance to flow in the fine capillaries, whereas in the present model internal resistance is neglected. It is also interesting to note that the theoretical drying time is less than the experimental drying time, and that the difference between the two grows as the thickness of the slab increases (Fig. 7). Therefore, an internal resistance exists, and it increases with the increase in slab thickness.

Examination of the deviation (18% for $\frac{3}{8}$-in. thickness) indicates that the internal resistance can be neglected for a thickness of $\frac{3}{8}$ in. or less.

Cheng (C5) studied the relation of the drying rate to drying conditions (balsa wood slabs were used for experiment). His data were utilized to show (Fig. 8) the effect of air velocity on the drying rate, and also to show the comparison between theoretical and experimental curves. Data from 15 runs were taken for comparison, but to avoid overcrowding of the curves, only three of these are shown. The value of the critical moisture content used was the same as determined by Kauh. Heat- and mass-transfer

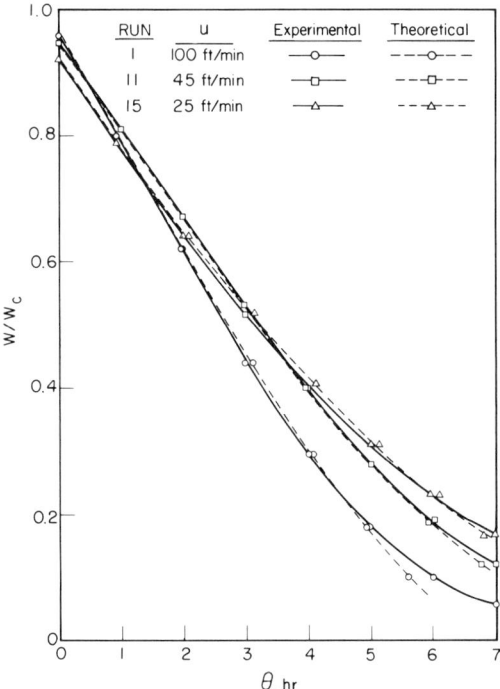

FIG. 8. Effect of air velocity on drying time.

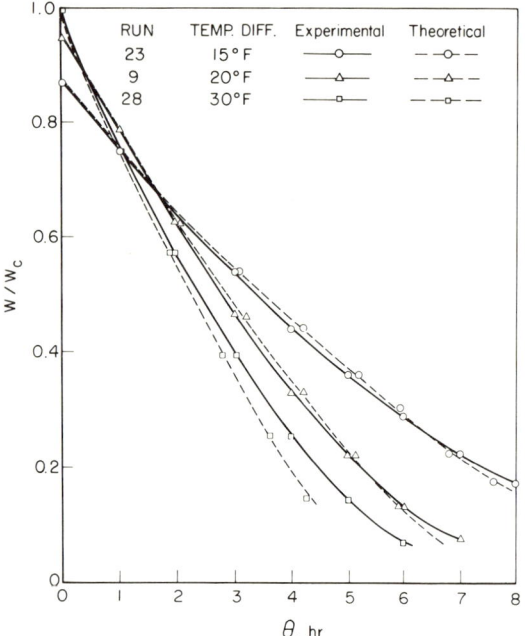

FIG. 9. Effect of initial moisture and humidity on drying time.

coefficients were determined from boundary-layer theory. (For details, see sample calculations.) The maximum deviation in the theoretical and experimental time in the first falling-rate period is about 5%.

This deviation is reasonable, considering the fact that the theoretical relations used to calculate heat- and mass-transfer coefficients are correct only within about 15%.

Figure 9 shows the effect of initial water concentration and humidity on the drying time.

Swanson (S16) took data on balsa wood, clay, and rutabaga slabs in order to determine the contribution of the film resistance. Comparison of theoretical and experimental curves for balsa wood shows (Fig. 10) a close agreement (within 18%) but for clay (which shrinks a little) deviation increases (up to 24%) due to the increase in internal resistance (Fig. 11). Although Eq. (38) is not very sensitive to critical moisture content, lack of its knowledge is another reason for a greater deviation in the case of clay. Data on rutabagas do not show good agreement because of shrinkage and solvent in the water.

Garud's (G4) data on the drying of welding electrodes show agreement within 15%, (Fig. 12) although the critical moisture content was not known accurately. Whenever data were not sufficient to calculate heat- and mass-transfer coefficients by boundary-layer theory, initial drying rate data was used for the purpose.

3. *Sample Calculations*

Cheng (C5) has recorded the following data in run 1:

Dry bulb temperature	120-F
Wet bulb temperature	100-F
Air velocity (u)	100 ft/min
Length of the slab (L)	4 in.
Width of slab	2 in.
Critical moisture content for balsa wood	3.16 lb/lb dry solid
Slab thickness (taped edges)	0.24 in.
Density of the solid	9.25 lb/ft^3
Latent heat (λ) (110-F)	1031 BTU/lb

Calculate the time to dry from 3.06 to 0.32 lb water/lb dry solid (a) using the generalized plots and (b) using the computer solution. Evaluating properties at 110°F in Eq. (4) gives

$$N_{\text{Re}} = \frac{Lu\rho}{\mu} = \frac{(100)(4)(0.070)}{(60)(12)(0.0188)(0.000672)} = 3070$$

$$h_m = 0.592 \frac{(0.016)(12)}{4} (N_{\text{Re}})^{1/2} = 1.7$$

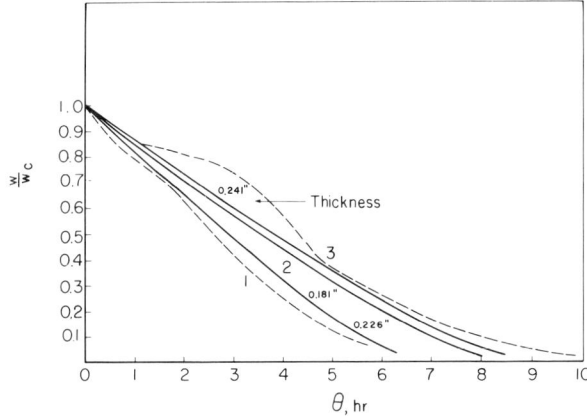

Fig. 10. Effect of thickness on drying time. ---- Experimental (S16); — theoretical; $H_r = 0.613$.

The average value of

$$h_r = \frac{0.17 \times 10^{-8} e(T_1^4 - T_2^4)}{T_1 - T_2}$$

$$= 0.17 \times 10^{-8} e(580^2 + 570^2)(580 + 570) = 1.29e$$

If e is about 0.8, then $h_r = 1.0$ and $h = 2.7$.

$$F = \theta h T_R A_0 / \lambda W_c$$

$$= (2.7)(580)(2)(12)\theta/1031(3.16)(9.25)(0.24) = 5.2\theta$$

(a) Time to dry from 3.06 to 0.32 lb water would be C' from 0.97 to 0.10. At $\Delta T = 20°$ and using curve 9-10, Fig. 3, $\Delta F = 29$ and $\theta = 29/5.2 = 5.6$ hr.

(b) Computer solution

$$J_{hm} = J_D = N_{Num}/N_{Re}(N_{Pr})^{1/3} = N_{Sh}/N_{Re}(N_{Sc})^{1/3}$$

$$N_{Num} = N_{Sh}(N_{Pr}/N_{sc})^{1/3}$$

Evaluate the properties at 110°F

$$h_m L/k = (k_g L R T/D)(C_p \rho_g D/k)^{1/3}$$

$$(k_g/h_m) = (D/kRT)(C_p \rho_g D/k)^{-1/3}$$

$$D = 1.09 \quad \text{ft}^2/\text{sec} = 0.282 \text{ cm}^2/\text{sec}$$

as $C = P/RT$ and $\partial C/\partial \theta = D \, \partial^2 C/\partial X^2 = (D/RT) \, \partial^2 P/\partial X^2$ and

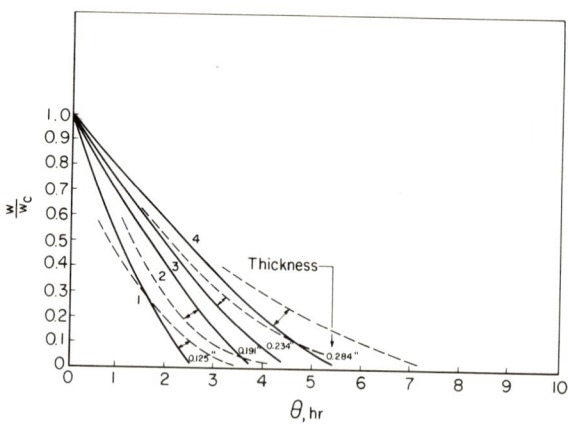

FIG. 11. Drying curves for clay. ---- Experimental (S16); — theoretical; $H_r = 0.345$.

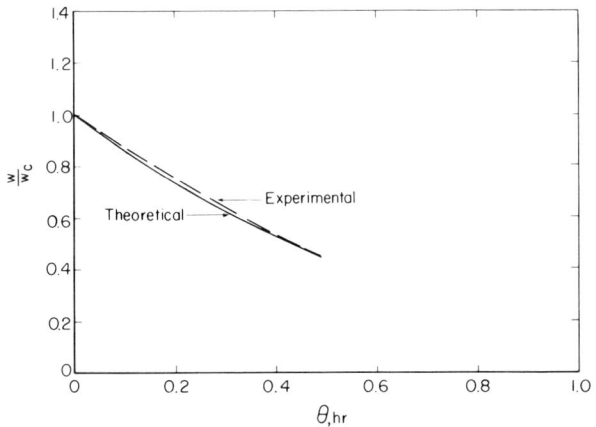

FIG. 12. Drying curves for welding flux.

$$D/RT = 1.09(18)(492)/(359)(1)(570) = 0.0474 \text{ lb/ft hr atm}$$
$$k = 0.0162 \text{ BTU/hr ft °F}$$
$$C_p = (0.24) \text{ BTU/lb °F}$$
$$\rho_g = 0.070 \text{ lb/ft}^3$$
$$k_g/h_m = [(0.0474)/(.0162)][(0.24)(.070)(1.091)/0.0162]^{-1/3}$$
$$= 2.93/1.04 = 2.9 \text{ °F lb/BTU atm}$$
$$h_m = 1.7; \quad k_g = 2.9(1.7) = 5.0; \quad h \cong 2.7$$
$$P_{R_s} = 1.693 \text{ PSIA}$$
$$G' = (k_g/h)(\lambda P_{R_s}/T_R) = 5(1031)(1.693)/(2.7)(570)(14.7) = 0.38$$
$$\phi = 18\lambda/RT_R = (18)(1031)/(2)(570) = 16.3$$

The dew point for the air in the drier is 96°F so the vapor pressure is 0.841

$$S = P_R/P_{R_s} = 0.841/1.693 = 0.50$$

The above three parameters can be used for the solution to Eqs. (23) and (24) and W/W_c will be solved in terms of F.

Since $F = 5.2\theta$, W/W_c can be obtained as a function of θ.

The computer solution gives $\theta = 5.7$ hr, while the experimental value was 5.9 hr (Fig. 8, run 1).

4. Effect of Air Velocity on Drying Times

In Cheng's run 1 as given above, the time to dry from 3.06 to 0.309 was 6 hr (C5). In run 15 the air velocity was 25 ft/sec and the initial water content was 2.90 lb/lb dry solid. All other conditions are the same as in run 1.

Estimate the time to dry this sample to a water content of 0.2 lb/lb dry solid. In run 1 C' went from 0.97 to 0.10 in 6 hr and Fig. 3 gives a value of ΔF of 29 where

$$\Delta F = \theta h T_R A_0 / \lambda W_c$$

$$h = \frac{\lambda W_c F}{\theta T_R A_0} = \frac{(1031)(3.16)(29)(9.25)(0.24)}{(6)(580)(2)(12)} = 2.5$$

$h_m = 1.5$ for run 1 if $h_r = 1.0$. For run 15, $h_m = 1.5(25/100)^{1/2} = 0.75$. $h = 1.75$ if $h_r = 1.0$. $F = 5.2 \times (1.75/2.5)\theta = 3.64\theta$. W/W_c goes from $2.9/3.16 = 0.92$ to $0.2/3.16 = 0.06$. $\Delta F = 33 - 2 = 31$ (Fig. 3, curve 9–10). $\theta = 31/3.64 = 8.5$ hr. The experimental time for the sample to reach 0.2 lb was 8.9 hr. Figure 13 shows the effect of air velocity on drying times in three runs for Cheng's data.

5. Effect of Humidity and Thickness on Drying Time

Sample Calculation. A $\frac{1}{8}$-in. piece of balsa wood dried to 0.316 lb water/lb dry solid in 1.65 hr. The dry bulb was at 91°F and the wet bulb was at 64°F^{K2} (run 11, Fig. 6).

A similar piece of balsa wood (K2) (run 23) with twice the thickness was dried at the same air velocity. The dry bulb was at 94°F and the wet bulb was at 74°F. Calculate the time for the $\frac{1}{4}$-in. wood to dry to 0.316 lb if both samples started at 3.16 lb water/lb solid (critical moisture).

An interpolation on Figs. 3 and 4 gives a value of $F = 23.3$ at $C = 0.1$ for $T_W = 64$ and $T_R = 91°F$.

$$F = \theta h T_R A_0 / \lambda W_c = (h T_R / \lambda W_c) \theta A_0$$

since the only variables are F, θ, and A_0 for these two runs.

$$(h T_R A_0 / \lambda W_c)_{\text{run 11}} = 23.3/1.65 = 14.1$$

and

$$(h T_R A_0 / \lambda W_c)_{\text{run 23}} = 14.1/2 = 7.0$$

At $T_R = 94$ and $T_W = 74$ curve 9 of Fig. 1 gives $F = 30$ at $C' = 0.1$

$$\theta = 30/7.0 = 4.2 \text{ hr}$$

The experimental value for $C' = 0.1$ in run 23, Fig. 7, was 4.8 hr.

When drying many materials there is a concentration of any solute at

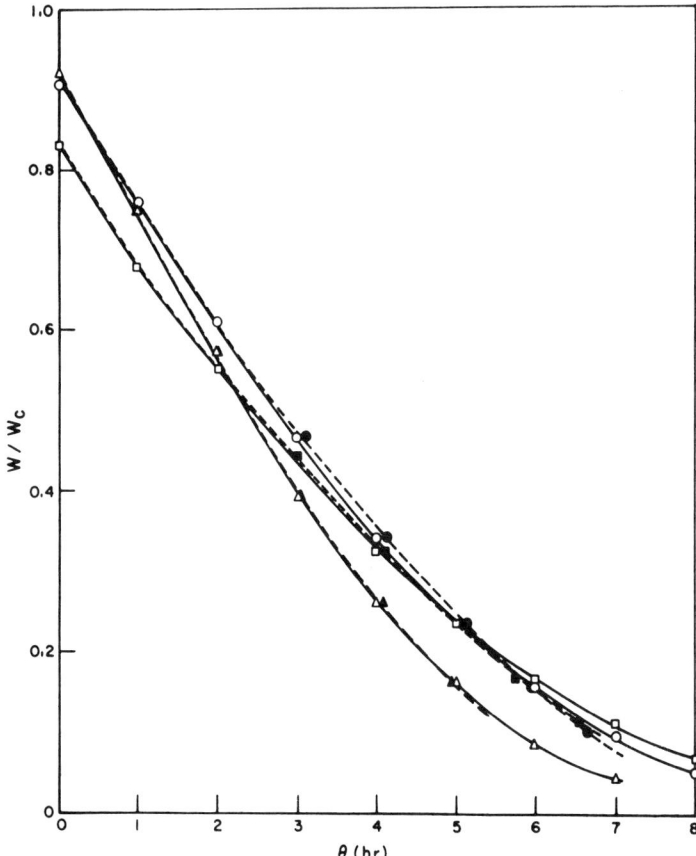

Fig. 13. Effect of air velocity and initial moisture on drying time.

Run	u (ft/min)	Experimental	Theoretical
2	96	—△—	-- ▲ --
10	50	—○—	-- ● --
14	31	—□—	-- ■ --

the surface and the vapor pressure of the solution is reduced. At low moisture concentrations the predicted times will be shorter than the actual time of drying. The higher the concentration of solids the greater will be this error. An examination of the data shows this effect. At high humidities this effect is more important than at low humidities.

When drying most food products such as rutabagas (S16) or macaroni, the solutions contain a great many soluble materials so that Eq. (17) would not be accurate. In drying welding rod (G4) the data indicate some of this error, and data on food products were even less accurate.

If water moves to the surface by capillarity, there cannot be any diffusion of water out from a receding water surface. The vapor pressure at the surface would be greater or equal to the water on the receding surface.

6. Evaluation of Parameters from Constant-Rate Data

In run 11, Fig. (18), the dry bulb temperature was 91°F; the wet bulb temperature was 64°F; the slab thickness $\frac{1}{8}$-in. (balsa wood); and the dry density was 9.25 lb dry wood ft.^{3} Calculate the time to dry to 0.316 from 3.16 lb water/lb dry solid.

$$\text{rate of drying} = 0.039 \frac{\text{lb water}}{\text{min lb dry solid}} \quad \text{(constant rate)}$$

$$W_c = 3.16 \text{ lb/lb dry solid}$$

$$A_0 = \frac{12}{\frac{1}{16} 9.25} = 20.8 \text{ ft}^2/\text{lb dry solid}$$

$$h = \frac{(0.039)(\lambda)}{A_0 \Delta T} = \frac{(0.039)(1057)(60)}{20.8(27)} = 4.4 \frac{\text{BTU}}{\text{hr ft}^2 \text{ °F}}$$

$$k_g = \frac{(0.039)}{A_0 \Delta P} = \frac{(0.039)(14.7)(60)}{20.8(0.30-0.15)} = 11.0 \frac{\text{lb}}{\text{ft}^2 \text{ hr atm}}$$

λ in the parameters G', ϕ, and F needs to be evaluated at $(91 + 64)/2 = 78°F$.

$$F = \frac{h T_R A_0 \theta}{\lambda W_c} = \frac{(4.4)(551)(20.8)\theta}{(1050)(3.16)} = 15.2\theta$$

$$G' = \frac{k_g}{h} \frac{\lambda P_{R_s}}{T_R} = \frac{11.0}{4.4} \frac{0.72}{14.7} \frac{1050}{551} = 0.23$$

$$\phi = \frac{18\lambda}{RT_R} = \frac{18(1050)}{(2)(551)} = 17.1$$

$$P_R = 0.15$$

$$S = \frac{0.153}{0.72} = 0.213$$

The above parameters can be used in Eqs. (23) and (24) to obtain a solution to the drying times. At $W/W_c = 0.1$ the answer is 1.5 hr, while the experimental value was 1.65 hr.

7. *Resistance in Both Phases*

Where there is appreciable resistance in the gas and solid phase, it is necessary to formulate a mathematical relation for the movement of water in this phase. The usual assumption for this case is that the moisture movement is proportional to the water gradient, or,

$$\partial W/\partial \theta = D_S \, \partial^2 W/\partial X^2 \tag{30}$$

Generalized plots for the drying times for thick material have been presented by Peck and Kauh (P2). Where the resistance in the solid is small compared to the surface resistance, Eq. (30) could give a fair answer.

Any model can be used to predict some drying data if enough parameters are used. However, no theory has been able to predict drying times where most of the resistance is in the solid phase.

8. *Conclusions*

Basic theory of drying for thin material under constant drying conditions can be solved by the use of Figs. 3 and 4. In an actual drier these drying conditions usually change. An approximate solution can be obtained by using the above figures in a stepwise fashion. However, a computer program for the solution is available for thin materials (A1) and for cases where there is some internal resistance to the flow of water (K2).

C. Drier Design for Packed-Bed Driers

In packed-bed drying, air is passed through a bed of wet material. The present work is applicable to a fixed bed with a constant moisture content at the initial time.

The air enters the bed at a given temperature and humidity. The temperature of the air falls and the humidity of the air increases as it passes up through the bed. The solids at the air entrance dry faster than solids at the air exit end of the drier.

Thodos *et al.* (T1–T4) presented data and calculated heat- and mass-transfer coefficients for packed beds.

Myklestad (M7) developed equations to predict the moisture content of granular material in fixed packed beds as a function of time and location.

Experimental data on the substance being dried or a similar substance are required to evaluate transfer parameters before the equations can be used. The theory was developed for materials drying in four stages: one stage at constant rate, one at increasing rate, and two at falling rate.

Kirkwood and Mitchell (K5) investigated the drying of porous ceramic granules, coke, and brewer's spent grain in packed beds by using factorial experiments to determine whether interactions exist between the effects of the operating variables on the drying time. Empirical expressions, specific for porous ceramic granules, coke, and brewer's spent grain, are presented and extensive experimental data for the drying of these materials are presented.

Sato (S2) presents complete and extensive experimental conditions and results for the drying of porous Al_2O_3 spheres in the constant rate, first falling rate and second falling rate periods.

1. *Theoretical Calculations*

A water balance around a small element gives

$$\rho_s(\partial W/\partial\theta)_x = G(\partial\bar{H}/\partial x)_\theta \tag{31}$$

if small terms are neglected. However,

$$\bar{H} = \tfrac{18}{29}[P_R/(\pi - P_R)] \tag{32}$$

or

$$(\partial\bar{H}/\partial x)_\theta = \tfrac{18}{29}(1 - P_R)^{-2}(\partial P_R/\partial x)_\theta \tag{33}$$

when π is 1 atm. Substituting Eq. (33) in (31) gives

$$\rho_s(\partial W/\partial\theta)_x = [18G/29(1 - P_R)^2](\partial P_R/\partial x)_\theta \tag{34}$$

A heat balance around a small element gives

$$GC_p(\partial T_R/\partial x)_\theta = -(\partial W/\partial\theta)_x \rho_s \lambda \tag{35}$$

The expressions for heat and mass transfer between the air and solid are

$$(\partial W/\partial\theta)_x = -k_g A(P_s - P_R) \tag{36}$$

and

$$(\partial W/\partial\theta)_x = -h A_0(T_R - T_s) \tag{37}$$

Substituting $A = A_0(W_c/W)^{2/3}$ for $W < W_c$ in Eq. (36),

$$(\partial W/\partial\theta)_x = -k_g A_0(W/W_c)^{2/3}(P_s - P_R) \tag{38}$$

The vapor pressure for water is given by the relation

$$P_s = P_{R_s} \exp[-(18\lambda/RT_R)(T_R/T_s - 1)] \tag{39}$$

A solution to Eqs. (34), (35), and (37)–(39) will give the complete history of the moisture distribution in the drying bed.

Values of heat and mass transfer from the air to the solids can be determined from correlations in the literature (T3). A computer program for the solution of the above equations was prepared by Max (M2).

2. *Comparison of Theoretical and Experimental Drying Curves*

The extensive experimental data given by Kirkwood and Mitchell (K5) for the drying of porous cylinderical ceramic granules and by Sato (S2) for the drying of porous spherical Al_2O_3 were used to test the ability of the model to predict drying curves. Experiments representing sharply different drying conditions were selected to test the model under a wide variety of drying conditions. In the experiments chosen, the bed height ranged from 0.5–3.94 in., the air temperature ranged from 120 to 212°F, the total drying time ranged from 15 min to 7 hr, the humidity of the inlet air ranged from 0.008 to 0.0298 lb H_2O/lb BDA, and the initial moisture content of the bed ranged from 0.315 to 0.70 lb H_2O/lb BDS. The porous cylindrical ceramic granules dried were $\frac{1}{4}$ in. in diameter by $\frac{1}{4}$ in. in length. Sato dried a variety of sizes of Al_2O_3 spheres, however the experiments using the 0.27- and 0.10-in. diameter size were chosen.

The comparison of the theory with the experimental data is shown in Figs. 14–16. Maximum deviation between the theoretical and experimental

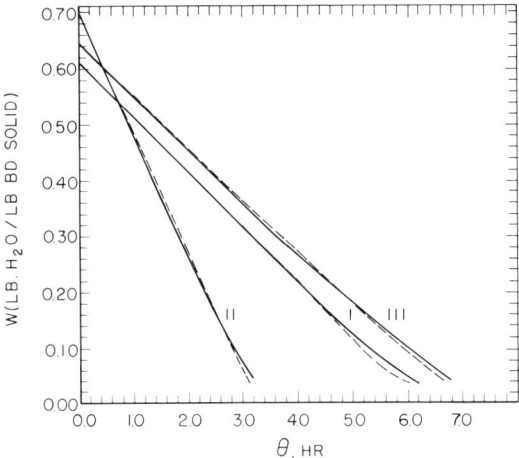

Fig. 14. Drying curves for porous Al_2O_3 spheres in a 3.94-in. bed. (I) Experiment no. 19, 0.10 in. spheres; (II) experiment no. 13, (III) experiment 14, 0.27-in. spheres. — Theoretical; – – – – experimental.

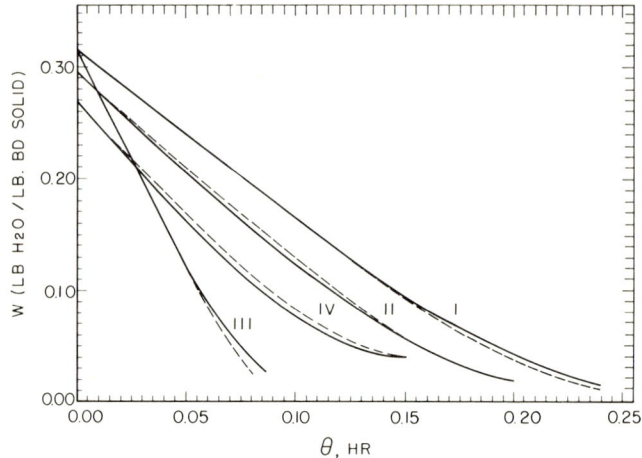

FIG. 15. Drying curves for ceramic granules and coke. (I) Experiment D3W3L1G1; (II) D3W1L1G1; (III) D3W1L1G3: 0.25-in. porous cylindrical ceramic granules in a 0.50-in. bed. (IV) Experiment D3W1L3G3: $\frac{3}{8}$ to $\frac{1}{4}$ mesh porous granular domestic grade coke in a 4.0-in. bed. — Theoretical; - - - - experimental.

drying curves was less than 5%. Deviation is defined as the ratio of the difference between the experimental and theoretical drying times to the experimental drying times. Figures 17 and 18 show bed moisture profiles for the beginning, midpoint, and end of the drying periods. Elements near the

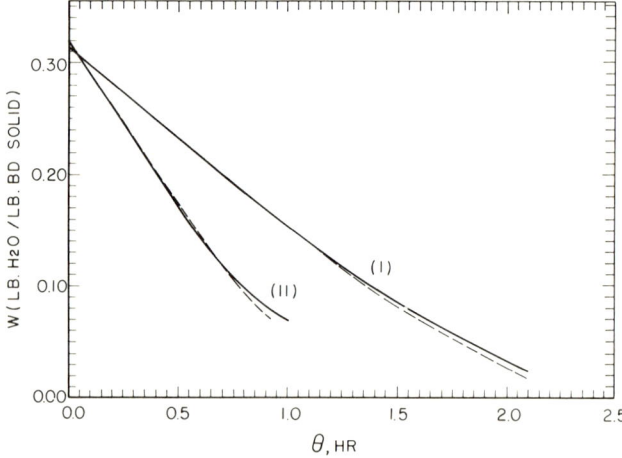

FIG. 16. Drying curves for ceramic granules and coke. (I) Experiment D1W1L3G1: 0.25-in. porous cylindrical ceramic granuler in a 2.0-in. bed. (II) Experiment D1W3L1G1 $\frac{3}{8}$ to $\frac{1}{4}$ mesh porous granular domestic grade coke in a 0.625-in. bed. — Theoretical; - - - - experimental.

DRYING OF SOLID PARTICLES AND SHEETS 277

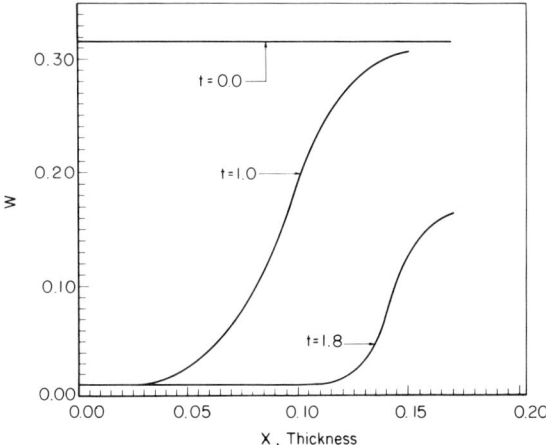

FIG. 17. Moisture distributions in a thick packed bed. Experiment D1W1L361 (S2).

bed inlet dried to equilibrium moisture content before elements near the bed outlet began to dry in all experiments studied except D3W3L1G1. Therefore it was not uncommon to have all drying periods represented in the bed at the same time.

Table I gives the basic computer input data for the nine runs which were presented. Equilibrium moisture for Al_2O_3 and coke was taken as 0.04 and ceramics was taken as 0.01. The sphericity of the coke and ceramics was taken as 0.866.

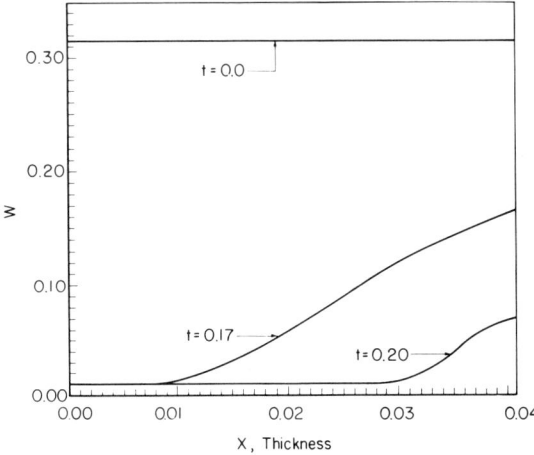

FIG. 18. Moisture distributions in a thin packed bed. Experiment D3W3L161 (S2).

TABLE I
BASIC COMPUTER INPUT DATA

RUN CODE	BH Height of the packed bed (FT)	AS Specific surface area of solid (ft²/ft³ of bed)	BDS Bulk density of bone dry solid (lb BD solid/ft³ bed)	T Bulk temp. inlet air (°K)	HUM Humidity of inlet (lb H₂O/lb BD Air)	G Air rate (air/min ft²)	X Bulk temp. of inlet air (°R)	C_s Initial uniform water content of bed (lb H₂O/lb BD Solid)	C_o Critical moisture concentration (lb H₂O/lb BD Solid)
Al₂O₃									
Exp. #13	0.328	160[a]	42[c]	371.0	0.008	1.76[d]	668.4	0.70	0.60[d]
Exp. #14	0.328	160[a]	42[c]	370.0	0.018	0.85[d]	666.6	0.645	0.60[d]
Exp. #19	0.328	470[a]	40[c]	370.0	0.019	0.84[d]	666.6	0.610	0.60[d]
Ceramic									
Exp. D3W1L1G1	0.042	120[a]	50[d]	366.0	0.0106[d]	4.0	660.0	0.295	0.30[d]
Exp. D3W1L1G3	0.042	120[a]	50[d]	366.0	0.0124[d]	12.0	660.0	0.315	0.30[d]
Exp. D3W3L1G1	0.042	120[a]	43[d]	366.0	0.0370[d]	4.0	660.0	0.315	0.30[d]
Exp. D1W1L3G1	0.167	120[a]	52[d]	320.0	0.0298[d]	4.0	578.0	0.315	0.30[d]
Coke									
Exp. D1W3L1G1	0.052	100[b]	24[d]	322.0	0.0568[d]	4.0	580.0	0.320	0.30[d]
Exp. D3W1L3G3	0.333	100[b]	26[c]	366.0	0.0104	12.0	660.0	0.270	0.30

[a] AS was estimated from the geometry of bed particles, an estimate of the void fraction and the relation: $AS = 6(1 - e)/D_p$.
[b] Particle geometry not given for coke. Multiple computer runs were made to find an AS which gave the best fit to experimental drying curves.
[c] Air from these beds is saturated and BDS was calculated from: $BDS = (60)(H_{out} - H_{in})(R)/(BH)$(Slope of drying curve). Consistancy between the variables in the foregoing equation reported in the references is thereby insured.
[d] Air from these beds is not saturated-guidance was obtained from approximate densities given in the reference, and densities calculated as explained in c above.

3. Conclusions

There are available computer programs (M2) that allow generation of drying curves and moisture profile curves for packed beds of nonshrinkable thin porous solids that have no lowering of the vapor pressure of water and which offer negligible resistance to the flow of liquid within the solid. These programs allow prediction of drying times within 5% provided physical properties, primarily the bulk density and specific solid surface area, and drying conditions are known precisely.

All the calculated results checked with the experimental data with a maximum deviation of 5%.

Care must be used when drying material with a high solute concentration. The solute moves to the surface and when the water is evaporated the vapor pressure is reduced below that predicted by Eq. (20).

V. Drying Porous Solids—Continuous Operations

A. Rotary Driers

1. General Description

Since rotary driers are relatively inexpensive, are easy to operate and clean, and require little maintenance, they are commonly used by the process industries to dry granular free-flowing solids. However, until a few years ago, little was known quantitatively about the factors influencing the operation and design of this equipment. It was difficult for the process engineer to estimate the size of such equipment or the conditions under which it will operate most efficiently. A suitable procedure should be developed that would permit the process engineer to estimate the size of a rotary drier required for a given job.

A rotary drier may be considered as having two distinct functions: first, that of conveyor, and second, that of heat and mass exchanger.

The progress of the charge through the drier is influenced by the following eleven variables in this system: (1) physical properties of the charge, (2) feed rate, (3) temperature of the charge, (4) rate of air flow, (5) temperature of air, (6) diameter of cylinder, (7) length of cylinder, (8) slope of cylinder, (9) rate of rotation of cylinder, (10) design of flights, and (11) number of flights in the drier.

The system has too many variables for exact analysis. A model which uses the more important variables and explains their effect on holdup times has been devised. It is felt that this model will give sufficiently accurate data for most engineering work.

In order to solve the relation for the residence time in the drier as a function of the experimental conditions, it is necessary to consider how the material moves through the drier.

2. Previous Work

Previous workers concerned with the conveying function of rotary driers have generally given primary consideration to determining the time of passage. It was believed that the holdup in the drier was of vital importance in the drier design. Work on this problem was presented by Sullivan et al. (S15) and Smith (S11, S12).

Prutton et al. (P5) were the first to propose a formula for the time of passage that incorporates an air-velocity factor.

The most extensive work on holdup and retention time in rotary driers has been conducted by Friedman and Marshall (F2). Unfortunately, much of their work was done on dry materials with no air flow. From an experimental standpoint, their results have a disadvantage because one of the conditions requires that the test be conducted at zero air velocity. With many materials the handling characteristics at zero air velocity are different from those with air flow in that the material may change considerably during drying.

Spraul (S14) modified Friedman and Marshall's work and made a comparison of holdup at two different air mass velocities. This relationship permits an evaluation of the effect of air velocity on holdup in the range where proper drying conditions are encountered. Care should be taken that extrapolation is not attempted over a wide range of air velocities.

Van Krevelen and Hoftijzer (V1) studied the time of passage of three different materials in a small drier 10 cm in diameter and 76 cm long. The materials studied were nitrochalk fertilizer granules, sand, and marl powder.

Miskell and Marshall (M6) studied retention time in driers and many other authors have written articles on the design of rotary driers (F1, S4, S5, T5) and presented empirical equations to correlate the results.

Saeman and Mitchell (S1) derived the retention time expression:

$$t = L/C^* D' R'(\alpha - U') \tag{40}$$

where t is retention time (min), L is drier length (ft), C^* is a constant between 2 and π (dimensionless), D' is drier diameter (ft), R' is rate of rotation (rpm), α is the slope of the drier (dimensionless), m is a constant (min/ft), and U' is air velocity (ft/min).

Hiraoka and Toei (H5) made experiments to study the volumetric heat-transfer coefficient.

Miller *et al.* (M5) discuss factors influencing the operation of rotary driers, such as diameter, length, mean temperature difference, mass velocity of air, flight size, and retention time. An excellent photographic study of flight action and showering is available. They also present graphs portraying the variation of temperatures and humidities with drier length.

3. *Development of the Equation*

When a particle moves forward along the X axis, the forces (Fig. 19) acting on the particle are given by

$$F_D - V_p'\rho_p g \sin \alpha = M(du/d\theta) \qquad (41)$$

where F_D is drag force, M is the mass of the particle, ρ_p is the density of the particle, α is the angle of inclination of the drier, u is particle velocity in the X direction, θ is time of fall, and V_p' is volume of the particle. The drag force is given by

$$F_D = V_p'K_0(U' - u)^n \qquad (42)$$

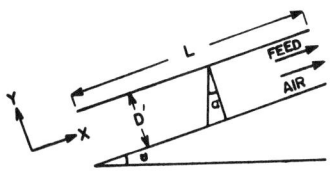

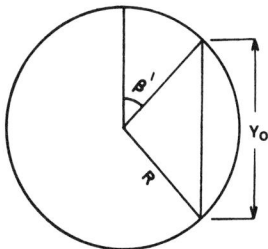

Fig. 19. Cross section of rotary drier.

where K_0 is a constant, U' is air velocity, u is particle velocity, and n is a constant.

The particle velocity in the X direction is usually negligible when compared to air velocity. Equation (42) now becomes

$$F_D = V_p' K_0 U'^n \tag{43}$$

Substituting Eq. (43) in (41),

$$V_p' K_0 U'^n - V_p' \rho_p g \sin \alpha = M(du/d\theta)$$

$$(1/\rho_p) K_0 U'^n - g \sin \alpha = du/d\theta$$

It is assumed that U' is constant across the drier as a first approximation, and so the above equation can be solved to give

$$u = (1/\rho_p)(K_0 U'^n - \rho_p g \sin \alpha)\theta + C_1$$

at $\theta = 0$, $u = 0$, therefore $C_1 = 0$, and thus

$$u = (1/\rho_p)(K_0 U'^n - \rho_p g \sin \alpha)\theta \tag{44}$$

Now

$$u = dX/d\theta \tag{45}$$

Substituting this in Eq. (44) and solving for X,

$$X = (1/\rho_p)(K_0 U'^n - \rho_p g \sin \alpha)\theta^2/2 + C_2$$

at $\theta = 0$, $X = 0$, therefore $C_2 = 0$, and thus

$$X = (1/\rho_p)(K_0 U'^n - \rho_p g \sin \alpha)\theta^2/2 \tag{46}$$

When the particle falls, forces acting on the particle are given by Eq. (41),

$$V_p' \rho_p g \cos \alpha - F_D' = M(dv_p/d\theta) \tag{47}$$

where F_D' is the drag force acting in the y direction when the particle falls and v_p is the particle velocity when it is falling.

Drag force F_D' can be neglected for the drying of some materials. This assumption is best for large dense particles in small driers. If F_D' is neglected Eq. (47) becomes

$$g \cos \alpha = dv_p/d\theta \tag{48}$$

Solving for v_p,

$$v_p = (g \cos \alpha)\theta + C_3$$

when $\theta = 0$, $v_p = 0$, therefore $C_3 = 0$, and thus

$$v_p = (g \cos \alpha)\theta \tag{49}$$

Now,

$$v_p = dy/d\theta \tag{50}$$

Substituting this in Eq. (49) and solving for y (distance of fall),

$$y = (g \cos \alpha)\theta^2/2 + C_4$$

at $\theta = 0$, $y = 0$, therefore $C_4 = 0$, and thus

$$y = (g \cos \alpha)\theta^2/2$$

or,

$$\theta^2/2 = y/g \cos \alpha \tag{51}$$

Now,

$$y_0 = D' \cos \beta' \tag{52}$$

where D' is the diameter of the drier; β' is the angle of showering, assumed to vary from $+\pi/2$ to $-\pi/2$; and y_0 is the total distance of fall. Substituting for y_0 in Eq. (51)

$$\theta^2/2 = (D' \cos \beta')/(g \cos \alpha) \tag{53}$$

Substituting Eq. (53) in (46)

$$X = \frac{(K_0 U'^n - \rho_p g \sin \alpha) D' \cos \beta'}{\rho_p g \cos \alpha} \tag{54}$$

This is the distance traveled in the X direction per fall.

The number of falls per revolution is approximately $2\pi/(\pi - 2\beta)$; therefore the distance traveled per revolution is $(X)[2\pi/(\pi - 2\beta')]$. Average distance the solid travels per revolution is

$$\bar{X} = \int_0^{W_F} \frac{(X)[2\pi/(\pi - 2\beta')]}{W_F} dW_F \tag{55}$$

where W_F is the total weight of the particles showered.

Assuming uniform showering,

$$dW_F/d\beta' = W_F/\pi \tag{56}$$

or,

$$dW_F = (W_F d_s')/\pi \tag{57}$$

Substituting Eq. (57) ii (55)

$$\bar{X} = \int_{-\pi/2}^{+\pi/2} \frac{(X)[2\pi/(\pi - 2\beta')]W_F}{W_F \pi} d\beta'$$

$$= 2 \int_{-\pi/2}^{+\pi/2} \frac{X}{\pi - 2\beta'} d\beta' \tag{58}$$

Substituting the value of X from Eq. (14) into (58)

$$\bar{X} = 2 \int_{-\pi/2}^{+\pi/2} \frac{(K_0 U'^n - \rho_p g \sin \alpha) D' \cos \beta'}{(\rho_p g \cos \alpha)(\pi - 2\beta')} d\beta'$$

$$= 2D' \left(\frac{K_0 U'^n}{\rho_p g \cos \alpha} - \tan \alpha \right) \int_{-\pi/2}^{+\pi/2} \frac{\cos \beta'}{\pi - 2\beta} d\beta' \quad (59)$$

$$= 2D' \left(\frac{K_0 U'^n}{\rho_p g \cos \alpha} - \tan \alpha \right) (0.924) \quad (60)$$

$\bar{X}$ is the average distance traveled per revolution. Let L be the length of the drier (ft), t the retention time (min), and R' the rate of rotation of the drier (rpm). Then,

$$L = 2D' \left(\frac{K_0 U'^n}{\rho_p g \cos \alpha} - \tan \alpha \right) (0.924)(R')(t)$$

$$= 1.85 D' R' t \left(\frac{K_0 U'^n}{\rho_p g \cos \alpha} - \tan \alpha \right) \quad (61)$$

or,

$$t = \frac{L}{1.85 D' R' [(K_0 U'^n / \rho_p g \cos \alpha) - \tan \alpha]} \quad (62)$$

or,

$$t = \frac{L}{1.85 D' R' [(K' U'^n / \cos \alpha) - \tan \alpha]} \quad (63)$$

where $K' \equiv K_0 / \rho_p g$, or in general,

$$t = \frac{L}{C^* D' R' [(K' U'^n / \cos \alpha) - \tan \alpha]} \quad (64)$$

The constant C^* would vary depending upon the limits of and also would depend upon the number of flights and their capacity. The constant evaluated here is for an infinite number of flights. It is believed that the major variation will occur with the first few flights (two to four) and that subsequent addition of flights will have minor effect. It should be evaluated by experimental conditions.

4. Rotary Drying Theory

The study of drying in rotary driers is based upon the following relationships derived from basic theory:

a. Mass Balance. The moisture lost by the solids is gained by the gas (humid air) flowing through the drier:

$$G \, d\bar{H} + W' \, dW = 0 \tag{65}$$

where $G \, d\bar{H}$ is the amount of water gained by the air, W gives pounds of water per pound of bone dry solid, and W' the number of pounds of bone dry solid fed per hour.

b. Heat Balance. The energy balance reduces to a heat balance when other slight energy variations, such as those of potential and kinetic energies are neglected:

$$\lambda_R G \bar{H} + G c_{P_{ha}}(T_g - T_{R'}) + W' c_{P_s}(T_s - T_{R'})$$
$$+ W'W(T_s - T_{R'}) + Q_{losses} = \text{constant} \tag{66}$$

where $\lambda_R G \bar{H}$ is the latent heat content of moisture in air, $G c_{P_{ha}}(T_g - T_{R'})$ is the sensible heat content of humid air, $W' c_{P_s}(T_s - T_{R'})$ is the sensible heat content of the dry solid, $W'W(T_s - T_{R'})$ is the sensible heat content of water in the solid, and Q_{losses} is the amount of heat lost through the drier shell by convection and radiation. Equation (66) may be written in differential form as follows:

$$G \, d(\lambda_R \bar{H}) + G \, d[c_{P_{ha}}(T_g - T_{R'})] + W' c_{P_s} dT_s$$
$$+ W' \, d[W(T_s - T_{R'})] + dQ_{losses} = 0 \tag{67}$$

where the specific heat content of the solids c_{P_s} was assumed constant. It is also assumed that the temperature distribution is uniform in the solid particle, i.e., all points are at the surface temperature.

The amount of heat lost is calculated by the relationship

$$dQ_{losses} = 0.15 (\Delta T)^{0.25} \pi D' (\Delta T) \, dL \tag{68}$$

c. Heat Transfer. The rate of heat transmitted by the gas to the solids is given by

$$hA_0(T_g - T_s) = c_{P_s} \frac{dT_s}{dt} + \frac{d}{dt}[W'W(T_s - T_{R'})] + \frac{G}{W'} \frac{d}{dt}(\lambda \bar{H})$$
$$+ 0.45 \frac{G}{W'} (T_g - T_s) \frac{d\bar{H}}{dt} \tag{69}$$

where $hA_0(T_g - T_s)$ is the total amount of heat transferred by the air to the solids; $c_{P_s} dT_s/dt$ is the variation of sensible heat content of the solids; $d[W'W(T_s - T_{R'})]/dt$ is the variation of the sensible heat content of water contained in the solids, the specific heat content of water being taken as equal to one and assumed constant; $(G/W') d(\lambda \bar{H})/dt$ is the variation of the latent heat content of moisture in air; and $0.45(G/W')(T_g - T_s) d\bar{H}/dt$ is the variation of the sensible heat of evaporated moisture. The factor 0.45 stands for the specific heat content of steam which is assumed to be constant.

d. Mass Transfer. The rate of mass transfer is given by

$$-W' dW/dt = G d\bar{H}/dt = W'k_g A_0 (p_s - p_R)(W/W_c)^{0.6} \quad (70)$$

The term $(W/W_c)^{0.6}$ accounts for the reduction of drying area on the solid particle's surface. This also implies that only the falling-rate period of drying is being considered, as most practical situations suggest.

The use of an expression such as (70) indicates that drying occurs by removal of water from the exposed areas on the external surface of the particle and in its capillaries. The treatment may not be applied for temperatures of the solids above the boiling point of water, when intense vapor streams completely alter the heat- and mass-transfer mechanism. The psychrometric ratios have little meaning if the temperature of the air is much above 250°F.

e. Drying Time. The drying time is related to the other variables by Eqs. (23) and (24).

Combining Eqs. (64) and (26) gives

$$\frac{dL}{dF} = \frac{C^* D' R' \lambda W_c}{h A_0 T_{R'}} \left[\frac{K'U'}{\cos \alpha} - \tan \alpha \right] \quad (71)$$

The constant C^* normally ranges from the value 2.8 to 3.5. K' is the drag coefficient that accounts for the drag effect of air on the falling solid particles. Both C^* and K' are obtained from experimental data for a given material. N in Eq. (64) can be taken as one for most cases. L is the drier length and F is evaluated by Eqs. (23), (24), (69), and (70) where T_R and P_R vary along the length of the drier.

Equation (64) has been shown to be in close agreement with the data published in the literature (B6, B7, C1, C2).

Many materials have different flow and tumble characteristics with water content. In these cases there may be considerable variation in h and A_0 with water content. Data on the laboratory unit must evaluate all these properties before trying to scale up to a large unit.

6. Sample Calculations

Parallel Flow. As an example, the case of run 2 in the set of data relating to the drying of cheese (D1) is considered: slope of drier, 0.031; water in feed, 0.458 (dry basis) (lb/lb); water in product, 0.25 (dry basis) (lb/lb); solid feed rate, 41.8 lbs/hr (dry); air rate, 1687 lb/hr (340 ft/min); initial temperature of solid, $T_0 = 90°F$; inlet air temperature, 130°F; humidity of inlet air, 0.008 lb water/lb dry air; drier diameter, 1.25 ft; speed of drier, 11.7 rpm (702 rph); $C^* = 2.8$; $K' = 1.45 \times 10^{-4}$ min/ft; $hA_0 = 25$.

Values of C^*, K', and hA_0 were taken from cheese data and are evaluated from this data (D1). Calculate the length of drier which would be needed.

Since temperatures do not vary considerably in the drier, it is acceptable to perform the calculations in one step. Assuming that the average temperature of the solids is 80°F, then $P_{s_1} = 0.51$.

$$\bar{H}(\text{average}) = 0.008 + \frac{0.458 - 0.25(41.8)}{(1687)(2)} = 0.011$$

and

$$P_R = 0.25, \quad S = 0.26/0.51 = 0.51 \tag{72}$$

The reference temperature T_R is now selected:

$$T_R = 120°F = 580°R \quad (\text{average for air}) \tag{73}$$

Therefore $P_{R_s} = 1.69$ psia and $\lambda = 1031$ Btu/lb. The psychrometric ratio is evaluated by the procedure described in the literature (K3). The value $(k_g/h) = 3$ is obtained. Equation (27) becomes

$$G' = \frac{k_g p_{R_s} \lambda}{h T_R} = 3 \frac{(1.69)(1031)}{(14.7)(580)} = 0.52 \tag{74}$$

Equation (28) becomes

$$\phi = \frac{18\lambda_0}{RT_R} = \frac{(18)(1031)}{(1.987)(580)} = 16.1 \tag{75}$$

A solution using Pimentel's (P3) program on Eqs. (23) (24), (69), and (70) gave a value of

$$L = 8.6 \text{ ft} \tag{76}$$

The actual drier was 8 ft long.

Where average humidities can be used it is possible to work this problem with the use of Fig. 4.

$$W_f = 0.546, \quad T_R = 120°, \quad S = 0.01 \tag{77}$$

and,

$$T_W = 77°F \tag{78}$$

$$\Delta T = 120 - 77 = 43° \tag{79}$$

The value read from Fig. 4 would be $F = 6.1$. The result would check that obtained by the above calculation.

When drying material to a low value of W it is necessary to divide the drier up into short sections and calculate ΔL for each section.

An approximate length can be obtained by the use of Fig. 4. The average wet and dry bulb temperatures can be estimated by making a heat balance for each section. ΔF can be calculated for each ΔW and Eq. (71) will then give the desired length of the drier.

6. *Conclusions*

A program on a computer has been prepared by Pimentel (P3) for both parallel and countercurrent flow rotary driers.

There is little basic data in the literature which give the desired parameters for a solution. For most problems, it is necessary to run some data on a small drier and use the above theory to scale up the size or predict the effect of changing operating conditions.

B. Tunnel Driers

Equations (21)–(28) and Eq. (65) are applicable to tunnel driers. The time (t) material spends in the drier can be controlled so that Eq. (64) is not needed.

Equations (23) and (24) can be used for the calculation of the exit water concentration. Since the humidity and temperature vary as the material moves through the tunnel, it is necessary to divide the tunnel into several sections so that the temperature and humidity do not vary too much in each section.

The program on the computer given by Pimentel (P3) could be used for both parallel and counterflow tunnel driers.

VI. Summary

It is our intention and hope that the material presented here will be of interest and use to the designer of driers.

There are many fields of drying which require much more work before proper design is possible. Very little work is available in most driers such as fluidized bed driers, steam tube driers, or almost any production unit.

Very little is known about the resistance to flow of moisture in solids. If this problem were to be solved, it would be possible to design driers for material where the major resistance is in the solid phase.

Little is known about drying of material that has a high solute concentration such as is the case for most foods. The field of high temperature or thick materials has little theory on which to base a design.

The unit operation of adsorption drying which is particularly suited for removing or eliminating small concentrations of substances, such as in the drying of air and gas and the removal of trace contaminants in gas streams has received considerable attention (B5)–(C2). This topic involves a different set of fundamental principles and has not been treated in this article.

Acknowledgment

Some of the work presented in this chapter has been supported by Chicago Bridge and Iron Company.

Nomenclature

A	Total area for mass transfer	h_m	Convective heat transfer coefficient
A_0	Total area of the solid surface	h_r	Radiation heat transfer coefficient
B	Psychrometric ratio		
C	Concentration of diffusing species in gas phase	j_D	j-factor for mass transfer
C^*	Constant in Eq. (64)	J_{hL}	j factor based on the heat-transfer coefficient at point L
C'	Moisture content of solids, W/W_c	j_{hm}	j factor for heat transfer based on the average heat transfer coefficient
C_p	Heat capacity of air		
C_{pg}	Heat capacity of fluid phase		
C_{pha}	Heat capacity of humid air	k_g	Gas-phase mass-transfer coefficient
C_{ps}	Heat capacity of dry solid		
d	Diameter of tube	k	Effective thermal conductivity
D	Molecular diffusivity	k_0	Constant
D'	Diameter of drier	K'	Drag coefficient $(K_0/\rho_p g)$
D_p	Particle diameter	L	Drier length, plate length
D_s	Coefficient in solid phase as defined by equation IV-19	m	Constant
		M	Mass of particle
e	Porosity	M_m	Mean molecular weight
E	$\theta h T_R A_0/\lambda W_c$	n	Constant
F_D	Drag Force	N_{Le}	Lewis number
$F_D{}^1$	Drag force acting in the y direction when the particle falls	N_{Re}	Reynolds number
		N_{Num}	Nusselt number based on the average heat transfer coefficient
f	Fanning friction factor		
G	Mass velocity		
G'	$k_g \lambda P_{RS}/h T_R$	N_{Nu}	Nusselt number
h	Total heat-transfer coefficient	N_{Pr}	Prandtl number
$h_c{}^*$	Convective heat transfer coefficient in the absence of mass transfer	N_{Sc}	Schmidt number
		N_{Sh}	Sherwood number
		ΔP_f	Pressure drop due to friction

P_{BM}	Log mean of the inert partial pressure	W^1	Mass flow of bone dry solid
P_{RS}	Vapor pressure of water at T_R	x	Distance measured from the surface into the material or length dimension
P_s	Vapor pressure of water at T_s		
P_R	Partial pressure of water in the drier	$\bar{x}$	Average distance the solids travel per revolution
Q_{losses}	Heat loss through the dryer shell by convection and radiation	y	Distance of fall
		y_0	Total distance of fall of particle
q	Heat-transfer rate		
R'	Rate of rotation of drier		
r_c	Radius of capillary with water surface at a distance X below the surface	GREEK LETTERS	
r_{cs}	Largest radius of capillary which can just bring water to the surface	α	Angle of inclination of dryer and angle between the surface and the horizontal
S	Moisture content of air	α_T	Thermal diffusivity
T_m	Mean temperature of gas stream	β	Contact angle between solid and liquid surfaces
T_R	Room temperature, also temperature in dryer		
		β'	Angle of showering
T_s	Temperature at the solid surface	γ	Dimensionless parameter which is a measure of the effect of concentration changes on temperature
T_W	Wet bulb temperature		
t	Drying time or retention time		
U'	Average linear velocity of the fluid or air		
		ϵ	Emissivity
U	Gas velocity or particle velocity in X direction	θ	Time
		λ	Latent heat of sublimation or vaporization
v_0	Free stream velocity		
v_p	Particle velocity when it is falling	μ	Viscosity of fluid
		ν	Kinematic viscosity of fluid
v_p'	Volume of particle	ρ	Density of fluid phase
w	Width of plate in direction of flow	ρ_g	Density of air
		ρ_p	Density of particle
W	Moisture content of the porous solid	ρ_s	Bulk density of solid or adsorbent supporter
W_c	Critical moisture content of the porous solid	σ	Surface tension
		ϕ	$18x/RT_R$
W_f	Total weight of particles showered	ψ	Particle shape factor
		∇	Difference

References

A1. Ahluwalia, M. S., Ph.D. Thesis, Illinois Institute of Technology, Chicago, 1969.
A2. Akbar, A., and Goerling, C., *Chem. Ing. Tech.* **33,** 619 (1961).
A3. Arzan, A. A., and Morgan, R. P., *62nd Nat. AIChE, 1967* Preprint no. 22-d (1967).
B1. Bedingfield, C. H., and Drew, T. B., *Ind. Eng. Chem.* **42,** 1164 (1950).
B2. Bird, R. B., Stewart, W. E., and Lightfoot, E. N., "Transport Phenomena." Wiley, New York, 1960.

B3. Bird, R. B., Stewart, W. E., and Lightfoot, E. N., "Transport Phenomena," pp. 411–412. Wiley, New York, 1965.
B4. Buckingham, E., U.S., Dep. Agr., Bur. Solids, Bull. **38,** (1907).
B5. Bulletin D-165. Honey Combe Dehumidifiers, Cargo-caire Engineering Corporation, New York, 1965.
B6. Bullock, C. E., Ph.D. Thesis, University of Minnesota, Minneapolis, 1965.
B7. Bullock, C. E., and Threlkeld, J. L., Prepr. ASHRAE Trans. (1966).
C1. Carter, J. W., Trans. Inst. Chem. Eng. **44,** T253 (1966).
C2. Carter, J. W., Trans. Inst. Chem. Eng. **45,** T213 (1968).
C3. Cassie, A. B., King, G., and Baxter, S., Trans. Faraday Soc. **36,** 445 (1940).
C4. Ceaglske, N. H., and Hougen, O. A., Trans. Amer. Inst. Chem. Eng. **33,** 283 (1937).
C5. Cheng, P. T., The effect of temperature, humidity and air velocity on the drying of wood. M.S. Thesis, Illinois Institute of Technology, Chicago, 1948.
C6. Childs, E. C., J. Agr. Sci. **26,** 114 and 527 (1936).
C7. Chilton, T. H., and Colburn, A. P., Ind. Eng. Chem. **26,** 1183 (1934).
C8. Corben, H., and Newitt, R., Trans. Inst. Chem. Eng. **33,** 52 (1955).
D1. Desai, T. P., M.S. Thesis, Chem. Eng. Dept., Illinois Institute of Technology, Chicago, 1959.
E1. Eckert, E. R. G., and Drake, R. M., Jr., "Heat and Mass Transfer." McGraw-Hill, New York, 1963.
F1. Fan, L. T., and Ahn, Y. K., Appl. Sci. Res., A **10,** 465 (1961).
F2. Friedman, S. J., and Marshall, W. R., Jr., Chem. Eng. Progr. **45,** 482 (1949).
F3. Fulford, G. D., Can. J. Chem. Eng. **47,** 378 (1969).
G1. Gardner, G. O., and Kestin, J., Int. J. Heat Mass Transfer **6,** 289 (1962).
G2. Gardner, W., Soil Sci. **7,** 313 (1919).
G3. Gardner, W., and Widtsoe, J. A., Soil Sci. **11,** 215 (1921).
G4. Garud, B. S., Drying schedules of welding electrodes. M.S. Thesis, Illinois Institute of Technology, Chicago, 1949.
G5. Goldstein, S., "Modern Developments in Fluid Dynamics," Vol. II. Oxford Univ. Press, London and New York, 1938.
G6. Gurr, C. G., Marshall, T. K., and Hutton, J. T., Soil Sci. **74,** 335 (1952).
H1. Haines, W. B., J. Agr. Sci. **17,** 264 (1927).
H2. Harmathy, T. I., Ind. Eng. Chem., Fundam. **8,** 92 (1969).
H3. Henry, H. C., and Epstein, N., Can. J. Chem. Eng. **48,** 595 (1970).
H4. Henry, P. S. H., Proc. Roy. Soc., Ser. A **171,** 215 (1939).
H5. Hiraoka, M., and Toei, R., Mem. Fac. Eng., Kyoto Univ. **25,** Port 1, 144 (1963).
H6. Hougen, O. A., McCauley, H. J., and Marshall, W. R., Jr. Trans. Amer. Inst. Chem. Eng. **36,** 183 (1940).
H7. Hutcheon, W. L., Highw. Res. Bd., Spec. Rep. **40,** 113 (1958).
J1. Jakob, M., "Heat Transfer," Vol. II, p. 261. Wiley, New York, 1957.
J2. Jakob, M., and Dow, W. M., Trans. Amer. Soc. Mech. Eng. **68,** 123 (1946).
K1. Kamei, S., J. Soc. Chem. Ind., Jap. **40,** 251, 257, 325, 366, and 374 (1937).
K2. Kauh, J. Y., Evaluation of drying schedules. Ph.D. Thesis, Illinois Institute of Technology, Chicago, 1966.
K3. Kauh, J. Y., Peck, R. E., and Wasan, D. T., Int. J. Heat Mass Transfer **10,** 1629 (1967).
K4. Kestin, J., and Persen, L., Int. J. Heat Mass Transfer **5,** 143 (1962).
K5. Kirkwood, K. C., and Mitchell, T. J., J. Appl. Chem. **15,** 256–280 (1965).
K6. Klausner, Y., and Kraft, R., Trans. Soc. Rheol. **10,** 603 (1966).

K7. Krischer, O., *Z. Ver Deut Ing., Beih.* **1**, 17 (1940).
K8. Ksenzhek, O. S., *Zh. Fiz. Khim.* **37**, 1297 (1963).
K9. Kuzmak, J. M., and Sereda, P. J., *Soil Sci.* **84**, 419 (1957).
L1. Latzko, H., *Z. Angew. Math. Mech.* **1**, 268 (1921).
L2. Lester, D. H., and Bartlett, J. W., A theory of bed drying of particulate solids including the role of capillary. Ph.D. Thesis, University of Rochester, Rochester, New York, 1969.
L3. Lewis, W. K., *Ind. Eng. Chem.* **13**, 427 (1921).
L4. Leverett, J., *Amer. Inst. Mining, Met. Eng., Tech. Publ.* **1223**, 17 (1940).
L5. Lynch, E. J., and Wilke, C. R., UCRL Report 8602. University of California, Berkeley, 1959.
M1. Markin, V. S., *Izv. Akad. Nauk SSSR, Ser. Khim.* **9**, 1523 (1965).
M2. Max, D. A., Packed bed drying. M.S. Thesis, Illinois Institute of Technology, Chicago, 1970.
M3. McCormick, P. Y., *Ind. Eng. Chem.* **60**, 52 (1968).
M4. McCormick, P. Y., *Ind. Eng. Chem.* **62**, 84 (1970).
M5. Miller, C. O., Smith, B. A., and Schuette, W. H., *Trans. Amer. Inst. Chem. Eng.* **38**, 123 and 841 (1942).
M6. Miskell, F., and Marshall, W. R., Jr., *Chem. Eng. Progr.* **52** (1956).
M7. Myklestad, O., *Int. J. Heat Mass Transfer* **11**, 675–687 (1968).
N1. Newman, A. B., *Trans. Amer. Inst. Chem. Eng.* **27**, 203 and 310 (1931).
P1. Peck, R. E., Ahluwalia, M. S., and Max, D., *Chem. Eng. Sci.* **26**, 389–403 (1971).
P2. Peck, R. E., and Kauh, J. Y., *AIChE J.* **15**, 85 (1969).
P3. Pimentel, L., Design of rotary driers. M.S. Thesis, Illinois Institute of Technology, Chicago, 1969.
P4. Pohlhausen, E., *Z. Angew. Math. Mech.* **1**, 115 (1921).
P5. Prutton, C. F., Miller, C. O., and Schuette, W. H., *Trans. Amer. Inst. Chem. Eng.* **38**, 123 (1942).
R1. Rai, C., Ph.D. Thesis, Illinois Institute of Technology, Chicago, 1960.
R2. Richards, L. A., *J. Agr. Res.* **37**, 719 (1928).
R3. Richards, L. A., *Physics* **1**, 318 (1931).
S1. Saeman, W. C., and Mitchell, T. R., Jr., *Chem. Eng. Progr.* **50**, 467 (1954).
S2. Sato, H., *Chem. Eng. Jap.* **28**, 585–589 (1964).
S3. Schlichting, H., "Boundary-Layer Theory." McGraw-Hill, New York, 1955.
S4. Schneider, P. G., *Trans. ASME* **89**, 765 (1957).
S5. Schofield, F. R., and Glikin, P. G., *Trans. Inst. Chem. Eng.* **40**, 183 (1962).
S6. Sherwood, T. K., *Ind. Eng. Chem.* **21**, 12 and 976 (1929); **22**, 132 (1930); **24**, 307 (1932).
S7. Sherwood, T. K., *Ind. Eng. Chem.* **26**, 1096 (1934); **25**, 1134 (1933).
S8. Sherwood, T. K., and Comings, E. W., *Ind. Eng. Chem.* **25**, 311 (1933).
S9. Sheth, H. P., M.S. Thesis, Illinois Institute of Technology, Chicago, 1971.
S10. Slitcher, C. S., *U.S., Geol. Surv., Annu. Rep.* **19**, Part II, 301 (1898).
S11. Smith, B. A., *Ind. Eng. Chem.* **30**, 993 (1938).
S12. Smith, B. A., *Trans. Amer. Inst. Chem. Eng.* **38**, 251 (1942).
S13. Spalding, D. B., *J. Appl. Mech.* **81**, 455 (1961).
S14. Spraul, J. R., *Ind. Eng. Chem.* **47**, 368 (1955).
S15. Sullivan, J. D., Mair, C. G., and Ralston, O. C., *U.S., Bur. Mines, Tech. Pap.* **384** (1927).

S16. Swanson, B. S., Air film resistance in drying thin slabs. Unpublished M.S. thesis, Illinois Institute of Technology, Chicago, 1944.
T1. Thodos, G., and Malling, G. F., *Int. J. Heat Mass Transfer* 489–498 (1967).
T2. Thodos, G., and Petrovic, L. J., *Ind. Eng. Chem., Fundam.* **7,** 274–280 (1968).
T3. Thodos, G., and Sen Gupta, A., *Chem. Eng. Progr.* **58,** 58–62 (1962).
T4. Thodos, G., and Sen Gupta, A., *AIChE J.* **8,** 608–610 (1962).
T5. Toei, R., and Hayashi, S., *Mem. Fac. Eng., Kyoto Univ.* **25,** 457 (1963).
T6. Toei, R., Hayashi, S., and Okazaki, S., *Mem. Fac. Eng., Kyoto Univ.* **25,** Part 1, 116 (1963).
T7. Tuttle, F., *J. Franklin Inst.* **200,** 609 (1925).
V1. van Krevelen, D. W., and Hoftijzer, P. J., *J. Soc. Chem. Ind., London* **68,** 91 (1949).
W1. Wakabayashi, M. M., *Kagaku Kogaku* **28,** 102 (1964).
W2. Walker, I. K., *N. Z. J. Sci.* **4,** 775 (1961).
W3. Wasan, D. T., and Wilke, C. R., *Int. J. Heat Mass Transfer* **7,** 87 (1964).
W4. Wilke, C. R., and Wasan, D. T., *Pap. Jt. Meet. AIChE and BIChE,* **1965** Vol. 6, p. 21 (1965).
W5. Wilsdon, B. H., *Mem. Dep. Agr. India, Chem. Ser.* **6,** Part I, 154 (1921).

AUTHOR INDEX

Numbers in parentheses are reference numbers and indicate that an author's work is referred to although his name is not cited in the text. Numbers in italics show the page on which the complete reference is listed.

A

Agers, D. W., 65, *109*
Ahluwalia, M. S., 273(A1), *290, 292*
Ahn, Y. K., 280(F1), *291*
Ainshtein, V. G., 132(G1), 136(G2), *189*
Akbar, A., 256, *290*
Ambrose, P. M., 23(A6), *99*
Anderson, A. E., 33(A6), *99*
Anderson, R., 16(B31), 17(B31), *100*
Arbstedt, P. G., 78(W5), *110*
Arden, T. V., 56, *99*
Arehart, T. A., 57, *99*
Arzan, A. A., 258, *290*
Ashbrook, A. W., 66(J5, R10), *105, 108*

B

Back, A. E., 25(R5), 26(R4), 76, 77, *99, 107*
Backer, L., 14, *104*
Baerns, M., 114(B1), *188*
Bailes, R. H., 28(M8), *106*
Baragwanath, J. G., 21, *99*
Baroch, C. J., 12(B5), *99*
Bartlett, J. W., 259, *292*
Barton, R. K., 112(B2), 160, 161, *188*
Bauer, D. J., 64(B6), *100*
Bautista, R. G., 31(B7), 62, 63(C10, H11, H12, I2), 67(C11), *100, 101, 104*
Baxter, S., 253(C3), *291*
Bays, C. A., 33(B9), *100*

Beck, J. V., 16(B10, B32), 17(B32), *100*
Becker, H. A., 121(B4), 127(B3), 142, 160, 174, 179, 183(B3), *188*
Bedingfield, C. H., 251(B1), 252, *290*
Begunova, T. G., 43(B11), *100*
Benedict, C. H., 21(B12, B13, B14, B15), *100*
Bennett, P. W., 83, *102*
Benson, B., 37(M15), *106*
Benz, T.W., 3(M5), 34(M5, M6), *105, 106*
Berquin, Y. F., 112(B5, B6), 176, *188*
Berti, L., 112(B7), *188*
Beyer, H. G., 72, *105*
Bhappau, R. B., 17(B17), 27, 29, *100, 105*
Bird, R. B., 249(B2), 250(B2, B3), *290, 291*
Bjerrum, J., 21, *100*
Bjorling, G., 34(B21), 50, *100*
Black, K. L., 64, *100*
Blake, C. A., 67(B30), *100*
Bochinski, J., 63(B24), *100*
Boldt, J. R., Jr., 2(B25), 3(B25), 17(B25), 35(B25), 49, 79, 82(B25), 90(B25), *100*
Bonsack, J. P., 65, *100*
Booth, R. B., 20(L16), *105*
Bowers, R. H., 112(B8), *188*
Bowles, K. C., 75(N1), 76(N1), *107*
Bradshaw, P., 218(B3), 219, 221, 223(B2, B3), 225, *245*
Braley, S. A., 16(L3, L5), 17(L4), *105*
Bramwell, P., 67(T15), *109*
Bratt, G. C., 22, *100*
Brennan, D., 167(H3), 168, *189*

Bremer, A., 83, *100*
Brereton, E., 20(J4), *105*
Bresse, J. C., 57(A8), *99*
Bridges, D. W., 3(B29), 52(B29), *100*
Broneer, P. T., 15(K10), *105*
Brown, K. B., 63(H34), 67(B30), *100, 104*
Brown, R. L., 138(B10), *189*
Bruce, R. W., 37(D15, D16), 45, *102*
Bryner, L. C., 16(B31, B32, B33), 17(B31, B32, B33, B43, B35), *100*
Buchanan, R. H., 112(B9), 160, *188*
Buchwalter, D. J., 20(R13), *108*
Buckingham, E., 254(B4), *291*
Bullock, C. E., 286(B6, B7), *291*
Burkin, A. R., 3(B36, B38), 82, 83, *100, 101*
Busse, F. H., 200, *245*
Butler, J. A., 15, *101*
Butler, J. N., 25, *101*
Byrne, J. B., 66(E6), *102*

C

Cahalan, M. J., 3(Z1), *110*
Caldwell, N. A., 33(C1), *101*
Callahan, J. R., 31(C2), *101*
Cameron, F. K., 33(A6), *99*
Carlson, C. W., 62, *101*
Carlson, E. T., 2(C4), 3(C5), 62, 94(C4), *101*
Carlson, W. J., Jr., 35, 42(I7), *104*
Carman, E. D., 60, *101*
Caron, M. H., 21, 81, *101*
Carter, J. W., 286(C1, C2), 289(C2), *291*
Cassie, A. B., 253, *291*
Casto, M. G., 63, 67, *101*
Ceaglske, N. H., 254(C4), 255, *291*
Cebeci, T., 206(C1), 207, *245*
Champagne, F. H., 201(C4), 219, 222(C4), 236(C4), 237(C4), 238(C4), 239(C4), 241(C4), *245*
Charlton, B. G., 123(C1), 126, *189*
Chatelain, J. B., 21, *99*
Chatterjee, A., 161, 162(C2), *189*
Cheng, P. T., 263(C5), 265, 267, 270(C5), *291*
Childs, E. C., 253, *291*
Chilton, C. H., 62(C18), *101*
Chilton, T. H. 251(C7), *291*
Cholette, A., 160(C3), *189*

Clark, J. B., 33, *101*
Clouse, R. J., 69(D30), *102*
Cochran, A. A., 63(89), *107*
Coffer, L. W., 3(C20), *101*
Cohen, E., 11, *101*
Colburn, A. P., 251(C7), *291*
Coleman, C. F., 67(B30), *100*
Colmer, A. R., 16(C23, T2), 17(C22), *101, 109*
Colombo, A. F., 20(C24), *101*
Conley, F. R., 33(S15), *108*
Conley, J. E., 13, 26(H22, H23), *104, 108*
Cook, W. R., 80, *108*
Corben, H., 256, *291*
Corrick, J. D., 16, 17(C26), *101*
Corrsin, S., 201(S4), 219(S4), 222(S4), 236(S4), 237(S4), 238(S4), 239(S4), 241(S4), *245*
Cowan, C. B., 179(C4), *189*
Crabtree, F. H., 63(L11), *105*
Crane, S. R., 5(G6), *103*
Crouse, D. J., 63(H34), 67(B30), *100, 104*

D

Daley, F. E., 68(R18), *108*
Daly, B. J., 232, 237, *245*
Danckwerts, P. V., 129, 159(D1), *189*
Darrah, R. M., 29(S14), *108*
Dasher, J. 4(G3), 54, *101, 103*
Davidson, J. F., 112(D2), 134, 135, 137(L2), 138, 145(L2), 148, 151, 157, 164, 165, 167(L2), 183, *189, 190*
Davis, F. T., 3(W2), *110*
Davis, J. B., 56(A7), *99*
Davis, J. G., 33(D2, D3), 34, *101*
Davis, M. W., Jr., 70, *101*
Davis, R. E., 215, *245*
Dean, R. S., 25(D8), 26, *101*
Deardorff, J. W., 199, 243(D3), *245*
de Bruyn, P., 37(R12), *108*
DeCuyper, J. A., 17(D9), *101*
Delchamps, E. W., 16(T3), 17(T3), *109*
Dement, E. R., 65, *101*
Denaro. A. R., 78(D11), *101*
Dennis, W. H., 78(D12), *101*
Desai, T. P., 287(D1), *291*
Dewey, J. C., 26(R4), *107*
Dewuff, A., 12(T13), *109*
Donald, M. B., 30, 33(D13), *101*

Donaldson, C. D., 232, *245*
Dorr, J. V. W., 20(D14), *102*
Douglas, W. D., 2(V2), 84(V2), *110*
Dow, W. M., 249, *291*
Downes, K. W., 37(D15, D16), 45, *102*
Drake, R. M., Jr., 249(E1), 250(E1), *291*
Dresher, W. H., 41, 47, 48, *102*
Drew, T. B., 251(B1), 252, *290*
Drobnick, J. L., 23(A5), 64(A3), 65(A4), 67(L12), *99, 105*
Dufour, M. F., 21, *102*
Duggan, E. J., 21(D20), *102*
Duncan, D. W., 15(T17), 17(D21, D22, D23, D24, T17), *102, 109*
Dunning, H. N., 64(S36), *109*
Dunstan, E. T., 20(D25), *102*
Durie, R. W., 33(D26), *102*
Dutrizae, J. E., 24(D27, D28, D29), *102*
Dykstra, J., 69, *102*

E

Ebner, M. J., 25(E1), *102*
Eckert, E. R. G., 249(E1), 250(E1), *291*
Eddy, L., 20(E2), *102*
Edwards, J. D., 13(E3), 34(E3), *102*
Ehrlich, H. L., 17(E4), *102*
Elkin, E. M., 83, *102*
Ellis, D. A., 66(E6), *102*
Elperin, I. T., 156(E1), 175(E1), *189*
Elsner, L., 18, *102*
Engel, A. L., 20(E8), *102*
Engel, G. T., 64(G13), *103*
Epstein, N., 154(E2), 169(E2), *189*, 251(H3), 252, *291*
Ergun, S., 175(E3), 180, *189*
Evans, D. J. I., 3(E10, M5), 34(M5, M6), 82, 87, 91, 92, 94(E10), 95, *102, 105, 106, 110*
Evans, L. G., 3(S10), 12(S10), 29(S10), 31(S10), *108*
Everest, D. A., 3(E12), 52(E12), 53, *102*

F

Falke, W. L., 26, *102*
Fan, L. T., 280(F1), *291*
Fassell, W. M., Jr., 41(D18), 47(D17), 48(D17), *102*

Ferriss, D. H., 218(B3), 219(B3), 221(B3), 223(B3), *245*
Fillus, H., 52(K7), *105*
Finkelstein, N. P., 2(V2), 84(V2), *110*
Fischer, R. E., 63(M42), *107*
Fisher, J. R., 17(F2), *102*
Fitzhugh, E. F., Jr., 46, 77, *102, 108*
Flanders, H. E., 75(N1), 76(N1), *107*
Fleming, R. J., 175(R1), 176(R1), 177(R1), 178(R1), 179(R1), 182(R1), 183(F2), *189, 190*
Fletcher, A. W., 63, *102*
Forrest, W., 18(M18, M19), *106*
Forward, F. A., 3(F14), 5(H8), 27, 28, 34(F8), 35, 37(F11, V8), 38, 39(F6), 41, 43(F12, V9), 90, *103, 110*
Foster, J. S., 76(N6), *107*
Fox, A. L., 26, *101*
Franklin, J. W., 73, 76, *103*
Frary, F. C., 13(E3), 34(E3), *102*
Friedman, S. J., 280, *291*
Frint, W. R., 33(C1), *101*
Frisch, N. W., 55, *103*
Fulford, G. D., 248(F3), 257, 259, *291*

G

Garaud, B. S., 263(G4), 267, 272(G4), *291*
Gardner, G. O., 250(G1), 251, *291*
Gardner, W., 254, *291*
Gaudin, A. M., 4(G3), 8(G1), 20(G2), 54(D1), *101, 103*
Gawain, T. H., 220, 242(G1), *245*
Gelperin, E. N., 132(G1), *189*
Gelperin, N. I., 132, 136, *189*
George, D. R., 5(G6), *103*
Gerlach, K. J., 38, *103*
Ghosh, B., 123(G3), 126, 176, *189*
Gibson, A., 169, *189*
Gilkin, P. G., 280(S5), *292*
Gishler, P. E., 112(M11), 113(M10), 114(M10), 115, 117(M10), 120(M10, T1), 123(M10), 124, 127(T1), 128(M10), 141(M10, T1), 142(M10, T1), 143(M10), 153(T1), 154(T1), 169(T1), 173(M10), 174(M10), 180(M10), 181, 183(M10), *189, 190, 191*
Glushko, G. S., 232, *245*
Goerling, C., 256, *290*
Goldstein, S., 249(G5), 250(G5), *291*

Goltsiker, A. D., 114(G7), 120, 123(G7), 130, 136, 167(G7), 172, 173, *189*
Golubev, L. G., 123(N1), 130, *190*
Gorshtein, A. E., 123(G8), 130(M16, M17), 136, 140, 145, 150, 151, 153, 165(M16), 169(M16, M17, 172, *189, 190*
Gosman, A. D., 212(G3), 231, *245*
Gow, W. A., 17(H13, H14), *104*
Gray, P. M., 34(G8), 36, *103*
Grimes, M. E., 64(G10), *103*
Grinstead, R. R., 66(G11), *103*
Griswold, G. G., Jr., 20(M41, S11), *106, 108*
Gruenfelder, J. G., 3(G12), *103*
Gruzensky, W. G., 64(G13), *103*
Gurr, C. G., 253, *291*
Gustafson, E. G., 52(K7), *105*

H

Habashi, F., 18, 80(H1, H4, H5), *103*
Haines, W. B., 255, *291*
Halpern, J., 3(H7), 5(H8), 34(F8), 35, 39(F6, P7), 45, *103, 106, 107*
Hancher, C. W., 57(A8), *99*
Hanjalic, K., 221, 235, 237, *245*
Hanson, C., 69, 71, *103*
Happel, J., 141(H1), *189*
Harada, T., 63(H11, H12), *104*
Hard, R. A., 62, *100*
Harlow, F. H., 221, 232, 233, 237, *245*
Harmarthy, T. I., 253, *291*
Harris, V. G., 201(C4), 219(C4), 222(C4), 236(C4), 237(C4), 238(C4), 239(C4), 241(C4), *245*
Harrison, D., 112(D2), *189*
Harrison, V. F., 17(H13, H14), *104*
Hatton, A. P., *245*
Hattori, H., 137(M6), 140, 142, *190*
Haver, F. P., 23, 24, *104*
Hayashi, S., 258(T5, T6), 259(T5, T6), 280(T5), *293*
Hayden, W. M., 87(H17), *104*
Hazen, W. C., 72, *105*
Hedley, N., 18, 19(H19), 20(H18, H20), *104*
Heindl, R. A., 26(H22, H23), *104*
Heinen, H. J., 20(E8, H24), *102, 104*
Heiser, A. L., 112(H2, S1), *189, 190*
Heister, N. K., 3(H25), 52(H25), *104*
Helbronner, A., 16, *108*

Helfferich, F., 3(H26), 52(H26), *104*
Hendriksson, S. T., 78(W5), *110*
Henrie, T. A., 20(S4), *108*
Henry, H. C., 251(H3), 252, *291*
Henry, P. S. H., 253, *291*
Herbst, W. A., 114(M8), *190*
Herring, A. P., 26, *104*
Herring, H. J., 201, 202, 210, 222, 225, 233, *245, 246*
Herwig, G. L., 56(A7), *99*
Herzog, E., 14, *104*
Hester, K. D., 63, *102*
Hiester, N. K., 56, *110*
Higbie, K. B., 5(G6), 64(B6), *100, 103*
Higgins, I. R., 57(H29), *104*
Hills, R. C., 21, *102*
Hinkle, M. E., 16(C23), 17(C22), *101*
Hinze, J. O., 233, *245*
Hiraoka, M., 281, *291*
Hirt, C. W., 233, *245*
Hoertel, F. W., 26, *105*
Hoftijzer, P. J., 280, *293*
Hollis, R. F., 59, *104*
Holmes, T. W., 89(M22), *106*
Holt, P. H., 114(M8), *190*
Holt, R. J. W., 3(M14), 52(M16), *106*
Hougen, O. A., 254, 255(C4), *291*
House, J. E., 64(S36), 65(A4), *99, 109*
House, S. E., 64(L13), *105*
Howard, E. V., 32, *104*
Howard, L. N., 200, *245*
Hudson, A. W., 29(H32), *104*
Hughson, M. R., 17(H13), *104*
Hunt, C. H., 167(H3), 168, *189*
Hurlbut, C. S., Jr., 10(H33), *104*
Hurst, F. J., 63(H34), *104*
Husband, W. H. W., 33(O3), *107*
Hussain, A. K., M. F., 214, *245*
Hussey, S. J., 59, *104*
Hutcheon, W. L., 253, *291*
Huttl, J., 30(H36), 36, *104*
Hutton, J. T., 253(G6), *291*

I

Ingraham, T. R., 24(D27, D28, D29), 75, *102, 104, 110*
Ioannou, T. R., 63(I2), *104*
Irving, J., 29(I3), *104*

Isaeoff, E. G., 52 (K7), *105*
Ivanov, M. V., 15(K10), *105*
Ivanov, V. I., 17(I4), *104*
Ivarson, K. C., 17(H14), *104*
Iverson, H. G., 5(I5), *104*
Iwasaki, I., 35, 42, *104*
Izzo, T. F., 59, *107*

J

Jackson, K. J., 23(J1), *105*
Jacobi, J. S., 12(J2), *105*
Jacobsen, F. M., 72, *105*
Jakob, M., 249, *291*
Jameson, A. K., 16(B33), 17(B33), *100*
Jarman, A., 20(J4), *105*
Jeffries, Z., 13(E3), 34(E3), *102*
Jelden, C. E., 75(N1), 76(N1), *107*
Jessen, F. W., 33(D26), *102*
Joe, E. G., 66(J5, R10), *105, 108*
Joffe, J. S., 16, *110*
Johnson, P. H., 29, 45, *105*
Johnson, P. W., 12(P8), *107*
Johnston, J. P., 209(J1), 225, *245*
Johnston, W. E., 19(J8), *105*
Jones, L. H., 20(P4), *107*
Jones, L. W., 17(B34), *100*
Jones, W. P., 221(H1), 235(H1), 237(H1), *245*
Joseph, T. L., 25(D8), *101*
Jury, S. H., 57(A8),

K

Kahata, H., 42(I7), *104*
Kajic, J. E., 17, *105*
Kamei, S., 253, *291*
Kaneko, T. M., 47 (D17, 48(D17), *102*
Kasahara, A., 215, *245*
Kauh, J. Y., 251, 258(P2), 263(K2), 270(K2), 272, 273(K2), 287(K3), *291, 292*
Kays, W. M., 202, 205(L2), 229(L2), *245, 246*
Kearney, D., 230, *245*
Kelsall, D. F., 72, *105*
Kenahan, C. B., 79, *105*
Kendall, J. M., 215, *245*

Kenny, H. C., 21(B15), *100*
Kentro, D. M., 20(H18), *104*
Kerby, R., 75, *104*
Kershner, K. K., 26, *105*
Kestin, J., 251, *291*
Kindig, J. K., 72, *105*
King, G., 253(C3), *291*
Kinzel, N. A., 16, *105*
Kirby, R. G., 12(P8), *107*
Kirkwood, K. C., 274, 275, *291*
Klassen, J., 120(T1), 127(T1), 141(T1), 142(T1), 153(T1), 154(T1), 169(T1), *190*
Klausner, Y., 256, *291*
Kline, S. J., 206(K5), 222(K5), *245*
Kolta, G. A., 50, *100*
Koslov, J., 64(B23), *100*
Koyanagi, M., 178(K1), *189*
Kraft, R., 256, *291*
Krischer, O., 254, *292*
Ksenzhek, O. S., 256, *292*
Kudryk, V., 42(F10), *103*
Kugo, M., 123(K2), 125, 135, 159, 161, *189*
Kunii, D., 177, *189*
Kunin, R., 52(K7, K8, P12), *105, 107*
Kurushima, H., 96 (K9), 97, *105*
Kuzmak, J. M., 253, *292*
Kuznetsov, S. I., 15(K10), *105*

L

Lama, R. F., 123(M1), 134, 135, *189, 190*
Lamborn, R. H., 2(L1), 74(L1), *105*
Latzko, H., 249, *292*
Launder, W. F., 221(H1), 235(H1), 237(H1), *245*
Lawrence, H. M., 21(L2), *105*
Leaphart, C., 29, *106*
Leathen, W. W., 16(L3, L5), 17(L4), *105*
Leaver, E. S., 18(L8), 20(L6, L7), 25(D8), *101, 105*
Lefroy, G. A., 134, 135, 137(L2), 138, 145(L2), 148, 151, 157, 164, 165, 167, 170, 183, *190*
LeGay, E., 20(J4), *105*
Leitch, H., 5(I5), *104*
Lemmon, R. J., 19(L10), 20(L9), *105*
Lessels, V., 20(R13), *108*
Lester, D. H., 259, *292*

Leva, M., 114(L3), *190*
Levenspiel, O., 160(L4), 177, *189, 190*
Leverett, J., 256, *292*
Lewis, C. J., 63(L11), 64(A3, L13), 67(L12), *99, 105*
Lewis, W. K., 253, *292*
Lightfoot, E. N., 249(B2), 250(B2, B3), *290, 291*
Lighthill, M. J., 219(L1), *245*
Lilge, E. O., 47, *105*
Lindstrom, R. E., 20(S4), 64(B6), *100, 108*
Link, R. F., 26(T14), *109*
Lodding, W., 47, *105*
Long, R. S., 66(E6, G11), 67, *102, 103, 108*
Lowe, E. A., 17(D24), *102*
Lowenthal, W., 112(H2), *189*
Lower, G. W., 20(L16), *105*
Lowrie, R. S., 68(R18), *108*
Loyd, R. J., 205, 229, *246*
Lu, B. C. Y., 134, 139(M4), 165(M4), 168(M3), 177, 178, 179, 183, 185(M4), *190*
Lumley, J. L., 219, 234, 235, 237, 242, *246*
Lundgren, D. G., 17(S17), *108*
Lundgren, T. S., 200(L4), *246*
Lundquist, R. V., 13, *105*
L'vova, S. D., 132(G1), *189*
Lyalikova, N. N., 15(K10), *105*
Lynch, E. J., 251(L5), *292*
Lyons, D. A., 23, *105*

M

McArthur, C. K., 59, *104*
McArthur, J. A., 29, *106*
McArthur, J. S., 18(M18, M19), *106*
McCabe, C. L., 6(M20), 7(M20), *106*
McCauley, H. J., 254(H6), 255(H6), *291*
McCormick, P. Y., 248(M3, M4), *292*
Macdonald, R. J., 54(D1), *101*
MacDonald, R. D., 14, 24(D27, D28, D29), *102, 109*
McGarvey, F. Z., 55, *103*
MacGregor, R. A., 17(M1), 18(M2), 33, *105*
McIntyre, L. D., 17(4), *105*
McKay, D. R., 45, *106*
Mackay, T. L., 39, *105*
McKinley, H. L., 89(M22), *106*
McKinney, W. A., 2(R14), *106, 108*

Mackiw, V. N., 2(M4), 3(M5, M25), 27, 34(M5, M6), 37(M7, P3, V5, V10), 41, 43(P3), 44(P3), 82, *103, 105, 106, 107, 110*
McNeill, R., 56, *106*
Madonna, L. A., 123(M1), 134, 135(M4), 139(M4), 165(M4), 168(M4), 185(M4), *190*
Magner, J. E., 28(M8), *106*
Mair, C. G., 280(S15), *292*
Malek, M. A., 123(M5), 126(M5), 134, 135, 139(M4), 165(M4), 168(M3), 177, 178, 179, 183, 185(M4), *190*
Malling, G. F., 273(T1), *293*
Malouf, E. E., 15(M9), 17(M9), 31(M10), 76(S20), *106, 108*
Mamuro, T., 137(M6), 140, 142, *190*
Mancantelli, R. W., 49, *106*
Mantell, C. L., 78(M12), *106*
Manurung, F., 112(B9), 115, 123(M7), 124, 125, 126(M7), 128, 133, 139(M7), 140, 160, 175, 176, 180(M7), 181, 185(M7), *188, 190*
Markim, V. S., 256, *292*
Marsden, D. D., 2(V2), 23, 84(V2), *106, 110*
Marshall, T. K., 253(G6), *291*
Marshall, W. R., Jr., 254(H6), 255(H6), 280, *291, 292*
Martin, F. S., 3(M14), 52(M14), *106*
Martin, W. L., 33(S15), *108*
Maschmeyer, S., 37(M15), *106*
Maslenitsky, N. W., 40, *106*
Massimilla, L., 167(V2), 168(V2), *191*
Matheson, G. I., 114(M8), *190*
Mathur, K. B., 112(M10), 113(M10, 114(M10), 115, 117(M10), 120(M10, T1, T2), 122(T1, T2), 123(M10), 124, 126(T1, T2), 127(T1, T2), 128(M10), 135(T2), 139(T2), 141(M10, T1, T2), 142(M10, T2), 143(M10, T2), 145(T2), 146(T2), 147(T2), 153(T1), 154(T1, T2), 157(T2), 165(M9, T2), 168(T2), 169(T1), 180(M10), 181(M10), 183, 185(M9), *189, 190, 191*
Matsen, J. M., 169(M12), *190*
Maurer, E. E., 63(M42), *107*
Max, D. A., 275, 279(M2), *292*
Meddings, B., 3(M25), 82, *106*
Mellor, G. L., 201, 202, 205, 207(M1), 209, 210, 222, 225, 233, *245, 246*

Mellor, J. W., 18(M26), *106*
Merenkov, K. V., 122(T5), 130(T5), 165(T3), *191*
Merigold, C. R., 65, *101*
Mikhlaik, V. D., 145(M13), 150, 156(E1), 163, 166(M13), 167(M13), 168, 175(E1), *189, 190*
Miller, A., 29(M27), *106*
Miller, C. O., 280(P5), 281, *292*
Miller, R. W., 12(T10), *109*
Mindler, A. B., 52(M28), *106*
Mioen, T., 78(W5), *110*
Miskell, F., 280, *292*
Mitchell, J. S., 35, 47, *106*
Mitchell, T. J., 274, 275, *291*
Mitchell, T. R., Jr., 280, *292*
Moffat, R. J., 205(L2), 206(S1), 229(L2), *246*
Moison, R. L., 56, *106*
Monninger, F. M., 75, *106*
Moore, J. D., 64(B23), *100*
Morgan, J. A., 6(M20), 7(M20), *106*
Morgan, R. P., 258(A3), *290*
Morris, J. B., 123(C1), 126(C1), *189*
Morrison, B. H., 4(M40), *106*
Morrow, B. S., 20(M41), *106*
Mosinskis, G., 207(C3), *245*
Mukhlenov, I. P., 123(G8), 130(M16, M17), 136, 140(M15), 145, 150, 151, 153, 165(M16), 169(M16, M17), 172, *189, 190*
Musgrove, R. E., 73(M42), *107*
Myers, R. Y., 52(K8), *105*
Myklestad, O., 273, *292*

N

Nabiev, M. N., 112(V4), 122(T5), 130(T5), 165(T3), *191*
Nadkarni, R. M., 75(N2, N3), 76(N2, N3), *107*
Nagirnyak, F. I., 17(I4), *104*
Nakayama, P. I., 221, *245*
Naot, M. M., 242, *246*
Napier, E., 53(E11), *102*
Nash, J. F., 224, *246*
Nash, W. G., 74(N4), *107*
Nashner, S., 49, *107*
Newitt, R., 256, *291*

Newman, A. B., 253, *292*
Ng, W. K., 11, *101*
Nielson, R. H., 62, *101*
Nikalaev, A. M., 123(N1), 130, *190*
Nolfi, F. V., 76, *107*
North, A. A., 66, *107*

O

O'Hern, H. A., Jr., 56, *106*
O'Kane, P. T., 37(V4), 50, *107, 110*
Okazaki, S., 258(T6), 259(T6), *293*
O'Leary, V. D., 12, *107*
Olson, E. H., 12(B5), *99*
Oppenheimer, C. H., 15(K10), *105*
Osberg, G. L., 118(T2), 120(T2), 122(T2), 124(T2), 126(T2), 135(T2), 139(T2), 141(T2), 142(T2), 143(T2), 145(T2), 146(T2), 147(T2), 154(T2), 157(T2), 165(T2), 167(T2), 168(T2), 179(C4), 180(T2), *189, 191*
Othmer, D. F., 129(Z2), 132(Z2), 161(Z2), 162(Z2), *191*
Ozsahin, S., 33(O3), *107*

P

Painter, L. A., 59, *107*
Palmer, R., 17(B35), *100*
Panlasigue, R. A., 22, *107*
Patankar, S. V., 207, *246*
Paulson, C. F., 52(M28), *106*
Pavlova, A. I., 112(V4), *191*
Pawlek, F. E., 38, *103*
Pearce, R. F., 37(P3), 43(P3), 44, *107*
Peck, R. E., 251(K3), 258(P2), 272, 287(K3), *291, 292*
Pellegrini, S., 37(V5), *110*
Penneman, R. A., 20(P4), *107*
Perlov, P. M., 40, *106*
Perkins, E. C., 14, *107*
Perry, J. H., 61(P6), *107,* 117(P1), *190*
Persen, L., 251, *291*
Peters, E., 15(M9), 17(M9), 39(P7), *106, 107*
Peters, F. A., 12(P8), *107*
Peterson, D. W., 26, *110*
Peterson, W. S., 112(P2), 177, 179(C4), *189, 190*

Petrovic, L. J., 273(T2), *293*
Pickering, R. W., 22, *100*
Pimentel, L., 287, 288, *292*
Polhausen, E., 248, *292*
Porter, B., 20(H24), *104*
Powell, H. E., 63(P9), *107*
Powell, J. E., 54, 55, *107*
Prater, J. D., 5(G5), 16(22), 17, 18(22), 31(M10), 76(S20), *103, 106, 108, 110*
Preuss, A., 52(P12), *107*
Prillig, E. B., 112(S1), *190*
Pritchard, J. W., 220, 242(G2), *245*
Prutton, C. F., 280, *292*
Pun, W. M., 212(G3), 231(G3), *245*

Q

Queneau, P., 2(B25), 3(Q1), 13(B25), 17(B25), 35(B25), 49, 79, 82(B25), 90(B25), *100, 107*
Quinlan, M. J., 160(Q1), *190*

R

Radcliffe, J. S., 160(Q1), *190*
Rahn, R. W., 67, *107*
Rai, C., 257, *292*
Ralston, O. C., 12(T10), 23(S12), *105, 108, 109*, 280(S15), *292*
Rampacek, C., 30, 43, 44(S22), *106, 108, 109*
Rashkovskaya, N. B., 112(R2), 113, 136(G6), 173(R4), 174(R4), 175(R4), *189, 190*
Raso, G., 120, 121, 122(V1, V3), 167(V2), 168(V2), *191*
Ratcliffe, J. S., 112(B2), 160(B2), 161(B2), *188*
Ravitz, S. F., 25(R5), 26(B3, R4), *99, 104, 107*
Razzell, W. E., 17, *107*
Read, F. O., 59(R7), *107*
Reddy, K. V. S., 123(S2), 125, 127(S2), 129(S2), 175, 176, 177, 178, 179(S2), 182, *190*
Reger, E. O., 112(R2), *190*
Ramirez, R., 86(R8), *107*
Reynolds, A. J., 236(T2), 239(T2), 240(T2), 241(T2), *246*
Reynolds, D. H., 17(B17), 27(B18, B19), *100*
Reynolds, W. C., 214, *245*
Richards, J. C., 138(B10), *189*
Richards, L. A., 256, *292*
Richardson, J. F., 148, 177(R3), *190*
Rigby, G. R., 112(B2), 100(B2), 161(B2), *188*
Ritcey, G. M., 3(R9), 52(R9), 66(J5), *105, 108*
Rizaev, N. U., 122(T5), 130(T5), 165(T3), *191*
Roberson, A. H., 74, *110*
Roberts, E. G., 3(R11), 31(R11), *108*
Roberts, R. T., 57(H29), *104*
Rodi, M. M., 220, 231, *246*
Roman, R. J., 17(B17), 27(B18), *100*
Romankiw, L. T., 37(R12), *108*
Romankov, P. G., 112(R2), 113, 136(G6), 173(R4), 174(R4), 175(R4), *189, 190*
Romero, N. C., 233, *245*
Rose, D. H., 20(R13), *108*
Rosenbaum, H., 232, *245*
Rosenbaum, J. B., 2(R14), *108*
Ross, A. H., 5(H8), *103*
Ross, J. R., 4(R15), 5(G5, R15), *103, 108*
Rotta, J. C., 220(R1), 231, *246*
Roy, T. K., 82, *108*
Rudolfs, W., 16, *108*
Runchal, A. K., 212(G3), 231(G3), *245*
Ruppert, J. A., 26(H22, H23), *104*
Russell, B., 200, *246*
Ryan, V. H., 87(R17), *108*
Ryon, A. D., 67(B30), 68, *100, 108*

S

Saccone, L., 122(V3), *191*
Saeman, W. C., 280, *292*
Sallans, H. R., 160, *188*
Samis, C. S., 42(F10), 44, *103, 108*
Sato, H., 274, 275, 277(S2), *292*
Saubestre, E. B., 83, *108*
Saunby, J. B., 118(T2), 120(T2), 122(T2), 124(T2), 126(T2), 135(T2), 139(T2), 141(T2), 142(T2), 143(T2), 145(T2), 146(T2), 147(T2), 154(T2), 157(T2), 165(T2), 167(T2), 168(T2), 180(T2), *191*
Schack, C. H., 4(R15), 5(R15), *108*

AUTHOR INDEX

Schaufelberger, F. A., 82, 90(S2), *108*
Scheiner, B. J., 20(S4), *108*
Schlain, D., 79, *105*
Schlichting, H., 249(S3), 250(S3), *292*
Schneider, P. G., 280(S4), *292*
Schofeld, F. R., 280(S5), *292*
Schuhmann, R., Jr., 4(G3), *103*
Schuette, W. H., 280(P5), 281(M5), *292*
Schwab, D. A., 27(B18), *100*
Sege, G., 69, *108*
Seidel, D. C., 30, 46, 77, *102, 108*
Sen Gupta, A., 273(T3), 275(T3), *293*
Seraphim, D. P., 44, *108*
Sereda, P. J., 253, *292*
Shavit, M. M., 242(N3), *246*
Shaw, K. G., 66(G11), 67, *103, 108*
Sheffer, H. W., 3(S10), 12(S10), 29(S10), 31(S10), *108*
Sheridan, G. E., 20(S11), *108*
Sherman, M. I., 23(S12, S13), *108*
Sherwood, T. K., 253, 256, *292*
Sheth, H. P., 259, *292*
Shewmon, P. G., 76(N6), *107*
Shock, D. A., 33(D3), 34, *101*
Shoemaker, R. S., 15(M9), 17(M9), 29(S14), *106, 108*
Shibata, T., 123(K2), 125(K2), 135(K2), 159(K2), 161(K2), *189*
Siebert, H., 47, *105*
Sievert, J. A., 33(S15), *108*
Silo, R. S., 27, *108*
Silverman, M. P., 17(S17), *108*
Simons, C. S., 2(C4), 3(C5), 94(C4), *101*
Simpson, R. L., 206(S1), *246*
Sims, C., 23(S18), *108*
Singiser, R. E., 112(H2, S1), *189, 190*
Skow, M. L., 13, 26(H22, H23), *104, 108*
Slitcher, C. S., 255, *292*
Smith, A. M. O., 207(C3), *245*
Smith, B. A., 280, 281(M5), *292*
Smith, J. W., 123(S2), 125, 127(S2), 129(S2), 175(R1), 176(R1), 177(R1), 178(R1), 179(R1, S2), 182(R1), *190*
Smith, L. L., 63(P9), *107*
Smith, S. E., 67(W12), *110*
Smith, W. A., *190*
Smutz, M., 12(B5), 63(B24, C10, H11, H12, I2), 67(C11), *99, 100, 101, 104, 107*
Sousa, L. E., 89(M22), *106*
Spalding, D. B., 207, 212(G3), 220, 231(S3), *245, 246,* 251(S13), *292*

Spedden, H. R., 76, *108*
Spedding, F. H., 54, 55, 63(B24), *100, 107*
Spence, W. W., 80, *108*
Spraul, J. R., 280, *292*
Stahmann, W. S., 27(B19), *100*
Stanczyk, M. H., 43, 44(S20), *108, 109*
St. Clair, H. W., 13(S24), *109*
Stepanov, B. A., 17(I4), *104*
Stephens, F. M., 14, *109*
Stevens, J. W., 112(B8), *188*
Stewart, R. H., 215, *246*
Stewart, R. M., 56(A7), *99*
Stewart, W. E., 249(B2), 250(B2, B3), *290, 291*
Stickney, W. A., 25(S27), 26(T14), 63, *109*
Stone, R. L., 40, *109*
Strickland, J. D. H., 23 (J1, S12, S13), *105, 108*
Suckling, R. D., 112(B8), *189*
Sullivan, J. D., 16, 29(S31), 30, *109,* 280, *292*
Sullivan, P. M., 22, *110*
Sulman, H. L., 20(S34), *109*
Sutton, J. A., 16, 17(C26), *101*
Swanson, B. S., 263(S16), 266, 267(S16), 268(S16), 272(S16), *293*
Swanson, R. R., 64(S36), 65(A4), *99, 109*
Swift, J. H., 26(T19), *109*
Swinton, E. A., 56(A7, M24), *99, 106*

T

Tabachnick, H., 18, 19(H19), 20(H20, H21), *104*
Takahashi, Y., 42(I7), *104*
Tame, K. E., 5(G6), 25(R5), 26(B3, R4), *99, 103, 107*
Taylor, J. H., 2(T1), 29(T1), 31, *109*
Temple, K. L., 16(C23, T2, T3), 17(T2, T3), *101, 109*
Thodos, G., 273, 275(T3), *293*
Thomas, R. W., 33(T4), *109*
Thompson, B. H., 69(D30), *102*
Thorley, B., 118, 120(T1, T2), 122(T2), 124, 126(T2), 127(T1), 135, 139(T2), 141(T1, T2), 142(T1, T2), 143, 145, 147, 154, 157, 165, 167, 168(T2), 169, 180, *190, 191*
Thornhill, E. B., 25, *109*
Tiemann, T. D., 14, 40(S29), *109*

AUTHOR INDEX

Tilley, G. S., 12(T10), *109*
Timokhova, L. P., 136(G2), *189*
Todd, D. B., 70, *109*
Toei, R., 258, 259(T5, T6), 280(T5), 281, *291, 293*
Tolun, R., 66(T12), *109*
Torres-Acuna, N., 80, *103*
Tougarinoff, B., 12(T13), *109*
Town, J. W., 25, 26, *109*
Townsend, A. A., 200, 216(T1), 219(T1), 236(T1), 244(T2), *246*
Tremblay, R., 67(T15), *109*
Treybal, R. E., 69, *109*
Trussell, P. C., 15(T17), 17(D22, D23, D24, T17), *102, 109*
Tschirner, H. J., 87(R17, T18), *108, 109*
Tsunoda, S., 96(K9), 97, *105*
Tsvik, M. Z., 122(T5), 130, 165(T3), *191*
Tucker, H. J., 236(T2), 239(T2), 240(T2), 241(T2), *246*
Tuttle, F. J., 253, *293*
Turner, T. L., 26(T19), *109*

U

Uchida, K., 23(H15), 24(H16), *104*
Uemaki, O., 123(K2), 125(K2), 135(K2), 159(K2), 161(K2), *189*

V

Van Arsdale, G. D., 3(V1), 29(V1), 33(V1), 84(V1), 85(V1), *109*
Van Goetsenhoven, F., 12(T13), *109*
van Krevelen, D. W., 280, *293*
van Zyl, J. J. E., 2(V2), 84(V2), *110*
Vedensky, D. N., 25(V3), 26(V3), *110*
Veltman, H., 27(F12), 37(F11, M7, V4, V5, V8, V10), 43(F12, V9), *103, 106, 110*
Vermeulen, T., 56, *110*
Vizsolyi, A., 37(V8, V10), 38, 43, *110*
Vizsolyi, H., 27(F12), 28, 43(F12), *103*
Volpicelli, G., 120, 121, 122(V1, V3), 167(V2), 168, *191*
von Hahn, E. A., 75, *110*
Vyzago, V. S., 112(V4), 122(T5), 130(T5), *191*

W

Wadia, D. R., 11, *110*
Wadsworth, M. E., 3(W2), 11, 39, 41(D18), 47(D17), 48(D17), 75(N1, N2, N3), 76(N1, N2, N3), *102, 105, 107, 110*
Wakabayashi, M. M., 254, *293*
Waksman, S. A., 16, *110*
Walden, C. C., 15(T17), 17(D22, D23, D24, T17), *102, 109*
Walden, S. J., 78, *110*
Walker, I. K., 253, *293*
Walker, R. B., 17(B35), *100*
Walsh, T. H., 125(M5), 126(M5), *190*
Warner, J. P., 37(P3), 43(P3), 44(P3), *107*
Warren, I. H., 3(F14), 37(V10, W6), *103, 110*
Wartman, F. S., 74, *110*
Wasan, D. T., 250(W3), 251(K3, W4), 252, 287(K3), *291, 293*
Watanabe, N., 123(K2), 125(K2), 135(K2), 159(K2), 161(K2), *189*
Weber, E. J., 70, *101*
Weed, R. C., 33, *110*
Weiss, D. E., 56(A7), *99*
Wells, R. A., 3(E12), 52(E12), 53(E11), 66, *102, 107*
Welsh, J. Y., 26(W9), *110*
Wen, C. Y., 182, *191*
Whatley, M. E., 67, *110*
Wheelock, T. D., 22, *107*
Whelan, P. F., 2(T1), 29(T1), 31, *109*
Whitaker, J. F., 87(W11), *110*
White, P. A. F., 67(W12), *110*
Whitten, D. G., 206(S1), *246*
Widtsoe, J. A., 254, *291*
Wilke, C. R., 250(W3), 251(L5, W4), 252, *292, 293*
Williams, G. H., 123(C1), 126(C1), *189*
Williams, L. A., 87(T18), *109*
Williams, L. M., 2(W13), *110*
Wilsdon, B. H., 254, *293*
Wilson, D. A., 22, *110*
Wilson, D. G., 16(Z2), 17, 18(Z2), *110*
Wilson, F., 2(W15), 94(W15), *110*
Winchell, H. V., 74(W16), *110*
Windolph, F. J., 88(W17), *110*
Wolfshtein, M., 212(G3), 231(G3), 242(N3), *245, 246*

Wong, M. M., 23(H15), 24(H16), *104*
Woodcock, J. T., 31(W18), *110*
Woodfield, F. W., 69, *108*
Woodward, J. R., 49, *106*
Woolf, J. A., 18(L8), 20(L6, L7), *105*
Wu, S. M., 40(S29), *109*
Wyman, W. F., 25(R5), 26(R4), *107*

Y

Yu, Y. H., 182, *191*

Yurko, W. J., 52(Y1), 84(Y1, Y2), *110*

Z

Zabrodsky, S. S., 114(Z1), 156(E1), 175(E1), *189, 191*
Zakarias, M. J., 3(Z1), *110*
Zaki, W. N., 148, 177(R3), *190*
Zenz, F. A., 129, 132(Z2), 161(22), *191*
Zimmerley, S. R., 16(Z2), 17, 18(Z2), *110*
Zubryckyi, N., 27, *110*

SUBJECT INDEX

A

Acid pressure leaching, 36–39
Acid solution, in elevated pressure leaching, 35–39
Air drying, of solids, 247–289
Air velocity, drying time and, 270–271
Alkaline pressure leaching, 39–41
Alumina, extraction of from silicates, 5
Aluminum, ore minerals of, 9
Ammonia leaching, 20–22
Ammonium acetate leaching, 44–45
Ammonium carbonate, in magnesium recovery, 26
AMSCO mineral spirit diluent, in solvent extraction, 68
Anaconda Company, 12
Anion exchangers, in hydrometallurgy, 64
Antimony, ore minerals of, 9
Aqueous ammonia solution, pressure leaching with, 41–42
Aqueous solutions
  displacement reactions in, 74–78
  metal reduction from, 72–83
Arizona Chemcopper Company, 84–85
Arsenic, ore minerals of, 9
Arsenical concentrates, cobalt extraction from, 35
Atmospheric pressure leaching, 11–34
  acid solution in, 11–13
  aqueous ammonia solution in, 20–22
  ammonium carbonate in, 26
  bacterial media in, 15–18
  basic solution in, 13–15
  cyanidation and, 18
  dissolution media in, 11–28
  hydrochloric acid in, 26
  methods used in, 29–34
  nonaqueous solvents in, 27–28
  solution mining and, 32–34
  tank or vat method in, 29–30
  water source in, 12
Autoclave, for pressure leaching, 48

B

Bacterial media, in atmospheric pressure leaching, 15–18
Bagdad Copper Company, 65
Bauxites, upgrading of, 13
Bedded resins, in resin ion exchange, 54–58
Bornite, in atmospheric pressure leaching, 24
Brine, in atmospheric pressure leaching, 23

C

Capillary theory, for porous solids, 255–257
Carnotite ores, leaching of, 15
Caron ammonia process, 42
Cement copper, 84
Centrifugal contactor, in solvent extraction, 70–71
Ceramic granules, drying curves for, 276
Chalcopyrite, in atmospheric leaching, 24
Chemical Construction Company, 90
Chemical reduction, hydrometallurgy and, 81–83
Chromium, ore minerals of, 9
Clay, drying curves for, 268
Climax Molybdenum Company, 88
Cobalt
  acid pressure leaching of, 37–38
  from arsenical concentrates, 35
  chemical processing of, 83
  electrorefining of, 81

hydrometallurgy of, 2
recovery flow diagram for, 92
Copper
   from acidic dump leach liquors, 65
   cementation launder for, 76
   cyanidation-precipitation of, 20
   electrolysis of, 86, 98
   feed materials in hydrometallurgy of, 84–88
   hydrometallurgy of, 84–88
   ore minerals of, 9
   precipitation from sulfate solutions by metallic iron, 74–75
   recovery of, 4–5
   scrap metal leaching in, 87
Copper electrolysis, 86
   in Kosaka process, 98
Copper industry, hydrometallurgical processing in, 83–88
Copper ores
   heap leaching of, 29
   hydrochloric acid leaching of, 27
Copper oxide ores, hydrometallurgy of, 86–87
Copper sulfide flotation concentrates, leaching of, 65
Copper sulfide minerals, ammonia leaching of, 43
Copper sulfide waste dumps, leaching of, 31–32
Copper-zinc concentrates, hydrometallurgy of, 96–98
Couette flow solution, in turbulent flows, 204–205
Cubanite, in atmospheric pressure leaching, 24–25
Cyanidation, in atmospheric pressure leaching, 18
Cyanide solutions, solubility of minerals in, 18–20

D

Diffusion theory, for porous solids, 253–254
Displacement reactions, in aqueous solutions, 74–78
Dissolution media, in elevated pressure leaching, 35–46

Dodecyl phosphoric acid, uranium recovery with, 64
Dowa Mining Company, 96
Driers
   packed-bed, 273–279
   rotary, 279–288
Drying
   defined, 247
   heat- and mass-transfer coefficients in, 248–252
   of porous solids, 253–258
   rotary driers for, 279–288
Drying curves
   for ceramic granules and coke, 276
   generalized, 262
   for porous aluminum oxide, 275
   theoretical vs. experimental, 263–267, 275–278
Drying time
   air velocity and, 265, 270–271
   humidity and, 264, 266
   initial moisture and, 266
   thickness and, 264, 267
Dump leaching, 31–32

E

Electrorefining, 80–81
Electrowinning, in hydrometallurgy, 78–80
Elevated pressure leaching, 34–50
   acid solution in, 35–39
   alkaline solution in, 39–41
   aqueous ammonia solution in, 41–44
   dissolution media in, 35–46
   equipment used in, 46–50
Ethylene glycol, in atmospheric leaching, 28
Evaporation-conduction theory, for porous solids, 253
Extractive metallurgy, defined, 1

F

Ferric sulfate, in atmospheric pressure leaching, 24
Ferrobacillus ferroxidans, 16
Ferrobacillus sulfoxidans, 16
Ferrous ores, bacterial oxidation and, 16
Fluctuating velocity field (FVF) closure, 199

Fluidization, spouting and, 111–112
Freeport Nickel Company, 2, 94–95

G

Galena, ammonium acetate oxidation of, 44–45
Gas flow rate, spouting and, 118–119
Gas-solids contacting spectrum, spouting in, 115–117
Gold
  in cyanide solutions, 20
  ore minerals of, 9, 20
  resin ion exchange recovery of, 59
Gold ore, cyanidation of, 20

H

Heap leaching, 29
Heat convection, in laminar and turbulent flow, 248–250
Heat transfer, mass transfer and, 250–251
Heat-transfer coefficient, in drying, 248–252
Homogeneous flows, new ideas for, 236–242
Humidity, drying time and, 263–266
Hydrochloric acid, in atmospheric pressure leaching, 26
Hydrogen reduction, in hydrometallurgy, 81–82
Hydrometallurgy, 1–99
  air/water pollution and, 3
  anion exchangers in, 64
  atmospheric pressure leaching in, 11–36
  chemical reduction in, 81–83
  of copper, 84–88
  of copper-zinc concentrates, 96–98
  current interest in, 2
  defined, 1–2
  electrolysis in, 78–81
  electrorefining in, 80–81
  electrowinning in, 78–80
  elevated pressure leaching in, 34–50
  and metal reduction from aqueous solutions, 72–83
  of molybdenum, 88–90
  ore concentration in, 4
  processing operations in, 83–99
  raw material preparation in, 3–7
  reduced-pressure leaching in, 50–51
  resin-in-pulp method in, 58–61
  resin ion exchange in, 52–61
  separation and concentration processes in, 51–72
  solid reductant in, 82
  solvent extraction in, 61–72
  solvent-in-pulp extraction in, 66–67

I

Ion-pair transfer, in solvent extraction system, 62–63
Iron, ore minerals of, 9
Iron laterites, nickel from, 46
Iron ore, silica removal from, 14
Iron oxides, magnetic vs. nonmagnetic, 14
Iron-oxidizing bacteria, 16–17
Isotropic disturbance, decay of, 236–238

K

Kennecott Copper Corp., 16

L

Laminar flow, forced heat convection in, 248–249
Lanthanum, separation of from monazite chlorides, 64
Laterite ores, hot water leaching of, 27
Leached ore slurries, extraction from, 66–67
Leaching
  at atmospheric pressure, 11–34
  bacteriological, 15–18
  dump method, 31–32
  elevated-pressure type, 34–50
  heap method, 29
  in hydrometallurgy, 4–5, 7–51
  metal concentration in, 7
  percolation type, 30
  at reduced pressure, 50–51
  solution mining and, 32–34
  tank and vat method in, 29–30
Lead ores, 9
  ammonia leaching of, 22
  brine leaching of, 23
Liquid ion exchangers
  in hydrometallurgy, 64
  in solvent extraction, 63–65

Liquid-liquid contactors, in solvent extraction, 67–68

M

Magnesium, ore minerals of, 9
Manganese
  atmospheric pressure leaching of, 25–26
  ore minerals of, 9
Mass transfer, heat transfer and, 250–251
Mass-transfer coefficient, in drying, 248–252
Mean Reynolds-stress (MRS) closure, 231–236
Mean turbulent energy (MTE), future importance of, 242
Mean turbulent energy closure, 199, 216–223
Mean turbulent energy Newtonian (MTEN), 219–243
Mean turbulent energy structure (MTES), 219–228
Mean-velocity field (MVF) calculations, 206–215
Mean-velocity field closure, in turbulent flows, 200–215
Mean-velocity field predictions, 242
Mean-velocity field Newtonian (MVFN)
  calculations in, 211–215
  closures in, 201–202, 219–231
Mercury ores, 9, 25
Metals, ore minerals of, 9–10
Minerals, solubility of in cyanide solutions, 18–20
Mixing characteristics, for spouted beds, 158–163
Moisture, drying time and, 266
Molybdenite, leaching of, 41
Molybdenum
  hydrometallurgy of, 88–90
  ore minerals of, 10
Molybdenum oxide, hydrometallurgy of, 88–89
Molybdenum sulfide, hydrometallurgy of, 89–90
MRS closure, see Mean Reynolds-stress closure
MTE, see Mean turbulent energy
MTEN, see Mean turbulent energy Newtonian

MTES, see Mean turbulent energy structure
Multiple-bomb leaching equipment, 47
MVF, see Mean-velocity field
MVFN, see Mean-velocity field Newtonian

N

National Center for Atmospheric Research, 215
Navier-Stokes equations, in turbulent flow, 198–199
Nickel
  chemical processing of, 83
  electrorefining of, 81
  hydrometallurgy of, 2, 90–96
  ore minerals of, 10
  precipitation from acidic solutions, 77
Nickel-cobalt matte, ammonia pressure leaching of, 43
Nickel-cobalt sulfides, hydrometallurgy of, 96
Nickel ores, ammonia leaching of, 21, 43
Nickel oxide, hydrometallurgy of, 94
Nickel sulfide, hydrometallurgy of, 90–91
Nickel sulfide flotation, in Sherritt-Gordon process, 90
Nitric acid, in silver recovery, 12–13

P

Packed bed
  forced convection through, 250
  moisture distributions in, 277
Packed-bed drying, 273
Particulate iron precipitant, 76
Perchlorate-copper system, 75
Percolation leaching, 30
Pitchblende ores, leaching rate for, 39–40
Platinum, ore minerals of, 10
PMC-Powdered Metals Corp., 86
Podbielniak centrifugal contactor, 70–71
Porous solids
  batch drying of, 258–270
  capillary theory for, 255–257
  continuous drying of, 279–288
  diffusion theory for, 253–254
  drier design for thin materials in, 259–273
  drying theory for, 257–258

evaporation-condensation theory for, 253
moisture movement through, 252–258
rotating driers for, 279–288
tunnel driers for, 288
Precipitation cone-type recovery system, 76
Pressure drop, in spouted beds, 131–140
Pressure leaching
 atmospheric, 11–34
 elevated, 34–50
 equipment for, 46–50
 reduced, 50–51
Psychrometric ratio, correlation for, 251–252
Pyrite, oxidation of, 45
Pyrometallurgy, defined, 1–2

R

Rare earths, ion exchange separation of, 54–55
Reduced-pressure leaching, 50–51
Resin-in-pulp techniques, in hydrometallurgy, 58–61
Resin ion exchange
 bedded resins in, 54–58
 fixed bed in, 54–56
 in hydrometallurgy, 52–61
 movable bed in, 56–58
Rotary driers, 279–288
 calculations for, 287
Rotary drying, theory of, 285–286

S

Scandium, recovery of, 4
Scrap iron, precipitation cone type recovery system for, 76
Selenium, pressure leaching of, 4
Sheets, drying of, 247–289
Sheritt-Gordon recovery process, 2, 49, 90, 92
Silica recovery, caustic digester in, 13
Siliceous bauxites, upgrading of, 13
Siliceous iron ore
 alkaline pressure leaching of, 40
 sintering and leaching of, 14
Silver
 nitric acid recovery of, 12–13
 ore minerals of, 10
 recovery of, 4, 12–13, 59
 resin ion exchange recovery of, 59
Silver ore
 cyanidation of, 20
 minerals in, 10
Sodium bicarbonate pressure leaching, of uranium alloys, 49
Solid particles, drying of, 247–289
Solids, porous, see Porous solids
Solution mining, in atmospheric pressure leaching, 32–34
Solvent extraction
 centrifugal contactor in, 70–71
 differential extractors in, 68–69
 in hydrometallurgy, 61–72
 ion-pair transfer in, 62–63
 liquid ion exchange in, 63–65
 liquid-liquid contactors in, 67–68
 mixer-settlers in, 68
 solvent-in-pulp process in, 66–67
Solvent-in-pulp extraction process, 66–67
Sponge iron, as iron precipitant, 76
Spout, voidage distribution through, 169–173
Spouted beds
 see also Spouting
 air distribution in, 143
 annulus, solids flow velocity in, 153–158
 bed structure in, 163–173
 comparison of calculations for, 184–186
 continuous operation in, 115
 defined, 112
 depth of, 113, 180–186
 dynamics of, 111–186
 flow patterns in, 140–163
 gas flow pattern in, 140–144
 maximum spoutable bed depth for, 180–186
 peak pressure-drop data for, 132–135
 piezoelectric technique in, 145–146, 171
 pressure drop across, 114, 132–135
 pressure gradient of, 114
 Reynolds number for, 172
 solids flow pattern in, 144–158
 solids mixing characteristics for, 158–163
 spout height and flow pattern in, 144–153
 spout shape in, 163–169
 total solids flow rates for, 155–156

SUBJECT INDEX 311

vertical profile in, 146–153
voidage distribution in, 169–173
Spouting
  see also Spouted beds
  in fluidization, 111–112
  gas flow rate and, 118–119
  location of in gas-solids contacting spectrum, 115–117
  mechanism of, 117–123
  minimum velocity in, 123–131
  orifice-to-column-diameter ratio in, 174
  phenomenon of, 111–115
  pressure drop in, 136–140
  stability of, 173–186
Spouting stability
  column geometry in, 174–176
  cone angle in, 174–175
  gas flow in, 179–180
  maximum spoutable bed depth and, 180–186
  particle size and, 176–178
  size distribution and, 178
  solids density and, 179
  solids properties and, 176–179
Spouting velocity
  for conical vessels, 130–131
  for cylindrical vessels, 123–127
  deviations from standard equation in, 124–131
  equation for, 123–125
Spouting vessel, 112–113
Spout pinching, 168
Spout shapes
  changes in, 166–167
  spouted bed structure and, 163–169
Stanford University conference (1968), 194–198
Sulfide ores
  leaching of, 36
  roasting of, 7
Sulfur, pyrrhotite leaching of, 45
Sulfuric acid, in elevated pressure leaching, 36

T

Taconite ores, leaching of, 40
Tellurium, pressure leaching of, 4
Thickness, drying time and, 264, 267
Thin material, drying of, 253–273

Thiobacillus concretivorus, 16
Thiobacillus ferroxidans, 16
Thiobacillus thioxidans, 16
Tin, ore minerals of, 10
Titanium, ore minerals of, 10
Tungsten ores
  ethylene glycol leaching of, 28
  minerals of, 10
Tunnel driers, 288
Turbulent boundary layer prediction calculation (TBLPC), test flows in, 196–198
Turbulent flow
  closure types in, 198–200
  computation of, 193–244
  Couette flow solution in, 204–205
  forced heat convection in, 249–250
  mean Reynolds-stress closure in, 231–236
  mean turbulent energy structure (MTES) in, 219–228
  mean velocity field (MVE) closure in, 200–215
  opportunities and outlook in, 236–244
  Stanford conference (1968) in, 194–198
  "transport theory" in, 243
  wall boundary condition in, 222
  wall-layer thickness parameters in, 203

U

Ultrasonic energy, in copper electrowinning, 79
Union Carbide Corp., 89
Universal Minerals and Metals, Inc., 87
Uraninite ore, leaching of, 36
Uranium
  for acid leach slurries, 66
  carbonate leaching and, 5
  dodecyl phosphoric acid recovery system for, 64
  leaching process in, 8
  ore minerals of, 10
  recovery of, 4
  resin-in-pulp process for, 60
  sulfuric acid extraction of, 35–36, 68
  from wet-process phosphoric acid, 63
Uranium compounds, hexavalent vs. tetravalent, 8

Uranium oxide, hydrometallurgical processing of, 83
Uranium ores
  atmospheric leaching of, 28
  roasting of, 6
  in sodium bicarbonate leaching solution, 49
Uranyl nitrate, liquid-liquid extraction of, 69

V

Vanadium
  carbonate leaching of, 5
  ore minerals of, 10
Vanadium ore, roasting of, 6

W

Waste streams, hydrometallurgy and, 5
Water, in atmospheric pressure leaching, 22–23
Welding flux, drying curves for, 269
Wet-process phosphoric acid, 63

Z

Zinc, ore minerals of, 10